About the Authors

Maggie Norris is a freelance science writer living in the San Francisco Bay area. As Fine Print Publication Services LLC, Maggie offers contract medical and technical writing services to clients in the pharmaceutical, biotech, and medical technology industries; patient care institutions; and research institutions.

Donna Rae Siegfried has written about pharmaceutical and medical topics for 15 years in publications including *Prevention, Runner's World, Men's Health,* and *Organic Gardening.* She has taught anatomy and physiology at the college level. She is also the coauthor of *Biology For Dummies,* 2nd Edition.

Dedication

To Susan.

—Maggie Norris

Publisher's Acknowledgments

We're proud of this book; please send us your comments at http://dummies.custhelp.com. For other comments, please contact our Customer Care Department within the U.S. at 877-762-2974, outside the U.S. at 317-572-3993, or fax 317-572-4002.

Some of the people who helped bring this book to market include the following:

Acquisitions, Editorial, and Media Development

Project Editor: Victoria M. Adang

Acquisitions Editor: Stacy Kennedy

Copy Editor: Todd Lothery

Assistant Editor: David Lutton

Technical Editors: Kristen Blake Bruzzini, PhD; Steve Dougherty

Editorial Manager: Michelle Hacker

Editorial Assistant: Rachelle Amick

Art Coordinator: Alicia B. South

Cover Photo: © iStockphoto.com/Max Delson Martins Santos

Cartoons: Rich Tennant (www.the5thwave.com)

Composition Services

Project Coordinator: Patrick Redmond

Layout and Graphics: Samantha K. Cherolis, Corrie Socolovitch

Proofreaders: Jessica Kramer, Evelyn Wellborn

Indexer: Becky Hornyak

Illustrator: Kathryn Born

Special Help: Megan Knoll

Publishing and Editorial for Consumer Dummies

    Diane Graves Steele, Vice President and Publisher, Consumer Dummies

    Kristin Ferguson-Wagstaffe, Product Development Director, Consumer Dummies

    Ensley Eikenburg, Associate Publisher, Travel

    Kelly Regan, Editorial Director, Travel

Publishing for Technology Dummies

    Andy Cummings, Vice President and Publisher, Dummies Technology/General User

Composition Services

    Debbie Stailey, Director of Composition Services

Contents at a Glance

Introduction 1

Part I: Locating Physiology on the Web of Knowledge 7
Chapter 1: Anatomy and Physiology: The Big Picture 9
Chapter 2: What Your Body Does All Day...27
Chapter 3: A Bit about Cell Biology 43

Part II: Sizing Up the Structural Layers...................... 71
Chapter 4: Scrutinizing the Skeletal System 73
Chapter 5: Muscles: Setting You in Motion 97
Chapter 6: Getting the Skinny on Skin, Hair, and Nails 119

Part III: Exploring the Inner Workings 133
Chapter 7: The Nervous System: Your Body's Circuit Board 135
Chapter 8: The Endocrine System: Releasing Chemical Messages 157
Chapter 9: The Circulatory System: Getting Your Blood Pumping 177
Chapter 10: The Respiratory System: Breathing Life into Your Body 199
Chapter 11: The Digestive System: Beginning the Breakdown 213
Chapter 12: The Urinary System: Cleaning Up the Act 233
Chapter 13: The Immune System: Living in a Microbe Jungle 249

Part IV: Life's Rich Pageant: Reproduction and Development 269
Chapter 14: The Reproductive System 271
Chapter 15: Change and Development over the Life Span 295

Part V: The Part of Tens 311
Chapter 16: Nearly Ten Chemistry Concepts Related to Anatomy and Physiology...313
Chapter 17: Ten Phabulous Physiology Phacts 319

Index 327

Table of Contents

Introduction 1

About This Book 1
Conventions Used in This Book 2
What You're Not to Read 2
Foolish Assumptions 3
How This Book Is Organized 3
    Part I: Locating Physiology on the Web of Knowledge 4
    Part II: Sizing Up the Structural Layers 4
    Part III: Exploring the Inner Workings 4
    Part IV: Life's Rich Pageant: Reproduction and Development 5
    Part V: The Part of Tens 5
Icons Used in This Book 5
Where to Go from Here 6

Part 1: Locating Physiology on the Web of Knowledge 7

Chapter 1: Anatomy and Physiology: The Big Picture 9

Scientifically Speaking 9
    How anatomy and physiology fit into science 10
    Anatomy, gross and otherwise 11
A Little Chat about Jargon 12
    Creating better communication 12
    Establishing precise terminology 13
Looking at the Body from the Proper Perspective 15
    Getting in position 15
    Dividing the anatomy 17
    Mapping out your regions 17
    Casing your cavities 21
Organizing Yourself on Many Levels 22
    Level I: The cellular level 24
    Level II: The tissue level 24
    Level III: The organ level 24
    Level IV: The organ system level 25
    Level V: The organism level 25

Chapter 2: What Your Body Does All Day .27
Transferring Energy: A Body's Place in the World.................................28
Building Up and Breaking Down: Metabolism.....................................28
Why your cells metabolize 29
How your cells metabolize 29
Staying in Range: Homeostasis 34
Maintaining a constant temperature: Thermoregulation 34
Swimming in H_2O: Fluid balance 35
Adjusting the fuel supply: Blood glucose concentration................36
Measuring important variables..37
Growing, Replacing, and Renewing 38
Growing 38
Replacing...38
Repairing parts 40
Healing wounds 40
Lasting parts..41

Chapter 3: A Bit about Cell Biology. 43
The Functions of Cells...43
Building themselves 43
Building tissues 44
Transforming energy 45
Making and transporting products.......................................45
Communicating 46
Seeing the Inside of Eukaryotic Cells 46
Containing the cell: Cell membrane.....................................48
Controlling the cell: Nucleus 51
Cytoplasm...52
Internal membranes 52
Powering the cell: Mitochondria..52
The protein factory...53
Lysosomes...54
Building Blocks That Build You...54
Joining together: The structure of macromolecules......................55
Nucleic acids and nucleotides 55
Polysaccharides 57
Proteins 57
Genes and Genetic Material 58
Traiting you right...59
Gene structure...59
Synthesizing protein..60

The Cell Cycle 62
Cells that divide, cells that don't 62
Interphase 63
DNA replication 63
Mitosis 65
Organizing Cells into Tissues 66
Connecting with connective tissue 67
Continuing with epithelial tissue 67
Mixing it up with muscle tissue 69
Getting nervous about nervous tissue? 69

Part II: Sizing Up the Structural Layers 71

Chapter 4: Scrutinizing the Skeletal System.................... 73

Reporting for Duty: The Jobs of Your Skeleton 74
Checking Out the Skeleton's Makeup 74
Caring about connective tissue.. 74
The structure of a bone 76
How the skeleton develops 77
Classifying bones 78
Joints and the Movements They Allow 79
Categorizing the types of joints 79
Knowing what your joints can do 80
The Axial Skeleton 81
Keeping your head up: The skull 82
Setting you straight on the curved spinal column 84
Being caged can be a good thing 86
The Appendicular Skeleton 88
Wearing girdles: Everybody has two 88
Going out on a limb: Arms and legs 90
Pathophysiology of the Skeletal System 93
Abnormal curvature 93
Osteoporosis 93
Cleft palate 94
Arthritis 94

Chapter 5: Muscles: Setting You in Motion.................... 97

Functions of the Muscular System 98
Supporting your structure.. 98
Moving you 98
Poised positioning 98
Maintaining body temperature 99
Pushing things around inside.. 99
Talking about Tissue Types 100
Defining unique features of muscle cells 101
Skeletal muscle... 102
Cardiac muscle 104
Smooth muscle 105

Getting a Grip on the Sliding Filament 105
   Assembling a sarcomere 105
   Contracting and releasing the sarcomere 106
Naming the Skeletal Muscles 107
   Starting at the top 108
   Twisting the torso 110
   Spreading your wings 113
   Getting a leg up 114
Pathophysiology of the Muscular System 116
   Muscular dystrophy 117
   Muscle spasms 117
   Fibromyalgia 118

Chapter 6: Getting the Skinny on Skin, Hair, and Nails 119
Functions of the Integument 120
Structure of the Integument 121
   Touching the epidermis 121
   Exploring the dermis 124
   Getting under your skin: The hypodermis 125
Accessorizing Your Skin 126
   Now hair this 126
   Nailing nails 127
   Nothing's bland about glands 127
Your Skin Saving You 129
   Controlling your internal temperature 129
   Your skin is sensational 130
Pathophysiology of the Integument 130
   Skin cancer 131
   Dermatitis 131
   Alopecia 131
   Nail problems as signs of possible medical conditions 132

Part III: Exploring the Inner Workings 133

Chapter 7: The Nervous System: Your Body's Circuit Board 135
Integrating the Input with the Output 136
Neural Tissues 136
   Neurons 136
   Neuroglial cells 137
   Nerves 138
   Ganglia and plexuses 138
Integrated Networks 139
   Central nervous system 139
   Peripheral nervous system 140

Thinking about Your Brain...142
    Keeping conscious: Your cerebrum 143
    Making your moves smooth: The cerebellum.....................144
    Coming up roses: Your brain stem 145
    Following fluid through the ventricles 146
    Regulating systems: The diencephalon 147
    Blood-brain barrier 147
Transmitting the Impulse 148
    Across the neuron 148
    Across the synapse...149
Making Sense of Your Senses 151
    Touch 152
    Hearing and balance...152
    Sight..153
    Olfaction...154
    Taste...155
Pathophysiology of the Nervous System.................................156
    Chronic pain syndrome..156
    Multiple sclerosis...156
    Macular degeneration 156

**Chapter 8: The Endocrine System: Releasing
Chemical Messages 157**
Homing In on Hormones 157
    Hormone chemistry..158
    Hormone sources..159
    Hormone receptors 161
Grouping the Glands 162
    The taskmasters: The hypothalamus and pituitary 163
    Controlling metabolism..165
    Getting the gonads going 168
    Enteric endocrine 170
    Other endocrine glands 171
Pathophysiology of the Endocrine System 172
    Abnormalities in insulin metabolism 172
    Thyroid disorders...173
    Androgen insensitivity 175

Chapter 9: The Circulatory System: Getting Your Blood Pumping... 177
Getting Substances from Here to There 177
Cardiac Anatomy 178
    Sizing up the heart's structure...178
    Examining the heart's tissues 179
    Supplying blood to the heart...180
Looking at Your Blood Vessels 181
    Starting with the arteries 182
    Cruising through the capillaries 183
    Visiting the veins..185

Carrying Cargo: Your Blood and What's in It 186
  Watering down your blood: Plasma 187
  Transporting oxygen and carbon dioxide: Red blood cells 187
  Plugging along with platelets 189
  Putting up a good fight: White blood cells 189
Physiology of Circulation 189
  Putting your finger on your pulse 189
  Generating electricity: The cardiac cycle 190
  On the beating path: The circuits of blood
    through the heart and body 192
  Going up, going down, holding steady: Blood pressure 193
  Not going with the flow 194
Pathophysiology of the Circulatory System 195
  Cardiac disorders 195
  Vascular disorders 196
  Blood disorders 197

Chapter 10: The Respiratory System: Breathing Life into Your Body 199

Functions of the Respiratory System 199
Nosing around Your Respiratory Anatomy 200
  Nose 200
  Pharynx 201
  Trachea 202
  Lungs 202
  Respiratory membrane 204
  Diaphragm 205
Breathing: Everybody's Doing It 205
  Normal breathing 206
  Breathing under stress 207
  Controlled breathing 208
Pathophysiology of the Respiratory System 210
  Hypoxemia 210
  Airway disorders 210
  Lungs 211

Chapter 11: The Digestive System: Beginning the Breakdown 213

Functions of the Digestive System 213
Structures of the Digestive Tract 214
  Examining the walls of the digestive tract 215
  Starting with the mighty mouth 216
  Pharynx and esophagus: Not Egyptian landmarks 217
  Stirring it up in your stomach 218
  Moving through the intestines 219
Doing the Chemical Breakdown 222
  The liver delivers 222
  Pancreas 224
  Digestive fluids, enzymes, and hormones 225

Pathophysiology of the Digestive System 227
Diseases of the oral cavity...227
Disorders of the stomach and intestines..........................227
Bowel syndromes...229
Diseases of the accessory organs...................................230

Chapter 12: The Urinary System: Cleaning Up the Act **233**
Functions of the Urinary System 233
Structures of the Urinary System 235
Putting out the trash: Kidneys 235
Holding and releasing...237
The Yellow River...239
Composition of urine...239
Filtering the blood 240
Selectively reabsorbing..240
Expelling urine...242
Maintaining Homeostasis 242
Fluid balance and blood pressure 243
Regulating blood pH 244
Pathophysiology of the Urinary System 245
Kidney pathologies 245
Urinary tract pathologies..246

Chapter 13: The Immune System: Living in a Microbe Jungle. **249**
Functions of the Immune System 249
Loving Your Lymphatic System.....................................251
Lymphing along...251
Structures of the lymphatic system 252
Identifying Immune System Cells....................................254
Looking at leukocytes...256
Lymphocytes..256
Phagocytizing leukocytes 257
Examining Immune System Molecules...............................258
Histamine...258
Complement system proteins 258
Antibodies..259
Immune System Mechanisms..260
Phagocytosis 260
Degranulation..261
Inflammation is swell..262
Immunity...262
Antibody-mediated immunity 263
Cell-mediated immunity...263
Immunization...263
Pathophysiology of the Immune System 264
The immune system and cancer.....................................264
Immune-mediated diseases 265
Infectious diseases...266

Part IV: Life's Rich Pageant: Reproduction and Development.. 269

Chapter 14: The Reproductive System 271

Functions of the Reproductive System..................................271
Producing Gametes 272
    Meiosis 272
    Female gametes: Ova...274
    Male gametes: Sperm...275
    Determining sex 276
The Female Reproductive System..276
    Organs of the female reproductive system 276
    Cycling approximately monthly..280
The Male Reproductive System...282
    The organs of the male reproductive system 283
    Seminal fluid and ejaculation 285
Pausing for Pregnancy 285
    Steps to fertilization 286
    Implantation 286
    Adapting to pregnancy...286
    Labor and delivery..288
Pathophysiology of the Reproductive System.......................290
    Infertility..290
    Cancer 290
    Sexually transmitted diseases...290
    Premenstrual syndromes...291
    Endometriosis 291
    Cryptorchidism 292
    Hypogonadism 292
    Klinefelter's syndrome 292
    Erectile dysfunction..292
    Pathophysiology of pregnancy 292
    Pregnancy loss 294

Chapter 15: Change and Development over the Life Span 295

Programming Development 295
    Stages of development 296
    Dimensions of development...296
Development before Birth...298
    Free-floating zygote to protected embryo 298
    Dividing development into trimesters 301
The Human Life Span 302
    Changes at birth...303
    Infancy and childhood..304
    Adolescence 305
    Young adulthood 306
    Middle age...307
    Growing creaky 307

Part V: The Part of Tens 311

Chapter 16: Nearly Ten Chemistry Concepts Related to Anatomy and Physiology 313

Energy Can Neither Be Created Nor Destroyed.......................313
Everything Falls Apart...314
Everything's in Motion..314
Probability Rules 315
Polarity Charges Life..315
Oil and Water Don't Mix 316
Fluids and Solids...316
Diffusion and Osmosis 317
Redox Reactions Transfer Electrons......................................318

Chapter 17: Ten Phabulous Physiology Phacts.................319

Unique to You: Hand, Hand, Fingers, Thumb 319
Nothing's Better than Mother's Milk.....................................320
It's Apparent: Your Hair Is Different......................................320
The Almonds of Emotion 321
You Smell Well! 322
What a Small Mouth You Have, Grandma 323
Microbes: We Are Their World 323
Oxygen Habitually Overreacts..324
Talkin' about Breath Control 325
Hanging Out with Hemoglobin...325

Index 327

Introduction

· ·

Congratulations on your decision to study human anatomy and physiology. The knowledge you gain from your study is of value in many aspects of your life.

Begin with the most obvious: the social value of this knowledge. Human anatomy and physiology is always a suitable topic of discussion in social situations because it allows people to talk about their favorite subject (themselves) in a not-too-personal way. Thus, some particularly interesting detail of anatomy and physiology is an ideal conversation opener with attractive strangers or horrifying shirt-tail relatives. (First, though, be completely clear in your mind about the boundary between scientific anatomy and physiology on the one hand and personal clinical details on the other.) Choose the specific topic carefully to be sure of having your intended effect. For example, telling a young boy that he has the same density of hair follicles on his body as a chimp does will probably please him. Telling his teenage sister the same thing may alienate her. Use this power carefully!

A little background in anatomy and physiology should be considered a valuable part of anyone's education. Health and medical matters are part of world events and people's daily lives. Basic knowledge of anatomy and physiology gets you started when trying to make sense of the news about epidemics, novel drugs and medical devices, and purported environmental hazards, to name just a few examples. Everyone has a problem with some aspect of his or her anatomy and physiology at some point, and this knowledge can help you be a better parent, spouse, care-giver, neighbor, friend, or colleague.

Knowledge of anatomy and physiology may also benefit your own health. Sometimes, comprehension of a particular fact or concept can help drive a good decision about long-term health matters, like the demonstrated benefits of exercise, or it may help you take appropriate action in the context of a specific medical problem, like an infection, an infarction, a cut, or a muscle strain. You may understand your doctors' instructions better during a course of treatment, which may give you a better medical outcome.

About This Book

This book guides you on a quick walk-through of human anatomy and physiology. It doesn't have the same degree of technical detail as a textbook. It contains relatively little in the way of lists of important anatomical structures, for instance.

We expect that most readers are using this book as a complementary resource for course work in anatomy and physiology at the high-school, college, or career-training level. Most of the information overlaps with the information available in your other resources. However, sometimes a slightly different presentation of a fact or of the relationship between facts can lead to a small "aha!" Some technical details in your more comprehensive resources may become easier to master after that.

The goals of this book are to be informal but not unscientific; brief but not sketchy; and information-rich but accessible to readers at many levels. We've tried to present a light but serious survey of human anatomy and physiology that you can enjoy for the sake of the information it imparts and that will help you perform well on your tests. As always, the reader is the judge of its success.

You won't find clinical information in this book. Chapters 4 through 15 have a pathophysiology section that uses disorders and disease states to explore the details of some physiological processes, but this book contains nothing related to patient care or self-care. It's also not a health and wellness manual or any kind of lifestyle book.

Conventions Used in This Book

We use the following conventions throughout the text to make the presentation of information consistent and easy to understand:

- New terms appear in *italic* and are closely followed by an easy-to-understand definition.
- **Bold** is used to highlight keywords in bulleted lists.

If you're using this book as a supplement to an assigned textbook, your course materials may name structures and physiological substances using a different nomenclature (naming system) than the one we use in this book. (Very little in biology goes by only one name.)

What You're Not to Read

As much as we'd like you to read every word we've written, we recognize that you may have limited time or interest to do so. If you need to make the most of your time with the text, here's what's safe to skip:

- **Text in sidebars:** Sidebars are the shaded boxes that provide a more in-depth look at some aspect of anatomy or physiology. In some instances, they connect real-world experiences with how your body responds.

✔ **Text marked with a Technical Stuff icon:** Sometimes we give you a nugget of information that's a little more advanced. We mark these sentences with a Technical Stuff icon. If reading these paragraphs makes your head hurt, skip to the next paragraph.

Foolish Assumptions

When we wrote this book, we tried to keep you in mind. We're guessing that you fall into one of these categories:

✔ **Formal student:** You're a high-school or college student enrolled in a basic anatomy and physiology course for credit, or a student in a career-training program for a certification or credential. You need to pass an exam or otherwise demonstrate understanding and retention of data, terminology, and concepts in human anatomy and physiology.

✔ **Informal student:** You're not enrolled in a credit course, but gaining some background in human anatomy and physiology is important to you for personal or professional reasons.

✔ **Casual reader:** Here you are with a book on your hands and a little time to spend reading it. And it's all about you!

How This Book Is Organized

This isn't a textbook, although it's organized somewhat like a textbook. We present general information first, and then we break down each subject for more detailed discussions of the various organ systems defined by anatomists and physiologists. The book comprises 17 chapters, grouped into five parts. The table of contents and the index help you find general or specific information.

The book is illustrated with more than 60 line drawings and process diagrams. It also has 16 color plates — scientific yet original artistic renderings. Many of the color plates, found in the center of the book, show the distribution of an organ system through the entire body. Others show particularly important organs in some detail.

The material in the chapters is, at the very least, factual and logically presented, thanks to our technical review team. Some readers may also find the text amusing from time to time, as well as informative. If some reader should be struck by the awesome complexity and precision of biology while reading this book, its authors, editors, and illustrator will be well pleased.

Part I: Locating Physiology on the Web of Knowledge

Part I lays out some facts and concepts essential for understanding any field of biology. In Chapter 1, we map the position of human anatomy and physiology on the interconnected web of all biological and medical knowledge. We introduce the concept of five levels of biological organization, and we discuss how anatomical and physiological information fits into the tissue, organ, and organ system levels. In Chapter 2, we lay out the basics of everyday physiology: metabolism, energy flow, and homeostasis.

Chapter 3 is a helicopter flight over cell biology, intended to excite and amaze you and thus prime you to believe the seemingly miraculous cellular events that underlie a lot of the physiology discussion in the rest of the book. Most cell biology is subject to several alternative interpretations, but anything in this book not clearly labeled as "speculative" has been tested many times. Cell biologists are discovering the minute mechanisms of more wonders and catastrophes all the time.

Part II: Sizing Up the Structural Layers

Part II is devoted to the three organ systems that form the physical bulk of your body: the skeletal system, the muscular system, and the integumentary system (the skin, along with its glands and accessories). Bones and muscles give your body mass, shape, mobility, and power in your habitat. Your skin is a huge area of contact with the environment, with all the dangers and opportunities that come with that exposure.

Together, these organ systems make up more than half the (dry) weight of the average human body. The musculoskeletal system (the skeleton plus the skeletal muscle) accounts for around 50 percent, and the skin accounts for another 12 percent to 15 percent. In addition to building and maintaining their own tissues, these organ systems contribute much to the physiology of other organ systems.

Part III: Exploring the Inner Workings

Anatomists haven't prescribed a sequence in which the various organ systems should be presented. Because the systems are all so interconnected, any sequence is at least partly arbitrary (although some decisions seem obvious, like placing the discussion of the urinary system after the discussion of the circulatory system and the digestive system).

In this book, we begin our focus on physiology with chapters about the body's two information networks: the nervous system, for electrical messaging, and the endocrine system, for chemical messaging. Then come chapters on the circulatory, respiratory, digestive, urinary, and immune systems. Rather than assume the reader's familiarity with previous chapters, the chapters in this part are built to be read in any order. We give cross-references to other chapters when these may be helpful.

Part IV: Life's Rich Pageant: Reproduction and Development

Part IV is devoted to reproduction and the developmental life cycle of the new individual. This account strives to be as simple as possible, but the topic is inherently complex. The human body invests an enormous proportion of its energy and resources in reproduction, beginning with the partial development of the reproductive organs in its fetal stage. During puberty, this development is completed in a growth surge that sometimes seems to transform a boy or girl overnight. This physiological demand may be followed, eventually, by the demands of carrying a pregnancy and giving birth to a child. With the birth of a child, an individual holds a ticket in the lottery of evolutionary success.

Part V: The Part of Tens

The Part of Tens is often a favorite of *For Dummies* readers. The first of two chapters in this part lists ten fundamental chemistry concepts related to life processes. The second chapter looks briefly at ten remarkable and sometimes obscure details of human anatomy and physiology. You should be able to find at least one sure-fire conversation-starter here.

Icons Used in This Book

The little round pictures that you see in the margins throughout this book are icons that alert you to several different kinds of information.

The bull's-eye symbol lets you know what you can do to improve your understanding of an anatomical structure.

This little icon serves to jog your memory. Sometimes, the text is information that we think you should permanently store in your anatomy and physiology file. Other times, the info here makes a connection between what you're reading and related information elsewhere in the book.

This icon flags extra information that takes your understanding of anatomy or physiology to a slightly deeper level, but the text isn't essential for understanding the organ system under discussion.

The information next to this icon provides you with interesting tidbits about the body, giving you some facts for impressing (or grossing out) people at parties.

Where to Go from Here

If you're a formal student (that is, one who's enrolled or planning to enroll in a formal course in human anatomy and physiology), you may get the most benefit by becoming familiar with this book a week or two before your course begins. Flip to the color plates in the center of the book to get started. The illustrations, charming as well as scientific, are arranged to follow the flow of the text, and the callouts indicate important technical terminology.

Then peruse the book as you would any science book; look at the table of contents and the index. Read the Introduction. (See, you've started already!) Then start reading chapters. Look at the figures, especially the color plates, as you read. You'll probably be able to get through the entire book in just a couple of sittings. Then go back and reread chapters you found particularly interesting, relevant, or puzzling. Study the illustrations carefully. The line drawings as well as the color plates are keyed closely to the text and often clarify important facts. Pay attention to technical terminology; your instructors will use it and expect you to use it, too.

If you're a casual reader (you're not enrolled in a formal course in anatomy and physiology and have little or no background in biology), the following approach may work well. Take some time with the color plates at the center of the book. They give you a good feel for the flow of information (and a good feeling about the human body). Then read the book straight through, beginning to end. Look at the figures, especially the color plates, as you read. After you've been through it all quickly once, go back and reread chapters you found particularly interesting, relevant, or puzzling. Make a habit of studying the illustrations while reading the related text. Don't sweat too much over terminology; for your purposes, saying "of my lungs" communicates as well as "pulmonary." (If you also enjoy word games, though, you can get started on a whole new vocabulary.) Keep the book handy for future reference the next time you wonder what the heck they're talking about in a TV drug ad. The color plates alone make it worth space on your bookshelf.

Part I

Locating Physiology on the Web of Knowledge

The 5th Wave   By Rich Tennant

AMERICA'S FUNNIEST CELLULAR VIDEOS

Unrecognized hormone tries to pass through a cell membrane

In this part . . .

Part I introduces the basics of human anatomy and physiology: the fundamental concepts of organismal biology and cell biology, some elementary terminology, and some hints about the scope and utility of anatomical and physiological knowledge. Unlike with other sciences, you don't always have to go to a lab to perform experiments. You may have one, or even a pair, of anatomical structures close by to investigate. You discover some of the body functions that have been happening right under your nose — and in some cases right *inside* your nose — all your life.

Chapter 1

Anatomy and Physiology: The Big Picture

In This Chapter

▶ Placing anatomy and physiology in a scientific framework

▶ Jawing about jargon

▶ Looking at anatomy: planes, regions, and cavities

▶ Delineating life's levels of organization

Human *anatomy* is the science of the human body's structures — things that can be touched, weighed, and analyzed. In this book, we glance at hundreds of structures: tissues, organs, organ systems, and points of contact between organ systems. We take a slightly closer look at a few specific tissues and organs. We also cover human *physiology,* which is the chemistry and physics of these structures and how they all work together to support the processes of life in each individual.

Scientifically Speaking

Human anatomy and physiology are closely related to *biology,* which is the science of living beings and their relationship with the rest of the universe, including all other living beings. If you've studied biology, you understand the basics of how organisms operate.

Anatomy and physiology narrow the science of biology by looking at the specifics of one species, *Homo sapiens.*

Anatomy is form; physiology is function. You can't talk about one without talking about the other.

The anatomy and physiology of everything else

Scientifically speaking, human biology isn't more or less complex, specialized, or cosmically significant than the biology of any other species, and all are interdependent. Every species of animal, plant, and fungus on the planet has both anatomy and physiology. So does each species of *protist* (one-celled creatures, like amoebae and the plasmodia that cause malaria). At the cellular level (see Chapter 3), all these groups are astoundingly similar. At the levels of tissues, organs, and organ systems (the provenance of anatomy and physiology), plants are very different from animals, and both plants and animals are equally dissimilar to fungi.

Each of these major groups, called a *kingdom,* has its own characteristic anatomy and physiology. It's evident at a glance to everyone at the beach that a starfish and a human are both animals, while the alga in the tide pool and the cedar tree on the shoreline are both plants. Obvious details of anatomy (the presence or absence of bright green tissue) and physiology (the presence or absence of locomotion) tell that story. The different forms within each kingdom have obvious differences as well: The cedar must stand on the shore but the alga would die there. The starfish can move from one place to another within a limited range, while humans can (theoretically) go anywhere on the planet and, with the appropriate accoutrements of culture (a human adaptation), survive there for at least a while. (That is, assuming the cedar and the alga keep on photosynthesizing.) Scientists use these differences to classify organisms into smaller and smaller groups within the kingdom, until each organism is classified into its own "specie-al" group.

Not that human anatomy and physiology aren't "specie-al." Humans' bipedal posture and style of locomotion are very specie-al. There's nothing like a human hand anywhere but at the end of a human arm. Most specie-al of all, possibly, is the anatomy and physiology that allows (or maybe compels) humans to engage in science: humankind's highly developed brain and nervous system. It's entirely within the norms of evolutionary theory that people would be most interested in their own specie-alties, so more humans find human anatomy and physiology more interesting than the anatomy and physiology of the alga. From here on, we're restricting our discussion to the anatomy and physiology of our own species.

How anatomy and physiology fit into science

Biologists take for granted that human anatomy and physiology evolved from the anatomy and physiology of ancient forms. These scientists base their work on the assumption that every structure and process, no matter how tiny in scope, must somehow contribute to the survival of the individual. So each process — and the structures within which the chemistry and physics of the process actually happen — must help keep the individual alive and meeting the relentless challenges of a continually changing environment. Evolution favors processes that work.

Human *pathophysiology* is the science of "human anatomy and physiology gone wrong." (The prefix *path-* is Greek for "suffering.") It's the interface of human biology and medical science. *Clinical medicine* is the application of medical science to alleviate an anatomical or physiological problem in an individual human.

Clinical medicine isn't the subject of this book. Many of the chapters do contain pathophysiology sections, but those sections have no relevant information on patient care. We chose the conditions that we briefly sketch in those sections to demonstrate some characteristic of the system under discussion, especially its interaction with other systems. However, we're guessing that a large proportion of readers are using this book to supplement instructional material in career training for a clinical environment, so the information throughout the book is slightly slanted in that direction.

Anatomy, gross and otherwise

Some biologists specialize in the anatomy and physiology of animals at various hierarchical levels (horses, fish, frogs) or particular organs (mammalian circulatory systems, olfaction in fish, insect hormones). Some focus solely on humans, others concentrate on other species, and still others examine the areas of overlap between humans and other animal species. These various areas of study contribute to human knowledge of biology in general and of clinical medicine in particular. The work of anatomists contributes to medical advances, such as improved surgical techniques and the development of bioengineered prostheses.

Throughout this book, you encounter some information from each major subset of anatomy, including

- **Gross anatomy:** The study of the large parts of an animal body — any animal body — that can be seen with the unaided eye. That's the aspect of anatomy we concentrate on in this book.

- **Histologic anatomy:** The study of different tissue types and the cells that comprise them. Histologic anatomists use a variety of microscopes to study these cells and tissues that make up the body.

- **Developmental anatomy:** The study of the life cycle of the individual, from fertilized egg through adulthood, senescence (aging), and death. Body parts change throughout the life span. For information about human developmental anatomy, see Chapter 15.

- **Comparative anatomy:** The study of the similarities and differences among the anatomical structures of different species, including extinct species. This subject is closely related to evolutionary biology. Information from comparative anatomy can help scientists understand the human body's structures and processes. For example, the comparative anatomy of humans and living and extinct apes can elucidate the structures in the human limbs that enable the bipedal posture.

Taxonomy of *Homo sapiens*

Taxonomy is the science of evolutionary relationships, expressed as a series of mutually exclusive categories. The highest (most inclusive) category is *kingdom,* of which there are five: Monera, Protista, Fungi, Plantae, and Animalia. Within each kingdom, the system classifies each organism into the hierarchical subgroups (and sometimes sub-subgroups) of phylum, class, order, family, genus, and species. Here's the breakdown of humankind:

Kingdom Animalia: All animals.

Phylum Chordata: Animals that have a number of structures in common, particularly the *notochord,* a rodlike structure that forms the body's supporting axis.

Subphylum Vertebrata: Animals with backbones.

Superclass Tetrapoda: Four-footed vertebrates.

Class Mammalia: Tetrapods with hair. Other classes of the vertebrata are Pisces (fish), Amphibia (frogs), Aves (birds), and Reptilia (scaly things).

Order Primates: Apes and monkeys.

Superfamily Hominoidea: Apes (chimpanzees, gorillas, orangutans, humans).

Family Hominidae: Great apes, including humans.

Genus Homo: The human species is the only surviving species of our genus, though this genus included several species in the evolutionary past.

Species Sapiens: All species are given a two-part Latin name, in which the genus name comes first and a species epithet comes second. The biologists who name species sometimes try to use a descriptor in the epithet. For humans, they could have chosen "bipedal" or "talking" or "hairless," but they chose "thinker."

Variety Sapiens: Some species get a "varietal" name, usually indicating a difference that's obvious but not necessarily important from an evolutionary point of view. The human species has one other variety, *Homo sapiens neanderthalensis,* which has been extinct for tens of thousands of years. All humans living since then are of one species variety, *Homo sapiens sapiens.* In the evolutionary classification of humans, there's no biologically valid category below species variety.

A Little Chat about Jargon

Why does science have so many funny words? Why can't scientists just say what they mean, in plain English? Good question, with short and long answers.

Creating better communication

The short answer is, scientists do say what they mean (most of them, most of the time, to the best of their ability), but what they mean can't be said in the English language that people use to talk about routine daily matters. Scientists develop vocabularies of technical terminology and other forms of jargon so they can communicate better with other scientists. It's important that the scientist sending the information and the scientist receiving the information both use the same words to refer to the same phenomenon. To communicate in science, you must know and use the same terminology, too.

Establishing precise terminology

The longer answer to the question of why scientists don't say what they mean starts with a little chat about jargon. Contrary to the belief of some, jargon is a good thing. *Jargon* is a set of words and phrases that people who know a lot about a particular subject use to talk together. There's jargon in every field (scientific or not), every workplace, every town, even every home. Families and close friends almost always use jargon in conversations with one another. Plumbers use jargon to communicate about plumbing. Anatomists and physiologists use jargon and technical terminology, much of which is shared with medicine and other fields of biology, especially human biology.

Scientists try to create terminology that's precise and easy to understand by developing it systematically. That is, they create new words by putting together existing and known elements. They use certain syllables or word fragments over and over to build new terms. With a little help from this book, you'll soon start to recognize some of these fragments. Then you can put the meanings of different fragments together and accurately guess the meaning of a term you've never seen before, just as you can understand a sentence you've never read before. Table 1-1 gets you started, listing some word fragments related to the organ systems we cover in this book.

Table 1-1	Technical Anatomical Word Fragments	
Body System	*Root or Word Fragment*	*Meaning*
Skeletal system	os-, oste- arth-	bone joint
Muscular system	myo- sarco-	muscle striated muscle
Integument	derm-	skin
Nervous system	neur-	nerve
Endocrine system	aden- estr-	gland steroid

(continued)

Table 1-1 *(continued)*

Circulatory system	card-	heart (muscle)
	angi-	vessel
	hema-	blood
	arter-	artery
	ven-	venous
	erythro-	red
Respiratory system	pulmon-	lung
	bronch-	windpipe
Digestive system	gastr-	stomach
	enter-	intestine
	dent-	teeth
	hepat-	liver
Urinary system	ren-	kidney
	neph-	kidney
	ur-	urinary
Immune system	lymph-	lymph
	leuk-	white
	-itis	inflammation
Reproductive system	vagin-	vagina
	uter-	uterine

But why do these terms have to be Latin and Greek syllables and word fragments? Why should you have to parse and put back together a term like *iliohypogastric?* One reason is the contrast between the preciseness with which scientists must name and describe the things they talk about in a scientific context and the relative vagueness and changeability of terms in plain English. Terms that people use in common speech are understood slightly differently by different people, and the meanings are always undergoing change. Not so long ago, for example, no one speaking plain English used the term *laptop* to refer to a computer or *hybrid* to talk about a car. It's possible that, not many years from now, almost no one will understand what people mean by those words. In contrast, scientific Greek and Latin stopped changing centuries ago: *ilio, hypo,* and *gastro* have the same meaning now as they did 200 years ago.

Problems can come up when the specialists who use the jargon want to communicate with someone outside their field. The specialists must translate their message into more common terms to communicate it. Problems can also come up when someone approaching a field, such as a student, fails to make progress understanding and speaking the field's jargon. This book aims to help you make the necessary progress.

Every time you come across an anatomical or physiological term that's new to you, pull it apart to see whether any of its fragments are familiar. Using this knowledge, go as far as you can in guessing the meaning of the whole term. After studying Table 1-1 and the other vocabulary lists in this chapter, you should be able to make some pretty good guesses.

Looking at the Body from the Proper Perspective

We want to make sure that you know where we're coming from when we use certain terms. If you don't look at the body from the correct perspective, you'll have your right and left confused. This section shows you the anatomical position, planes, regions, and cavities, as well as the main membranes that line the body and divide it into major sections.

Getting in position

Stop reading for a minute. Stand up straight. Look forward. Let your arms hang down at your sides with your palms facing forward. You are now in *anatomical position* (see Figure 1-1). Whenever you see an anatomical drawing, the body is in this position. Using this position as the standard removes confusion.

The following list of common anatomical descriptive terms that appear throughout this and every other anatomy book may come in handy:

- **Anterior:** Front or toward the front of the body
- **Ventral:** Front or toward the front of the body
- **Posterior:** Back or toward the back of the body
- **Dorsal:** Back or toward the back of the body
- **Caudal:** Near or toward the tail
- **Prone:** Lying on the stomach, face down
- **Supine:** Lying on the back, face up
- **Lateral:** On the side or toward the side of the body
- **Medial or median:** In the middle or toward the middle of the body
- **Proximal:** Nearer to the point of attachment or the trunk of the body
- **Distal:** Farther from the point of attachment or the trunk of the body (think "distance")

✔ **Superficial:** Nearer to the surface of the body

✔ **Deep:** Farther from the surface of the body

✔ **Superior:** Above or higher than another part

✔ **Inferior:** Below or lower than another part

✔ **Central:** Near the center (median) of the body or middle of an organ

✔ **Peripheral:** Away from the center (midline) of the body or an organ

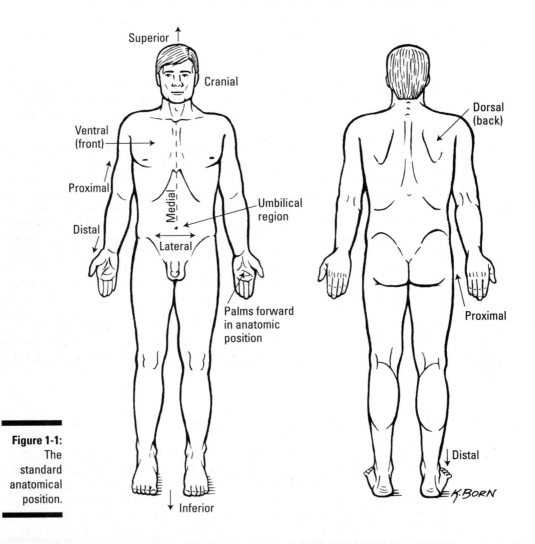

Figure 1-1: The standard anatomical position.

Dividing the anatomy

If you've taken geometry, you know that a *plane* is a flat surface and that a straight line can run between two points on that flat surface. Geometric planes can be positioned at any angle. In anatomy, usually three planes separate the body into sections. Figure 1-2 shows you what each plane looks like. The reason for separating the body with imaginary lines — or by making actual *cuts* referred to as *sections* — is so that you know which *half* or *portion* of the body or organ is being discussed. The anatomical planes are as follows:

- **Frontal plane:** Divides the body or organ into a front (anterior) portion and a rear (posterior) portion.

- **Sagittal plane:** Divides the body or organ lengthwise into right and left sections. If the vertical plane runs exactly down the middle of the body, it's referred to as the *midsagittal plane*. Otherwise, a sagittal plane can run vertically down through the body at any point, creating a *longitudinal section*.

- **Transverse plane:** Divides the body or organ horizontally, into top (superior) and bottom (inferior) portions. Dividing horizontally doesn't necessarily yield two equal divisions; that is, a transverse plane doesn't always go through the waist area to separate the body into top and bottom. Transverse planes can go anywhere to create *cross sections*. When looking at a cross section of a body part, imagine that the body is sectioned horizontally. Or think of a music box that has a top that opens on a hinge. The transverse plane is where the music box top separates from the bottom of the box. Imagine that you open the box by lifting the lid, and you look at the material lining the lid. That's the vantage point that you have when looking at a cross section.

Anatomical planes can "pass through" the body at any angle. The planes are arbitrary for the convenience of anatomists. Don't expect the structures of the body, and especially the joints, to line up or move along the standard planes and axes.

Mapping out your regions

Three types of planes divide the human body, but *regions* compartmentalize the body's surface. Just like on a map, a region refers to a certain area. The body is divided into two major portions: axial and appendicular. The *axial body* runs right down the center (axis) and consists of everything except the limbs, meaning the head, neck, thorax (chest and back), abdomen, and pelvis. The *appendicular body* consists of appendages, otherwise known as *upper and lower extremities*.

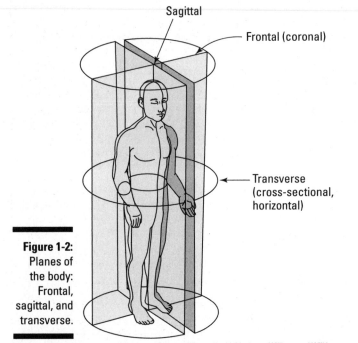

Figure 1-2:
Planes of
the body:
Frontal,
sagittal, and
transverse.

LifeArt Image©Wolters Kluwer Health/Lippincott Williams and Wilkins

Taking pictures of your insides

For anatomists and physiologists from Aristotle to Charles Darwin, the images they had were the sketches they made for themselves. Some of them were pretty good at it: Darwin's sketches of beaks of the finches of the Galapagos Islands were both beautiful and scientifically valuable.

Darwin made scientific history in his own way, of course, but it was a German physicist named Wilhelm Conrad Roentgen who's remembered as "the father of medical imaging." In 1895, Roentgen recorded the first image of the internal parts of a living human: an X-ray image of his wife's hand. By 1900, X-rays were in widespread use for the early detection of tuberculosis, at that time a common cause of death. *X-rays* are beams of radiation emitted from a machine toward the patient's body, and X-ray

images show details only of hard tissues, like bone, that reflect the radiation. In this way, they're similar to photographs. Refinements and enhancements of X-ray techniques were developed all through the 20th century, with extensive use and major advances during World War II. The X-ray is still a widely used method for medical diagnosis screening for signs of disease, usually tumors.

In the 1970s, computer technology took off, taking medical imaging technology with it. Digital imaging techniques began to be applied to convert multiple flat-slice images into one three-dimensional image. The first technology of this sort was called *computed axial tomography* (commonly called a CAT or CT scan), in which multiple X-ray images are combined into cross-sectional pictures of structures inside the body. These detailed and extensive images were unlike anything that had been available to anatomists before. CT technology is still an active area of development. Major technology manufacturers maintain robust product lines of CT instruments and accoutrements for use in clinics and clinical research.

Another class of imaging technologies uses radiation from within the body to create the images that show bodily processes — physiology as well as anatomy. A substance called a *radiopharmaceutical,* which combines a radioactive isotope and a drug, is administered to the patient. It travels to and concentrates in the anatomical structure of interest, and the drug is metabolized there. The isotope emits radiation continuously from within the body, allowing the metabolism of the drug to be traced by radiation detectors and digitally converted to images. One of these *nuclear medicine* techniques, called *positron emission tomography* (PET), can show precisely how some cells use sugar. Nuclear medicine techniques have been in use since the 1950s and, like other imaging technologies, continue to be developed and applied clinically and in research.

Ultrasound imaging technology uses the echoes of sound waves sent into the body to generate a signal that a computer turns into a real-time image of anatomy and physiology. Ultrasound can also produce audible sounds, so the anatomist or physiologist can, for example, watch the pulsations of an artery while hearing the sound of the blood flowing through it. Although all these technologies are considered noninvasive, ultrasound is the least invasive of all, and so is used more freely, especially in sensitive situations like pregnancy.

Since the early 1990s, neuroscientists have been using a type of specialized *magnetic resonance imaging* (MRI) scan, called *functional MRI* (fMRI), to acquire images of the brain. Functional imaging enables scientists to watch a patient's or research subject's thoughts as he or she is thinking them! This aspect of medical imaging has profound implications.

Digital imaging technologies produce images that are extremely clear and detailed. The images can be produced much more quickly and cheaply than older technologies allowed for, and the images can be easily duplicated, transmitted, and stored. The amount of anatomical and physiological knowledge that digital imaging technologies have helped generate over the past 30 years has transformed biological and medical science.

Here's a list of the axial body's regions:

✔ **Head and neck**

- Cephalic (head)

- Cervical (neck)

- Cranial (skull)

- Frontal (forehead)

- Nasal (nose)

- Occipital (back of head)

- Ophthalmic (orbital, eyes)

- Oral (mouth)

✔ **Thorax**

- Axillary (armpit)
- Costal (ribs)
- Mammary (breast)
- Pectoral (chest)
- Sternal (breastbone)
- Vertebral (backbone)

✔ **Abdomen**

- Celiac (abdomen)
- Gluteal (buttocks)
- Groin (area of pelvis near thigh)
- Inguinal (bend of hip)
- Lumbar (lower back)
- Pelvic (area between hipbones)
- Perineal (area between anus and external genitalia)
- Sacral (end of vertebral column)

Here's a list of the appendicular body's regions:

✔ **Upper extremity**

- Antebrachial (forearm)
- Brachial (arm)
- Carpal (wrist)
- Cubital (elbow)
- Palmar (palm)

✔ **Lower extremity**

- Crural (leg, from knee to ankle)
- Femoral (thigh)
- Patellar (front of knee)
- Pedal (foot)
- Popliteal (back of knee)
- Tarsal (ankle)

Casing your cavities

If you remove all the internal organs, the body is empty except for the bones and tissues that form the space where the organs were. Just as a dental cavity is a hole in a tooth, the body's cavities are "holes" where organs are held (see Figure 1-3). The two main cavities are the *dorsal cavity* and the *ventral cavity*.

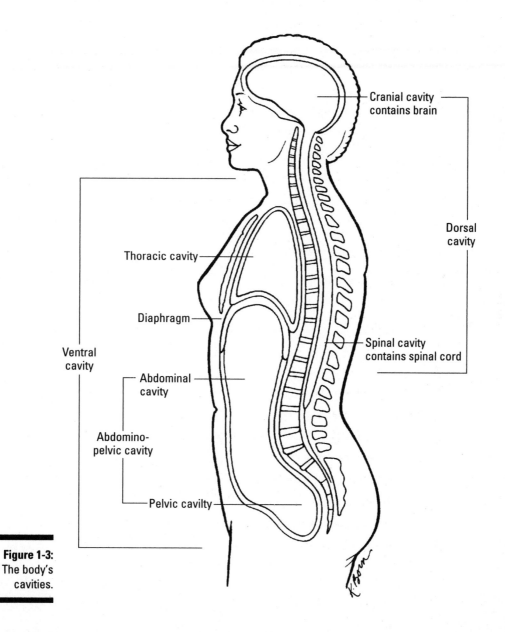

Cranial cavity contains brain

Dorsal cavity

Thoracic cavity

Diaphragm

Spinal cavity contains spinal cord

Ventral cavity

Abdominal cavity

Abdomino-pelvic cavity

Pelvic cavilty

Figure 1-3:
The body's cavities.

The dorsal cavity consists of two cavities that contain the central nervous system. The first is the *cranial cavity,* the space within the skull that holds your brain. The second is the *spinal cavity,* the space within the vertebrae where the spinal cord runs through your body.

The ventral cavity is much larger and contains all the organs not contained in the dorsal cavity. The ventral cavity is divided by the diaphragm into smaller cavities: the *thoracic cavity,* which contains the heart and lungs, and the *abdomino pelvic cavity,* which contains the organs of the abdomen and the pelvis. The abdominal organs are the stomach, liver, gallbladder, spleen, and most of the intestines. The pelvic cavity contains the reproductive organs, the bladder, the rectum, and the lower portion of the intestines.

Additionally, the abdomen is divided into quadrants and regions. The mid-sagittal plane and a transverse plane intersect at an imaginary axis passing through the body at the navel (belly button). This axis divides the abdomen into *quadrants* (four sections). Putting an imaginary cross on the abdomen creates the right upper quadrant, left upper quadrant, right lower quadrant, and left lower quadrant. Physicians take note of these areas when a patient describes symptoms of abdominal pain.

The regions of the abdominopelvic cavity include the following:

- **Epigastric:** Above the stomach and in the central part of the abdomen, just above the navel

- **Hypochondriac:** Doesn't moan about every little ache and illness but lies to the right and left of the epigastric region and just below the cartilage of the rib cage (*chondral* means "cartilage," and *hypo-* means "below")

- **Hypogastric:** Below the stomach and in the central part of the abdomen, just below the navel

- **Iliac:** Lies to the right and left of the hypogastric regions near the hipbones

- **Umbilical:** The area around the navel (the umbilicus)

- **Lumbar:** Forms the region of the lower back to the right and left of the umbilical region

Organizing Yourself on Many Levels

Anatomy and physiology focus on the level of the individual body, what scientists call the *organism.* The life processes of the organism are built and maintained at several physical levels, which biologists call *levels of organization:* the cellular level, the tissue level, the organ level, the organ system level, and the organism level (see Figure 1-4). In this section, we review these levels, starting at the bottom.

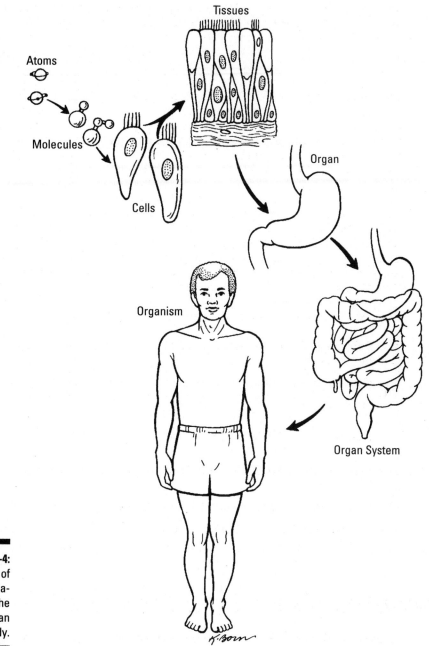

Figure 1-4:
Levels of organiza-
tion in the
human
body.

Level 1: The cellular level

If you examine a sample of any human tissue under a microscope, you see cells, possibly millions of cells. All living things are made of cells. In fact, "having a cellular level of organization" is inherent in any definition of "organism." We discuss the cellular level of organization in some detail in Chapter 3.

Level 11: The tissue level

A *tissue* is a structure made of many cells — usually several different kinds of cells — that performs a specific function. Tissues are divided into four classes:

- **Connective tissue** serves to support body parts and bind them together. Tissues as different as bone and blood are classified as connective tissue.

- **Epithelial tissue (epithelium)** lines the inside of organs within the body and covers the body. The outer layer of the skin is made up of epithelial tissue.

- **Muscle tissue** — surprise! — is found in the muscles, which allow your body parts to move; in the walls of hollow organs to help move their contents along; and in the heart to move blood along via the acts of contraction and relaxation. (Find out more about muscles in Chapter 5.)

- **Nervous tissue** transmits impulses and forms nerves. Brain tissue is nervous tissue. (We talk about the nervous system in Chapter 7.)

Level 111: The organ level

An *organ* is a part of the body that performs a specialized physiological function. For example, the stomach is an organ that has the specific physiological function of breaking down food. By definition, an organ is made up of at least two different tissue types; many organs contain tissues of all four types. Although we can name and describe all four tissue types that make up all organs, as we do in the preceding section, listing all the organs in the body wouldn't be so easy.

The organs that "belong" to one system can have functions integral to another system. In fact, most organs contribute to more than one system. The blood vessels are an excellent example: They serve as a transportation network, delivering nutrients produced by the digestive system to the skeletal muscles to provide energy for locomotion and to the uterus to support the developing fetus. They remove the byproducts of the energy consumed in locomotion and by the fetus in development and carry them to the organs of the urinary system for excretion.

Level IV: The organ system level

Human anatomists and physiologists have divided the human body into *organ systems,* groups of organs that work together to meet a major physiological need. For example, the digestive system is one of the organ systems responsible for obtaining energy from the environment. Other organ systems include the musculoskeletal system, the integument, the nervous system, and on down the list. The chapter structure of this book is based on the definition of organ systems.

Level V: The organism level

The whole enchilada. The real "you." As we study organ systems, organs, tissues, and cells, we're always looking from the organism level.

Chapter 2

What Your Body Does All Day

In This Chapter

▶ Seeing what your body does automatically every day

▶ Finding out what goes on inside of every cell

▶ Discovering the importance of homeostasis

▶ Building and maintaining your parts

This chapter is about your life as an organism. As Chapter 1 explains, *organism* is the fifth of five levels of organization in living things. Although the word *organism* has many possible definitions, for the purposes of this chapter, an organism is a living unit that metabolizes and maintains its own existence.

In this chapter, you see why your to-do list, crowded as it is, doesn't include items such as *Take ten breaths every minute* or *At 11:30 a.m., open sweat glands.* The processes that your body must carry out minute by minute to sustain life, not to mention the biochemical reactions that happen thousands or millions of times a second, can't be left to the immature and distractible frontal lobes (the conscious, planning part of your brain). Instead, your organs and organ systems, coordinated by the older parts of your brain, function together smoothly to carry out these processes and reactions automatically, without the activity ever coming to your conscious attention. All day and all night, year in and year out, your body builds, maintains, and sustains every part of you; keeps your temperature and your fluid content within some fairly precisely defined ranges; and transfers substances from outside itself to inside, and then back out again. These are the processes of *metabolism* and *homeostasis*.

Transferring Energy: A Body's Place in the World

The laws of thermodynamics are the foundation of how the physics and chemistry of the universe are understood. They're at the "we hold these truths to be self-evident" level for chemists and physicists of all specialties, including all biologists. The first law of thermodynamics states that energy can be neither created nor destroyed — it can only change form. (Turn to Chapter 16 for a brief look at the first law and other basic laws of chemistry and physics.) Energy changes form continuously — within stars, within engines of all kinds, and, in some very special ways, within organisms.

The most basic function of the organism that is you on this planet is to take part in this continuous flow of energy. As a *heterotroph* (an organism that doesn't photosynthesize), you ingest (take in) energy in the form of matter — that is, you eat the bodies of other organisms. You use the energy stored in the chemical bonds of that matter to fuel the processes of your *metabolism* and *homeostasis*. That energy is thereby transformed into matter called "you" (the material in your cells), matter that's "not you" (the material in your exhaled breath and in your urine), and some heat radiated from your body to the environment.

Hetero means "other," and *tropho* means "nourishment." A *heterotroph* gets its nourishment from others, as opposed to an *autotroph,* which makes its own nourishment, as a plant does.

Plants convert light energy from the sun into the chemical energy in carbohydrates, which comprise most of the matter of the plant bodies, recycling the waste matter (carbon dioxide) of your metabolic processes. Energy goes around and around, and some of it is always flowing through your body, being transformed constantly as it does so. You, my friend, are part of a cycle of cosmic dimensions!

Building Up and Breaking Down: Metabolism

The word *metabolism* describes all the chemical reactions that happen in the body. These reactions are of two kinds — *anabolic reactions* make things (molecules), and *catabolic reactions* break things down.

To keep the meanings of anabolic and catabolic clear in your mind, associate the word *catabolic* with the word *catastrophic* to remember that catabolic reactions break down products. Then you'll know that anabolic reactions create products.

Your body performs both anabolic and catabolic reactions at the same time and around the clock to keep you alive and functioning. Even when you're sleeping, your cells are busy. You just never get to rest (until you're dead).

Chapter 11 gives you the details on how the digestive system breaks down food into nutrients and gets them into your bloodstream. Chapter 9 explains how the bloodstream carries nutrients around the body to every cell and carries waste products to the urinary system. Chapter 12 shows you how the urinary system filters the blood and removes waste from the body. This chapter describes the reactions that your cells undergo to convert fuel to usable energy. Ready?

Why your cells metabolize

Even when your outside is staying still, your insides are moving. Day and night, your muscles twitch and contract and maintain "tone." Your heart beats. Your blood circulates. Your diaphragm moves up and down with every breath. Nerve impulses travel. Your brain keeps tabs on everything. You think. Even when you're asleep, you dream (a form of thinking). Your intestines push the food you ate hours ago along your alimentary canal. Your kidneys filter your blood and make urine. Your sweat glands open and close. Your eyes blink, and even during sleep, they move. Men produce sperm. Women move through the menstrual cycle. The processes that keep you alive are always active.

Every cell in your body is like a tiny factory, converting raw materials to useful molecules such as proteins and thousands of other products, many of which we discuss throughout this book. The raw materials (nutrients) come from the food you eat, and the cells use the nutrients in metabolic reactions. During these reactions, some of the energy from catabolized nutrients is used to generate a compound called *adenosine triphosphate* (ATP). Whenever ATP is catabolized it releases energy that the cell can use.

So, nutrients are catabolized (broken down), ATP is formed (anabolized), and when needed, ATP is catabolized. This principle of linked anabolic and catabolic reactions is one of the cornerstones of human physiology and is required to maintain life. Cellular metabolism also makes waste products that must be removed (exported) from the cell and ultimately from the body.

How your cells metabolize

The reactions that convert fuel to usable energy (ATP molecules) include glycolysis, the Krebs cycle (aerobic respiration) and anaerobic respiration, and oxidative phosphorylation. Together, these reactions are referred to as *cellular respiration.* These are complex pathways, so expect to take some time to understand them. See Figure 2-1 and refer to it as many times as necessary to understand what happens in cellular respiration.

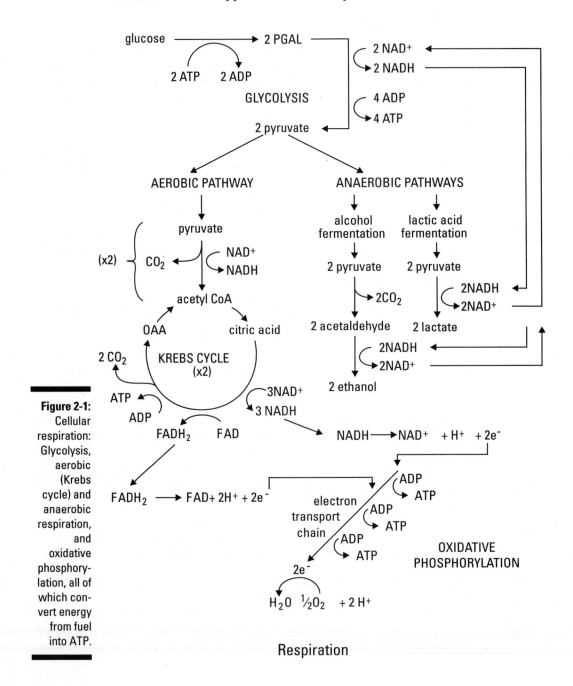

Figure 2-1: Cellular respiration: Glycolysis, aerobic (Krebs cycle) and anaerobic respiration, and oxidative phosphorylation, all of which convert energy from fuel into ATP.

Respiration

Glycolysis, the process that breaks down glucose, occurs in the *cytoplasm* (fluid portion) of every cell. Pyruvic acid, the product of glycolysis, moves from the cytoplasm into the cellular organelle called the *mitochondrion,* the cell's "powerhouse." The *Krebs cycle,* also called the *tricarboxylic acid cycle* or the *citric acid cycle,* takes place in the mitochondrion.

At the completion of the Krebs cycle, the high-energy molecules that are created during the cycle move into the *membrane* of the mitochondrion, where they're passed down the electron transport chain. At the end of that chain, the molecules are used to form ATP from adenosine diphosphate (ADP) and inorganic phosphate (P_i), and water is released.

ATP is the cell's "energy currency." Just as you can't keep spending money without earning some money to replenish your supply, your body can't keep expending energy without taking in more fuel. When the cell needs energy to fuel its metabolism, it "pays" with ATP molecules. See Figure 2-2 for the chemical structure of the ATP and related ADP molecules.

Glycolytic pathway (glycolysis)

Starting at the top of Figure 2-1, you can see that glucose — the smallest molecule that a carbohydrate can be broken into during digestion — goes through the process of *glycolysis,* which starts cellular respiration and uses some energy (ATP) itself. Glycolysis occurs in the cytoplasm and doesn't require oxygen. Two molecules of ATP are required to start each molecule of glucose rolling down the glycolytic pathway; although four molecules of ATP are generated during glycolysis, the net production of ATP is two molecules. In addition to the two ATPs, two molecules of *pyruvic acid* (also called *pyruvate*) are generated. They move into a mitochondrion and enter the Krebs cycle.

Krebs cycle

The *Krebs cycle* is a major biological pathway in the metabolism of every multicellular organism. It's an *aerobic pathway,* requiring oxygen.

As the pyruvate enters the mitochondrion, a molecule of a compound called *nicotinamide adenine dinucleotide* (NAD+) joins it. NAD+ is an electron carrier (that is, it carries energy), and it gets the process moving by bringing some energy into the pathway. The NAD+ provides enough energy that when it joins with pyruvate, carbon dioxide is released, and the high-energy molecule NADH is formed. The product of the overall reaction is *acetyl coenzyme A* (acetyl CoA), which is a carbohydrate molecule that puts the Krebs cycle in motion.

Cycles are endless. Products of some reactions in the cycle are used to keep the cycle going. An example is acetyl CoA: It's a product of the Krebs cycle, yet it also helps initiate the cycle. With the addition of water and acetyl CoA, *oxaloacetic acid* (OAA) is converted to *citric acid.* Then, a series of reactions proceeds throughout the cycle. Refer to "Digging deeper into the Krebs cycle" in this chapter.

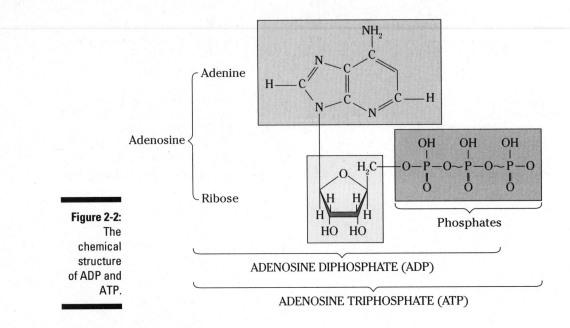

Figure 2-2:
The chemical structure of ADP and ATP.

ADENOSINE DIPHOSPHATE (ADP)

ADENOSINE TRIPHOSPHATE (ATP)

Oxidative phosphorylation

Oxidative phosphorylation is also called the *respiratory chain* and the *electron transport chain.* The electron carriers produced during the Krebs cycle — NADH and $FADH_2$ — are created when NAD+ and FAD, respectively, are "reduced." When a substance is *reduced,* it gains electrons; when it's *oxidized,* it loses electrons. (Turn to Chapter 16 for more information about such "redox reactions.") So NADH and $FADH_2$ are compounds that have gained electrons, and therefore, energy. In the respiratory chain, oxidation and reduction reactions occur repeatedly as a way of transporting energy. At the end of the chain, oxygen atoms accept the electrons, producing water. (Water from metabolic reactions isn't a significant contributor to the water needs of the body.)

As NADH and $FADH_2$ pass down the respiratory (or electron transport) chain, they lose energy as they become oxidized and reduced, oxidized and reduced, oxidized and . . . . It sounds exhausting, doesn't it? Well, their energy supplies become exhausted for a good cause. The energy that these electron carriers lose is used to add a molecule of phosphorus to adenosine *di*phosphate (ADP) to make it adenosine *tri*phosphate — the coveted ATP. And ATP is the goal for converting the energy in food to energy that the cells in the body can use. For each NADH molecule that's produced in the Krebs cycle, three molecules of ATP can be generated. For each molecule of $FADH_2$ that's produced in the Krebs cycle, two molecules of ATP are made.

Digging deeper into the Krebs cycle

The Krebs cycle is a complex set of chemical reactions that take place within the mitochondria of all eukaryotic cells. With the loss of water, citric acid changes to cis-aconitic acid. More water is taken in, and cis-aconitic acid becomes iso-citric acid. At this point, NAD^+ joins in, converting iso-citric acid to α-ketoglutarate; the reaction gives off carbon dioxide and NADH. The α-ketoglutarate converts to succinyl-coenzyme A when NAD^+ and coenzyme A are added. Carbon dioxide and NADH are given off in this reaction. Succinyl CoA is joined by guanosine diphosphate (GDP) and an inorganic phosphate molecule (Pi) to form succinic acid. Coenzyme A and guanosine triphosphate (GTP) are given off. Succinic acid (or succinate) is converted to fumaric acid (fumarate) when oxidized flavin-adenine dinucleotide (FAD) is added. FAD is an electron carrier like NAD^+, and it is also considered to be a nonprotein enzyme. That means it helps to pass on the energy to keep the reactions moving so that the ultimate goal can be reached. FAD is reduced to $FADH_2$ in this reaction. At this point in the cycle, more water is added to fumarate (see why you have to take in water?), which converts the fumarate to malic acid (malate). NAD+ joins the cycle again and converts malic acid to OAA. NADH is given off. After one spin of the Krebs cycle, you have the following amounts of energy-laden molecules:

- Three molecules of NADH (reduced NAD)

- One molecule of $FADH_2$ (reduced flavin adenine dinucleotide)

- One molecule of ATP

Okay. Understanding that one molecule of ATP equals one molecule of ATP is pretty easy. But if ATP is the only energy molecule the body can use, how many ATP molecules do you get out of NADH and $FADH_2$? (Hint: NADH and $FADH_2$ are used to synthesize ATP from ADP and inorganic phosphate during oxidative phosphorylation, which we cover elsewhere in this chapter.)

Theoretically, the entire process of aerobic *cellular respiration* — glycolysis, Krebs cycle, and oxidative phosphorylation — generates a total of 38 ATP molecules from the energy in one molecule of glucose: 2 from glycolysis, 2 from the Krebs cycle, and 34 from oxidative phosphorylation. However, this theoretical yield is never quite reached because processes, especially biological processes, are never 100 percent efficient. In the real world, usually around 29 to 30 ATP molecules per glucose molecule are expected.

Anaerobic respiration

Sometimes oxygen isn't present, but your body still needs energy. During these rare times, a backup system, an *anaerobic pathway* (called anaerobic because it proceeds in the absence of oxygen) exists. Lactic acid fermentation generates NAD^+ so that glycolysis, which results in the production of two molecules of ATP, can continue. However, if the supply of NAD^+ runs out, glycolysis can't occur, and ATP can't be generated.

Staying in Range: Homeostasis

Chemical reactions are not random events. Any reaction takes place only when all the conditions are right for it: All the required reagents and catalysts are close together in the right quantities; the fuel for the reaction is present, in sufficient amount and in the right form; and the environmental variables are all within the right range, including the temperature, salinity, and pH. The complicated chemistry of life is extremely sensitive to the environmental conditions; the environment is the body itself. *Homeostasis* is the term physiologists use to mean the subset of metabolic reactions that keep the internal environment of the body in a state conducive to the chemical reactions that maintain your life.

The following sections look at a few important physiological variables and how the mechanisms of homeostasis keep them in the optimum range in common, every-day situations.

As metabolic reactions, homeostatic reactions require energy.

Maintaining a constant temperature: Thermoregulation

All metabolic reactions in all organisms require that the temperature of the body be within a certain range. For animals that live in the sea, there's no problem: the constant temperature of the sea itself maintains the animal's temperature in the optimum range. (The exception here is marine mammals.) Animals that live on land have evolved different responses to the large and sudden temperature changes in their environment. One solution evolved by birds and mammals is called *homeothermy* or "warm-bloodedness": the maintenance of body temperature at a relatively constant level regardless of the ambient temperature. They do this by regulating their metabolic rate. (This applies also to marine mammals, who took homeothermy with them to their ocean habitat and evolved mechanisms to retain their body heat in that cold environment.) Warm-blooded animals have a large number of mitochondria per cell. This enables a high rate of metabolism, which generates a lot of heat. Warm-blooded animals must ingest a large quantity of food frequently to fuel their higher metabolism.

Regulating body temperature requires a steady supply of fuel (glucose) to the mitochondrial furnaces.

The opposite of homeothermy is *poikilothermy*, or "cold-bloodedness." Comparative anatomists and physiologists consider that homeothermy and poikilothermy are two ends of a spectrum of adaptations to the fluctuations of environmental temperature on land. However, all humans, and all primates, are true homeotherms, whose life processes require that they maintain their body temperature within a narrow range at all times.

Another way warm-blooded animals control their body temperature is by employing adaptations that conserve the heat generated by metabolism within the body in cold conditions or dissipate that heat out of the body in overly warm conditions. A few of the specific adaptations humans use to hold their internal temperatures constant are these:

- **Sweating:** Sweat glands in the skin open to dissipate heat by evaporative cooling of water from the skin. They close to conserve heat. Sweat glands are opened and closed by the action of muscles at the base of the gland, deep under the skin. Refer to Chapter 6.

- **Blood circulation:** Blood vessels close to the skin dilate (enlarge) to dissipate heat in the blood through the skin. They constrict (narrow) to conserve heat. That's why your skin flushes (reddens) when you're hot: That's the color of your blood visible at the surface of your skin. Refer to Chapter 9.

- **Muscle contraction:** When sweating and blood vessel constriction are not enough to conserve heat in cold conditions, your muscles will begin to contract automatically to generate more heat. This reaction is familiar as "shivering."

- **Insulation:** Mammals and birds evolved insulating structures on the body surface (hair and feathers, respectively) and regions of fatty tissue under the skin. Humans alone employ the cultural adaptation of clothing.

Swimming in H_2O: Fluid balance

A watery environment is part of the requirements for a great proportion of metabolic reactions. (The rest need a lipid, or fatty, environment.) The body contains a lot of water: in your blood, in your cells, in the spaces between your cells, in your digestive organs, here, there, and everywhere. Not pure water, though. The water in your body is a solvent for thousands of different ions and molecules (solutes). The quantity and quality of the solutes change the character of the solution. Because solutes are constantly entering and leaving the solution as they participate in or are generated by metabolic reactions, the characteristics of the watery solution must remain within certain bounds for the reactions to continue happening, fluid-balance homeostasis mechanisms have evolved.

✔ **The thirst reflex:** Water passes through your body constantly: mainly, in through your mouth and out through various organ systems, including the skin, the digestive system, and the urinary system. If the volume of water falls below the optimum level (dehydration), the mechanisms of homeostasis intrude on your conscious brain to make you uncomfortable. You feel thirsty. You ingest something watery. Your fluid balance is restored and your thirst reflex leaves you alone.

✔ **Changes in the composition of urine:** The kidney is a complex organ that has the ability to measure the concentration of many solutes in the blood, including sodium, potassium, and calcium. Very importantly, the kidney can measure the volume of water in the body by sensing the pressure of the blood as it flows through. (The greater the volume of water, the higher the blood pressure.) If changes must be made to bring the volume and composition of the blood back into the ideal range, the various structures of the kidney incorporate more or less water, sodium, potassium, et cetera into the urine. That's why your urine is paler or darker at different times. This and other functions of the urinary system are discussed in Chapter 12.

Adjusting the fuel supply: Blood glucose concentration

Glucose, the fuel of all cellular processes, is distributed to all cells dissolved in the blood. The concentration of glucose in the blood must be high enough to ensure that the cells have enough fuel. However, extra glucose beyond the immediate needs of the cells can harm many important organs and tissues, especially where the vessels are tiny, as in the retina of the eye, the extremities (hands and, especially, feet), and the kidneys. Diabetes is a disease in which there is a chronic overconcentration of glucose in the blood.

The amount of glucose in the blood is controlled mainly by the intestines (refer to Chapter 11) and by the hormone insulin. Insulin is a hormone released from the pancreas, an endocrine gland, into the blood in response to increased blood glucose levels. Turn to Chapter 8 for more details on hormones and the endocrine system. Most cells have receptors that bind the insulin, which increases the activity of glucose transporters in the cell membrane. Glucose is removed from the blood and into storage, mostly within the cells of the liver, the muscles (where it is stored as glycogen, the form of fuel your muscles use), and to the fat cells of the adipose tissue. At times when your intestines are not releasing much glucose, such as some hours after a meal, the production of insulin is suppressed and the stored glucose is released into the blood again. Refer to Chapter 8 for more information about insulin and the pancreas.

Measuring important variables

How does the pancreas know when to release insulin and how much is enough? How does the kidney know when the salt content of the blood is too high or the volume of the blood is too low? What tells the sweat glands to open and close to cool the body or retain heat? The answer in these and many other situations related to homeostasis is that the detection of threats to homeostasis and the response of organs to counter the threats involve an intricate system of communications between parts of the nervous, circulatory, and endocrine systems.

Receptors (sensors) in the blood vessels detect the state of the blood: Some detect temperature, some pressure (volume), some the concentration of glucose, and many others detect different variables. These receptors send their "data" through the nervous system to the brain, where an endocrine gland called the *hypothalamus* resides. The endocrine system makes and releases hormones, substances of great power and subtlety, that travel through the blood to the tissues and organs and cause them to "change their behavior" in ways that restore the variables to their optimum physiological range. The hypothalamus is sometimes called the "master gland" because it controls homeostasis by acting on other glands, notably the *pituitary* gland. Refer to Chapter 8.

Feedback in physiology

In biology and other sciences, *feedback* is information that a system generates about itself or its effects that influences how its processes continue. Feedback mechanisms can be *negative* or *positive*. These terms do not mean that one is harmful and the other beneficial, and they are not opposites — that is, they do not counteract one another in the same system or process. Organisms use both types of feedback mechanisms to control different aspects of their physiology.

A *negative feedback mechanism* functions to keep things within a range. It tells the system to stop, slow down, decrease its output when the optimum quantity or range has been achieved or to speed up or increase its output when the quantity is below the optimum range. In other words, it tells a process to start doing the opposite of what it's doing now. Negative feedback mechanisms maintain or regulate physiological conditions within a set and narrow range. Homeostasis depends on a vast array of negative feedback mechanisms.

A *positive feedback mechanism* tells a process to continue or increase its output. Positive feedback says, "Some is good; more would be better." It accelerates or enhances the output created by a stimulus that has already been activated. A positive feedback mechanism is usually a cascading process that increases the effect of the stimulus and pushes levels out of normal ranges, usually for a specific and temporary purpose. Because positive feedback can get out of control (think fire), evolution has favored relatively few positive feedback mechanisms. One example is the "clotting cascade" that occurs in response to a cut in a blood vessel, described in this chapter. Another is the release of oxytocin to intensify uterine contractions during childbirth.

Growing, Replacing, and Renewing

My, how you've changed, and are still changing! Growing up, growing old, and just living every day, you're building new parts and replacing old ones. From conception to early adulthood, your body was busy making itself: everything from scratch.

But the job wasn't finished when you were fully grown. Complex living tissues and organs almost all require replacement parts at some time, and many require them all the time. This necessity is one of the defining characteristics of organisms — the ability to organize matter into the structures that compose themselves and to replace and renew those structures as required, as we describe in the following sections.

As we discuss in the "Building Up and Breaking Down: Metabolism" section earlier in the chapter, making new cells and tissues is *anabolic metabolism,* and breaking up and eliminating old cells and tissues is *catabolic metabolism.*

Growing

You began life as a single cell and built yourself from there, with some help from your mom to get started. Your body developed along a plan, building a backbone with a head at the top and a tail at the bottom. (Somehow, you lost the tail.) Now look at you: 100 trillion cells, almost every one with its own special structure and job to do. Good work! Find more about the processes of development in Chapter 15.

Replacing

Just like the organism they're a part of, many kinds of cells have a life cycle: They're born, they develop, they work, they get worn out, and they die. For an organism to continue its life cycle, these cells must be replaced continuously, usually by the division and differentiation of *stem cells.* These relatively undifferentiated cells wait patiently until they're called upon to divide. Some of the daughter cells differentiate into their specific programmed type while others remain stem cells and wait to be called upon the next time. Stem cells are an active area of research in physiology and in the field of regenerative medicine.

Some of the cell and tissue types that must be continuously replaced are these:

- ✔ **Red blood cells:** The life cycle of a red blood cell is about 120 days. That means you replace all your red blood cells three times a year. New ones come from the red marrow of the bones, and old ones are scavenged for iron in the spleen and then eliminated in the feces. Old, dead, red blood cells give feces their characteristic color.

- ✔ **Epidermis:** The cells of the epidermis, the outer layer of the skin, are constantly shed from the surface and replaced from below. Your body replaces the entire epidermis about every six weeks. This process is discussed in Chapter 6.

- ✔ **Intestinal lining:** The epithelial cells of the intestinal lining are replaced about every week. You realize what a feat this is when you read about the intestine in Chapter 11.

- ✔ **Respiratory membrane:** Your body replaces the epithelial cells that line the alveolar wall and the pulmonary capillary vessels about every week. Refer to Chapter 10 for a description of the respiratory membrane.

- ✔ **Sperm:** The process of *spermatogenesis* (making sperm) is continuous, beginning in a male's puberty and ending with his death. The quantity and quality varies with his age and health. Turn to Chapter 14 for more details.

- ✔ **Bone:** As you can read about in Chapter 5, bone is living tissue and very active in a number of ways. The bones bear the body's weight and the stress of impact. Tiny cracks develop in bone all the time and are repaired quickly and constantly, a process known as *remodeling*. Bones serve as a storage depot for metal ions, especially calcium, which flows in and out of the bone constantly. For more info on bones, flip to Chapter 4.

Some other types of tissue replace their cells at a very slow rate, such as the following:

- ✔ **Brain cells:** Scientists thought for many decades that brain cells that died weren't replaced, and, in general, that no new cells were developed in the brain during adulthood. Brain researchers have now shown that this isn't true. The processes whereby new cells are born in the adult brain have attracted much research interest. See Chapter 7.

- ✔ **Cardiac muscle:** Until recently, physiologists believed that cardiac muscle cells couldn't regenerate, but that belief has recently been called into question. In 2009, researchers in Sweden reported evidence that, in healthy hearts, cardiac muscle cells do indeed divide, but slowly. The researchers estimated that a 20-year-old renews about 1 percent of heart muscle cells per year and that about 45 percent of the cardiac muscle cells of a 50-year-old are generated after birth. Research published in the early 2000s showed evidence that cardiac muscle cells regenerate to some extent after a heart attack.

Repairing parts

Your body repairs some tissues as necessary, such as after an injury:

- ✔ **Skeletal muscle:** Mature skeletal muscle cells, called *fibers,* don't divide and aren't replaced unless they're damaged. After they're formed, skeletal muscle fibers generally survive for your entire lifetime. But wait, you say that you've been working out and your biceps are twice as big as they were last year? Congratulations, but you didn't add cells. The cells you had just got bigger.

- ✔ **Smooth muscle:** Like skeletal muscle fibers, smooth muscle fibers are replaced when they're injured.

- ✔ **Skin fibroblasts:** These are different from epidermal cells. These cells proliferate rapidly to repair damage from a cut or wound and are responsible for generating scar tissue, as discussed in the next section.

- ✔ **Liver cells:** Normally, these cells divide only rarely. However, if large numbers of liver cells are removed — by surgical removal of part of the liver, for example — the remaining cells proliferate rapidly to replace the missing tissue. This makes it possible to transplant part of the liver of a living donor to a recipient, or to split a single liver from a nonliving donor to two recipients. In these cases, when all goes well, both parts regenerate into a complete and functioning liver.

Healing wounds

When you have a tiny, superficial surface wound (a little scratch), the epidermis simply replaces the damaged cells. In a few days, the scratch is gone. But when the wound is deep enough that blood vessels are damaged, the healing process is a little bit more involved. Turn to Chapter 9 for information on blood and blood vessels.

The immediate rush of blood washes debris and microbes out of the wound. Then, the vessels around the wound constrict to slow down the blood flow. A type of "formed element" in the blood called *platelets* sticks to the collagen fibers that make up the vessel wall, forming a natural band-aid called a *platelet plug.*

After the platelet plug forms, a complex chain of events results in the formation of a *clot* that stops blood loss altogether. This chain of events is called the *clotting cascade* or *coagulation cascade*. Enzymes called *clotting factors* initiate the cascade. Here's a rundown of what happens, focusing on the most important steps:

- ✔ **Prothrombin:** This clotting factor converts to thrombin. Calcium is required for this reaction.

- ✔ **Thrombin:** This factor acts as an enzyme and causes the plasma protein *fibrinogen* to form long threads called *fibrin*.

- ✔ **Fibrin threads:** Wrapping around the platelet plug, these threads form a meshlike template for a clot.

- ✔ **Clot:** The meshlike structure traps the red blood cells and forms a clot. As the red blood cells that are trapped on the outside of the clot dry out (or the air oxidizes the iron in them, like rust), they turn a brownish-red color, and a scab forms.

Underneath the scab, the blood vessels regenerate and repair themselves, and in the dermis, cells called *fibroblasts* spur the creation of new cells to regenerate the tissues in the damaged layers. Scars are created to provide extra strength to skin areas that are deeply wounded. Scar tissue has many interwoven collagen fibers, but no hair follicles, nails, or glands. Feeling is usually lost in the area covered with scar tissue because nerves are damaged.

Lasting parts

As we mention earlier in the chapter, *almost* all tissues and organs require replacement parts at some time. However, here are some exceptions:

- ✔ **Central nervous system:** For the most part, the cells and tissues of the central nervous system are incapable of self-repair and regeneration. Thus the poor prognosis in cases of spinal cord injury.

- ✔ **Peripheral nerves:** These are the nerve cells that transmit sensation or motor messages between the central nervous system and the skin and skeletal muscles (see Chapter 7). Many types of peripheral neurons don't undergo regular replacement in normal functioning. They are therefore some of the oldest cells in your body. Unfortunately, they're not regenerated when they die from injury, so some kinds of nerve damage are permanent. Because they're not replaced when they die, the number of such nerve cells declines throughout life.

- ✔ **Ova:** A woman has all the eggs she's ever going to have in her ovaries at birth. In most women, that's about half a million more than they'll ever need. Most eggs die before puberty. Only a few mature and participate in the monthly events of the ovarian (menstrual) cycle. And only a very, very few go on to participate in the events of reproduction described in Chapter 14.

Chapter 3

A Bit about Cell Biology

. . . .

In This Chapter

▶ Finding out what cells do

▶ Taking a close look at cell structure

▶ Expressing your genome

▶ Seeing what kinds of tissues form your body

. . . .

*B*iologists see life as existing at five "levels of organization," of which the cellular level is the first (see Chapter 1 for more on the levels of organization). A basic principle of biology says that all organisms are made up of cells and that anything that has even one cell is an organism. Understanding the basics of cell biology is necessary for understanding any aspect of biology, including human anatomy and physiology.

The first thing to know is that cell biology is astoundingly complex. You can gain a detailed understanding of this complexity only from years of hard study. This chapter aims only to give you some idea of how complex cell biology is, so as to provide context for the various physiological miracles we describe in later chapters.

The Functions of Cells

Almost all the structures of anatomy are built of cells, and almost all the functions of physiology are carried out in cells. A comprehensive list of cell functions would be impossible, but we can group cell functions into a few main categories, which we do in the following sections.

Building themselves

Cells arise from other cells and nowhere else. Once in an organism's lifetime, at the beginning, two cells fuse to form a new cell. Ever after in that organism's lifetime, two cells arise from the division of one cell, and all the cells ultimately derive from the first one. This process is how an organism builds itself from one single generic cell, called the *zygote,* to a complex organism comprising trillions of highly differentiated, highly specialized, and highly efficient cells all working together in a coordinated way. Here's a look at how a cell goes from one to many.

Fusing: The zygote

The organism's first cell is the *zygote,* made by the fusion of sex cells: an *ovum* (egg cell) from the female parent and a *sperm* cell from the male parent. (See Chapter 14 for more information about sex cells, and see Chapter 15 for more about the zygote.) The zygote has two complete copies of its species' DNA: one from its male parent and one from its female parent. The two copies combine in the zygote's nucleus. The zygote is said to be *diploid* — having a complete double-set of DNA.

Dividing: Mitosis

In the form of cell division called *mitosis,* one cell divides into two *daughter cells,* each of them complete but smaller than the original cell.

The process of mitosis occurs only in diploid cells — cells that have two copies of the DNA. That is, all cells except the mature sex cells, which are *haploid* — they have only one copy of the organism's DNA.

Differentiating

After mitosis is complete, each daughter cell goes on to its own separate life. One or both may start or continue down a path of *differentiation,* the name for processes that give cells their particular structures and functions. A cell destined to become a nerve cell starts down one path of differentiation; a cell destined to become a muscle cell starts down another path.

A variation on this mechanism involves a special kind of cell called a *stem cell.* A stem cell divides by mitosis, and one daughter cell remains a stem cell and goes on dividing again and again, while the other daughter cell goes on to differentiate into a specific type of cell in a particular tissue. Only some tissues have their own special stem cells, such as the skin and the blood.

The details of cellular differentiation are beyond the scope of this book. Its complexity is beyond imagining. This is the only explanation you really need: It's under genetic control.

Building tissues

All tissues are built of and by cells and are maintained by them, too. Cells in a tissue are to one degree or another *differentiated* or *specialized* for their anatomical or physiological function in the tissue.

Tissues are built largely of and supported by *structural proteins*. The tissue's chemistry is carried out with the assistance of *enzymes*, which are another type of protein (more on proteins and enzymes later in the chapter). Differentiated cells produce different proteins: Some produce only a few different proteins, and some produce many different proteins in response to signals they receive from other cells. The process of protein construction is basically the same in every cell and for every protein.

For the purposes of anatomy and physiology, remember that all cells have certain very important features in common, but they differentiate into a vast array of shapes and sizes, containing vastly diverse structures, and having different functions and life cycles.

Transforming energy

Most cells make ATP to fuel their own metabolism. Flip to Chapter 2 for a description of how cells make ATP.

Some cells release the glucose molecules that the process of ATP production requires and ship them out for use in other cells (see Chapter 11). Sometimes, the glucose made available by the digestive system is more than the body can use at that moment. Thus, some cells function to corral the extra glucose molecules and store them for later. When later comes and the body needs some glucose, some cells release glucose molecules into the blood, and these become available for the Krebs cycle in other cells (check out Chapter 2 for info on the Krebs cycle).

Making and transporting products

A great many types of cells make special chemicals that are incorporated into tissues and participate in metabolic reactions. Cellular products include thousands of specific proteins and polypeptides, signaling chemicals like neurotransmitters and hormones, small molecules and ions, lipids of many kinds, and cellular matrices of many kinds.

Some specialized cells do essentially nothing else but make and export one product for use by other cells; others make products and perform other functions.

Some cells specialize in transporting the products of other cells around the body, or in transporting metabolic waste products out of the body. Some of these transporting cells have other functions as well. Others do nothing but that one job through their entire life cycle. Red blood cells are an extreme example of the one-job model. They lose their nuclei during differentiation and thereafter do nothing but transport gas molecules from one place to another. They don't divide, don't produce ATP, and don't maintain themselves. When their gas-transporting structures wear out, RBCs have nothing to do. They're removed from circulation and disposed of through the intestine.

Communicating

Some cells transmit signals of various kinds while remaining in one place in the body. Some nerve cells perform only the functions of generating and conducting electrical signals and maintaining themselves, and they typically live for years, or even until the death of the organism itself. Other cells produce various kinds of signaling molecules, like hormones and neurotransmitters, or receive and react to those signaling molecules.

Seeing the Inside of Eukaryotic Cells

Although they're astoundingly varied, cells are also remarkably alike. (This is a theme of cell biology.) It's not just a matter that all the cells of one organism or even one species are a lot alike. All cells, at least all *eukaryotic* cells, are alike. Plants, animals, and fungi are *eukaryotes* (organisms made up of eukaryotic cells), and all their cells, in all their enormous complexity and variation, are fundamentally alike. Yes, your skin cells, your kidney cells, and your bone cells are fundamentally similar to the leaf cells and root cells of a carrot; the cells of a mold, mushroom, or yeast; and the single cell of microorganisms called *protists* that live in water and soil.

Here's a simplistic description of a eukaryotic cell: It's a membrane-bound sac containing smaller but distinctive structures, called *organelles* ("little organs"), suspended in a gel-like matrix called the *cytoplasm.* As their name suggests, organelles are functional subunits of a cell, as organs are functional subunits of an organism. One of the largest and most prominent organelles, the *nucleus,* controls a cell's functioning, similar to the way the nervous system controls an organism's functioning. The term *eukaryote* is derived from the Greek term *karyos,* meaning "nut" or "kernel," which early biologists used to refer to the nucleus. Figure 3-1 shows the general structure of a eukaryotic cell. Refer to this figure as you read about the various cellular structures in the following sections. Table 3-1 gives you an overview of the structures found within a eukaryotic cell.

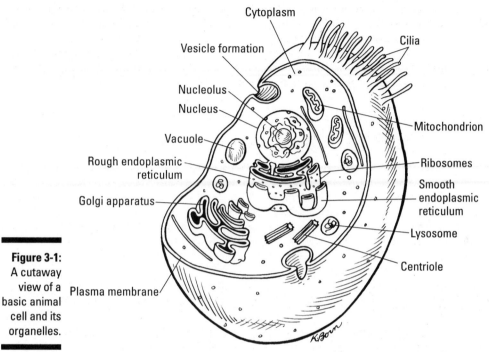

Figure 3-1:
A cutaway
view of a
basic animal
cell and its
organelles.

Table 3-1	Organelles of Animal Cells (Including Humans)
Organelle	**Function**
Nucleus	Controls the cell; houses the genetic material
Mitochondrion	Cell "powerhouse"
Endoplasmic reticulum	Plays an important role in protein synthesis; participates in transporting cell products; involved in metabolizing fats
Ribosome	Binds amino acids together under the direction of mRNA to make protein
Golgi apparatus	"Packages" cellular products in sacs called *vesicles* so some of the products can cross the cell membrane to exit the cell
Vacuoles	Membrane-bound spaces in the cytoplasm that sometimes serve in the active transport of materials to the cell membrane for discharge to the outside of the cell
Lysosomes	Contain digestive enzymes that break down harmful cell products and waste materials and actively transport them out of the cell

The organisms called *bacteria* (singular, *bacterium*) are made up of *prokaryotic* cells. Prokaryotic cells are very different from and much simpler than eukaryotic cells in their basic structure and organization. This difference between eukaryotic and prokaryotic organisms is the great divide in biology. At the cellular level, differences among animals, plants, fungi, and protists are almost negligible compared with the differences between these groups on the one hand and prokaryotes (bacteria) on the other.

Containing the cell: Cell membrane

A cell is bound by a membrane, the *cell membrane,* also called the *plasma membrane* or the *plasmalemma.* The cell membrane of all eukaryotes is made of *phospholipid* molecules. The molecules are made by cells, a process that requires energy. The molecules assemble spontaneously (without input of energy) into the membrane, obeying the forces of *polarity.* Turn to Chapter 16 for a discussion of polarity and how the membrane takes its distinctive form, often called the *phospholipid bilayer.*

Don't confuse *cell membranes* with *cell walls.* Every cell has a membrane, and a cell membrane is a fundamental characteristic of a cell. Some cells also have cell walls outside and separate from the cell membrane. No animal cells have cell walls, but some plant cells and fungal cells have them. They're dissimilar to cell membranes in structure and function.

Permeating the membrane: The fluid-mosaic model

The phospholipid bilayer is embedded with structures of many different kinds. Though the bilayer itself is essentially similar in all cells, the embedded structures are as various and specialized as the cells themselves. Some identify the cell to other cells (very important in immune system functioning); some control the movement of certain substances in or out of the cell across the membrane. Figure 3-2 is a diagrammatic representation of the phospholipid bilayer and embedded structures. This model of the cell membrane is called the *fluid-mosaic model.* "Fluid" describes the ability of molecules in the bilayer to move; "mosaic" pertains to the embedded structures.

The chemical properties of the phospholipid bilayer and the embedded structures contribute to a very important feature of the membrane: It's able to control which substances pass through it and which do not. This means the membrane is *semipermeable.*

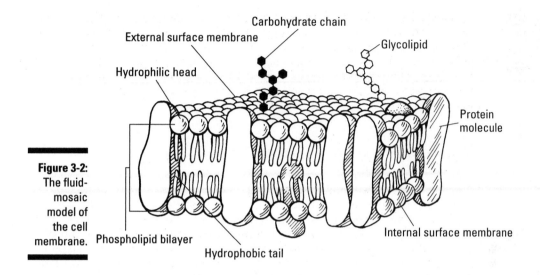

External surface membrane

Carbohydrate chain

Glycolipid

Hydrophilic head

Protein molecule

Figure 3-2: The fluid-mosaic model of the cell membrane.

Phospholipid bilayer

Hydrophobic tail

Internal surface membrane

Crossing the membrane passively

Some substances, mainly small molecules and ions, cross the membrane by a *passive transport* mechanism, meaning they more or less flow unimpeded across the bilayer, driven by the forces of "ordinary" chemistry, such as concentration gradients, random molecular movement, and polarity. Here are some ways that substances cross a membrane passively:

✓ **Diffusion:** A substance moves spontaneously down a *concentration gradient* (from an area where it's highly concentrated to an area where it's less concentrated). If you drop a teaspoon of salt into a jar of water, the dissolved sodium and chloride ions will, in time, *diffuse* (spread themselves evenly through the water). You can measure the time in seconds if you stir the solution or in days if you keep the solution perfectly still at room temperature. (To find out why, see Chapter 16.) Cellular and extracellular fluids are constantly being stirred and are at temperatures between 95 and 100 degrees Fahrenheit. Ions and molecules to which the cell membrane is permeable may diffuse into or out of the cell, constantly attempting to reach equilibrium.

✓ **Osmosis:** The diffusion of water molecules across a selectively permeable membrane gets a special name: *osmosis.* As with diffusion, a concentration gradient drives the mechanism. The pressure at which the movement of water across a membrane stops (that is, when the concentration of the solutions on either side of the membrane is equal) is termed the *osmotic pressure* of the system.

✔ **Filtration:** This form of passive transport occurs during capillary exchange. (*Capillaries* are the smallest blood vessels — they bridge arterioles and venules; see Chapter 9 for more on the circulatory system.) Capillaries are only one cell layer thick, and the capillary cell membrane acts as a filter, controlling the entrance and exit of small molecules. Small molecules dissolved in tissue fluid, such as carbon dioxide and water, diffuse through the capillary cell and into the blood, while substances dissolved in the blood, such as glucose and oxygen, diffuse into tissue fluid across the capillary cell membrane. The pulsating force of blood flow provides a steady agitation.

The blood pressure in the capillaries is highest at the arterial end and lowest at the venous end. At the arterial end, blood pressure pushes substances through the capillary membrane and into the tissue fluid. At the venous end, lower blood pressure (thus higher net osmotic pressure) pushes waste products out of the cell and pulls water from the extracellular fluid into the capillary.

Does passive transport contradict the idea that the cell controls what comes in and out through the membrane? No. The substances that move by passive mechanisms are "ordinary" small molecules and ions that are always present in abundance within and between every cell and kept within a physiologically healthy concentration range by the forces of homeostasis, the first line of defense against physiological abnormality. If at any time the physiological levels get too high or too low, the cell has protein pumps that can counteract the passive transport. You could say that the cell hasn't wasted energy evolving mechanisms to control things that are unlikely to get out of control and for which other remedies are available when they do.

Crossing the membrane actively

Active transport allows a cell to control which big, active, biological molecules move in and out of the cytoplasm. Active transport is a fundamental characteristic of living cells (whereas you can set up a system for diffusion, as we note earlier, in a jar of water).

Like many matters in cell biology, active transport mechanisms are numerous and widely varied. A simple active import mechanism has a molecule outside the cell that the cell needs for its functioning, as well as a membrane-embedded structure that can identify that molecule with unerring specificity, frequently using a kind of lock-and-key mechanism, and can communicate its presence to another membrane-embedded structure. The second structure then opens a channel that only that molecule can pass through, and the channel closes until the structure gets another reliable message to open up. A slightly more complex variation involves a transport molecule that brings the molecule from the cell where it was made to the cell where it's used.

The products made in a cell, whether *anabolites* (built by the cell for a useful purpose) or *catabolites* (wastes and byproducts of anabolic reactions), may be pushed out of the cell across the membrane by active transport mechanisms involving transport molecules.

Controlling the cell: Nucleus

As we mention previously, the defining characteristic of a eukaryotic cell is the presence of a nucleus (plural, *nuclei*) that directs the cell's activity. The largest organelle, the nucleus is oval or round and is plainly visible under a microscope. Refer to Figure 3-1, earlier in the chapter, to see the relationship of the nucleus to the cell; Figure 3-7, later in the chapter, shows a closer view of the nucleus's structure.

All cells have one nucleus, at least at the beginning of their life cycle. As a cell develops, it may lose its nucleus, as do red blood cells and the keratinocytes of the integument; or the cell may merge with other cells, with the merged cell retaining the nuclei of all the cells, such as the fibers of skeletal muscle. This type of cell is called a *syncytium.*

Just as all cells arise from other cells, each nucleus arises from the division of a nucleus. The nucleus contains one complete (diploid) copy of the organism's *genome* — the DNA that embodies the organism's unique genetic material. Every nucleus of every cell in an organism has its own complete and exact copy of the entire genome. It's bound by a semipermeable membrane called the *nuclear envelope.*

The cells produced from this identical DNA are unimaginably varied in structure, in function, and in the substances they produce (proteins, hormones, and so on). The differentiation of the cell (the structure it takes on) and everything about its products are directed by the nucleus, which controls *gene expression,* the selective activation of individual genes.

Keep in mind the relationship between the *genome* and *gene expression.* The genome (DNA) is identical in each cell and remains the same through the organism's life. Within any one cell, only a very few genes are ever expressed. Every cell contains the genes ("instructions") for making a hair, but only a few cells ever express those genes — that is, make a hair. The expression of individual genes within each of the trillions of active cells in an organism changes constantly, moment to moment and throughout the organism's life span. Gene expression is the fundamental process of metabolism.

Cytoplasm

Within the cell membrane, between and around the organelles, is a fluid matrix called *cytoplasm* or *cytosol* and an internal scaffolding made up of *microfilaments* and *microtubules* that support the cell, give processes the space they need, and protect the organelles. The organelles are suspended in the cytoplasm.

The cytoplasm is gelatinous in texture because of dissolved proteins. These are the enzymes that break glucose down into *pyruvate molecules* in the first steps of cellular respiration (see Chapter 2). Other dissolved substances are fatty acids and amino acids. Waste products of respiration and protein construction are first ejected into the cytoplasm and then enclosed by vacuoles and expelled from the cell.

Organelles — including the nucleus, the mitochondria, the endoplasmic reticulum, and the Golgi body — contain a fluid with a particular composition, similar to the cytosol and to one another but each suited to the particular organelle's needs.

Internal membranes

The plasma membrane isn't the only membrane in a cell. Phospholipid bilayer membranes (not bedizened with a "mosaic" of embedded structures) are present all through the cell, encapsulating each organelle and floating around, waiting to be useful. The network of membranes is sometimes called the *endomembrane system*. When an organelle makes a substance that must be expelled from the cell, a piece of bilayer moves in and encapsulates (surrounds) the material and moves toward the cell membrane. Arriving there, it merges with the membrane (which is fluid, remember), opens up, and releases the formerly encapsulated substance into the extracellular fluid. The Golgi apparatus, lysosomes, and vacuoles all make use of this transport mechanism.

Similarly, a fragment of the bilayer in the cell membrane may encapsulate a molecule in the extracellular fluid, extend into the cytoplasm, pinch itself off into a separate *vesicle* (membrane-bound subcellular structure), and then release the substance into the cytoplasm or within the membrane of an organelle.

Powering the cell: Mitochondria

A *mitochondrion* (plural, *mitochondria*) is an organelle that transforms energy into a form that can be used to fuel the cell's metabolism and functions. It's often called the cell's "powerhouse." We discuss the role of the mitochondrion in cellular respiration in Chapter 2.

The number of mitochondria in a cell depends on the cell's function. Cells whose function requires only a little energy, like nerve cells, have relatively few mitochondria; muscle cells may contain several thousand individual mitochondria because of their function in using energy to do "work." A mitochondrion can divide, like a cell, to produce more mitochondria, and it can grow, move, and combine with other mitochondria, all to support the cell's need for energy.

Mitochondria are very small, usually rod-shaped organelles (see Figure 3-3). A mitochondrion has an outer membrane that covers and contains it. The fluid inside the mitochondrion, called the *mitchondrial matrix,* is filled with water and enzymes that catalyze the oxidation of glucose to ATP. A highly convoluted (folded) inner membrane sits within the matrix, increasing the surface area for the chemical reactions.

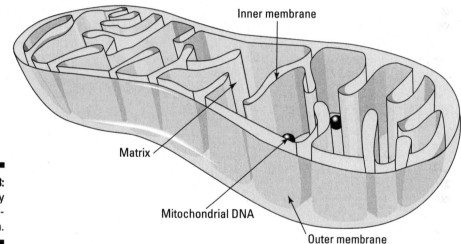

Inner membrane

Matrix

Mitochondrial DNA

Outer membrane

Figure 3-3:
A mighty
mitochon-
drion.

Unique among organelles, the mitochondrion contains a small amount of DNA in a separate chromosome. This DNA behaves separately and independently from the chromosomes in the nucleus. It duplicates and divides to give birth to new mitochondria within the cell, a separate event from mitosis. For more about the phenomenon of mitochondrial DNA, turn to Chapter 17.

The mitochondrion's two membranes aren't the same as the bilayer membrane of the nucleus.

The protein factory

The process of protein construction is a truly elegant system, as you see in the "Synthesizing protein" section later in the chapter. Here we look at the structures of the protein-construction system: the organelles and other intra-cellular structures and their relationships with one another.

The process of protein construction begins in the nucleus. In response to many different kinds of signals, certain genes become active, setting off the production of a specific protein molecule (gene expression). Think of the nucleus as a factory's administrative department.

The *endoplasmic reticulum* (or ER; literally, "within-cell network") is a chain of membrane-bound canals and cavities that run in a convoluted path, con-necting the cell membrane with the nuclear envelope. The ER brings all the components required for protein synthesis together. The *ribosomes,* another organelle involved in protein synthesis, adhere to the outer surface of some parts of the membrane, sticking out into the cytoplasm. These areas are called *rough ER* in contrast to *smooth ER,* where no ribosomes adhere. Think of the ER as the factory's logistics function.

The ribosome is the site of protein synthesis, where the binding reactions that build a chain of amino acids are performed. Ribosomes may float in the cytoplasm or attach to the ER. Ribosomes are tiny, even by the standard of organelles, but they're highly energetic, and a typical cell contains thousands of them. Think of the ribosomes as the production machinery.

The *Golgi body* forms a part of the cellular endomembrane system. It func-tions in the storage, modification, and secretion of proteins and lipids. Think of it as the shipping department.

Lysosomes

Old, worn-out cell parts need to be removed from the cells; if they aren't, they can become sources of toxins or severe energy drains. *Lysosomes* are organelles that do the dirty work of *autodigestion.* Lysosomal enzymes destroy another part of the cell, say an old mitochondrion, through a diges-tive process. Molecules that can be recovered from the mitochondrion are recycled in that cell or in another cell. Waste products are excreted from the cell in a membrane-bound vacuole.

Building Blocks That Build You

Though the processes of life may appear to be miraculous, biology always fol-lows the laws of chemistry and physics. Biochemical processes are much more varied and more complex than other types of chemistry, and they happen among molecules that exist only in living cells. Molecules many thousands of times larger than water or carbon dioxide are constructed in cells and react together in seemingly miraculous ways. This section talks about these large molecules, called *macromolecules,* and their complex interactions. The amazing chemistry of these polymeric macromolecules is the chemistry of life.

Joining together: The structure of macromolecules

The exceptionally large molecules of biology are the *nucleic acids* (DNA and RNA), *polysaccharides,* and *proteins.* All are made mainly of carbon, with varying proportions of oxygen, hydrogen, nitrogen, and phosphorus. Many incorporate other elements, such as magnesium, sulphur, or copper.

Macromolecules, as the term suggests, are huge. Like a lot of huge things, they're made up of smaller things, as a class is made up of students. A student is a subunit of the class, almost identical in many important ways to the other students but unique in other important ways. Macromolecules are made up of molecular subunits called, generically, *monomers* ("one piece"). Each type of macromolecule has its own kind of monomer. Macromolecules are, therefore, *polymers* ("many pieces").

In popular usage, *polymer* suggests plastics. Well, plastics are polymers, but not all polymers are plastics. The term refers to molecules that are made up of repeating subunits — its monomers. The chemical behavior of polymers is different from that of their constituent monomers.

The chemistry of macromolecules is like an infinite Lego set. Any block (monomer) can connect to another block if the shape of their connectors match. With enough blocks, some special connectors, and the energy to do so, you can eventually create a complex structure with bells, whistles, and wheels that turn. Then, you can do that 1,000 times more, and then connect all the complex structures together into one very large, very complex, highly functioning structure. (What's that? You don't have the materials or the energy to do that, and anyway, you wouldn't know how to make a structure fit for the Lego museum? That's okay. Your cells build much more complex things all day every day. All you need to do is keep supplying them with fuel.)

Nucleic acids and nucleotides

The nucleic acids *DNA* (deoxyribonucleic acid) and *RNA* (ribonucleic acid) are polymers made of monomers called *nucleotides* and arranged in a chain, one after another. And another, and another. DNA molecules are thousands of nucleotides long (see Figure 3-4). The functioning of genes is inseparable from the chemical structure of the nucleic acid monomers.

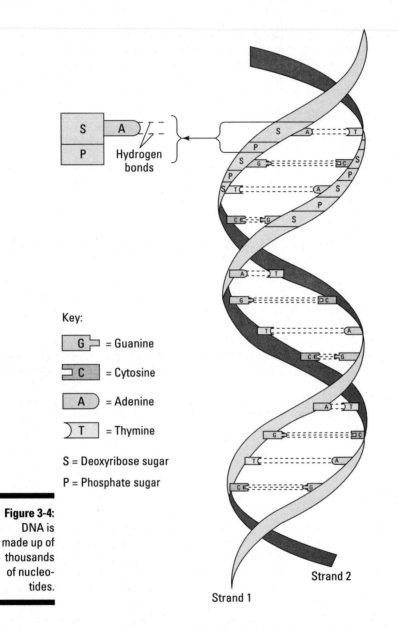

Key:

G = Guanine

C = Cytosine

A = Adenine

T = Thymine

S = Deoxyribose sugar

P = Phosphate sugar

Hydrogen bonds

Strand 2

Strand 1

Figure 3-4: DNA is made up of thousands of nucleo-tides.

A nucleotide is made up of a sugar molecule and a phosphate group attached to a nitrogenous base. The sugar molecule is either deoxyribose (in DNA) or ribose (in RNA). The nitrogenous base is one of four:

✔ Cytosine (C), guanine (G), thymine (T), or adenosine (A) in DNA

✔ Cytosine (C), guanine (G), thymine (T), or uracil (U) in RNA

The bases connect with each other in specific pairs. (Refer to the section "Gene structure" later in the chapter for a discussion of the biological significance of this complementary pairing.) The *complementary pairs* are:

- ✔ C with G
- ✔ T with A in DNA
- ✔ T with U in RNA

The structural similarities and differences between DNA and RNA allow them to work together to produce proteins within cells. The DNA molecule remains stable in the nucleus during normal cell functioning, protected from damage by the nuclear envelope. An RNA molecule is built on demand to transmit a gene's coded instructions for building proteins, and then it disintegrates. Some of its nucleotide subunits remain intact and are recycled into new RNA molecules.

Polysaccharides

The simple carbohydrate molecule glucose is the main energy molecule in physiology. *Polysaccharides* (polymers of carbohydrate monomers) are useful in animal physiology, including human physiology, as fuel storage (glycogen).

Polysaccharides play more roles in plant, fungal, and bacterial physiology, forming much of the (connective) structural tissue, a role taken largely by proteins in animals. However, polysaccharides are as vital (required for life) for animals as they are for plants.

Proteins

Proteins, also called *polypeptides,* are polymers of *amino acids.* The amino acid monomers are arranged in a linear chain and may be folded and refolded into a globular form. Structural proteins comprise about 75 percent of your body's material. The integument, the muscles, the joints, and the other kinds of connective tissue are made mostly of structural proteins like collagen, keratin, actin, and myosin. In addition, the *enzymes* that catalyze all the complex chemical reactions of life in all organisms (plants, fungi, bacteria, as well as animals) are also proteins.

Amino acids

Twenty different amino acids exist in nature. Amino acids themselves are, by the standards of nonliving chemistry, huge and complex. A typical protein comprises hundreds of amino acid monomers that must be attached in exactly the right order for the protein to function properly.

The binding proclivities among all the amino acids result in the structural precision that makes proteins functional for the exacting processes of biology.

Enzymes

Enzymes are protein molecules that *catalyze* the chemical reactions of life. (Flip to Chapter 16 for a brief description of catalysis in chemistry.) Enzymes can only speed up a reaction that is otherwise chemically possible. How effective are enzymes in speeding up reactions? Well, a reaction that may take a century or more to happen spontaneously happens in a fraction of a second with the right enzyme. Enzymes are involved in every physiological process, and each enzyme is extremely specific to one or a very few individual reactions. Your body has tens of thousands of different enzymes.

Anytime an enzyme is discovered, it is named, usually with some physiology shorthand for its function, frequently with the suffix *–ase*. An enzyme named *pyruvate dehydrogenase* removes hydrogen atoms from pyruvate molecules in the processes of cellular respiration (see Chapter 2).

The recipe for an enzyme — that is, the specific amino acids and their exact order in the chain — is a good working definition of a *gene*.

The malfunctioning of a single enzyme, which can be caused by a single nucleotide being out of order, is responsible for some very nasty, sometimes fatal diseases. *Phenylketonuria* (PKU) is an inherited metabolic disease caused by faulty *phenylalanine hydroxylase*. The inability to properly metabolize the amino acid phenylalanine, found in many foods, results in mental retardation, organ damage, unusual posture, and even death, unless the afflicted person can limit the ingestion of foods containing phenylalanine, which is a very difficult thing to do.

Genes and Genetic Material

Your anatomical structures are specified in detail, and all your physiological processes are controlled by your very own unique set of *genes*. Unless you're an identical twin, this particular set of genes, called your *genome,* is yours alone, created at the moment your mother's ovum and your father's sperm fused. The genome itself (all your genes) is incorporated in the DNA in the nucleus of each and every one of your cells.

The genome consists of many thousands of genes. Early research into the number of individual genes in the human genome has produced varying estimates, from about 23,000 to around 75,000, with estimates toward the lower end of the range predominating. Within any given cell, only a very few of these genes is ever *expressed* — that is, activated and used in protein formation.

Traiting you right

Your genes are responsible for your *traits.* If your genes specify that you'll grow to 6 feet tall (a trait), they cause bone and tissue to grow until your body reaches that height, and they maintain it thereafter (assuming a favorable environment), until the cells the genes work through age and die. If you have the genes for brown eyes (a trait), your genes direct the production of pigments that color the eyes. And if your genes include those for abundant production of low-density cholesterol, you have a tendency toward atherosclerosis, another trait. In an environment where the available nutritional resources include abundant red meat, this trait could give you trouble. Otherwise, this trait is harmless.

Calling a trait "harmless" may be technically incorrect. Assuming that any trait has some survival-enhancing purpose, at least under some environmental conditions, is probably more correct, even if you don't know what the purpose is.

Gene structure

The physical structure of the DNA molecule, called the *DNA double helix,* is key to the functioning of genes (refer to Figure 3-4). The DNA double helix can be compared with a ladder or a zipper. In function, it's more like a system of code.

Code "you"

The nucleotides are the symbols in the genetic code. In modern English, a written word is a code made of certain subunits (letters) set down in a specific order. For example, *this, than, then,* and *ten* are different words with different meanings and functions in expressing a thought. A *gene* is made of certain nucleotides in a certain order. A gene may be a few or many nucleotides in length (imagine a single word 20 pages long), but a given gene is always exactly the same nucleotides in exactly the same order along one strand of DNA. The order of the nucleotides is absolutely crucial: "ACTTAGGCT" is not the same as "ACTAAGGCT." According to the prevailing theory, each gene specifies the construction of one enzyme, a type of protein molecule. This is called the *one gene, one protein* model. If the nucleotides are out of order, the enzyme molecule they make will probably be useless, and the organism's functioning will likely be impaired, a little or a lot.

Every model used to explain cell biology, including the *one gene, one protein* model, is subject to change as more information becomes available. But the changes are likely to be minor for understanding basic anatomy and physiology.

Pairing at the molecular level

Remember nucleotide complementary pairs? (If not, see the section "Nucleic acids and nucleotides" earlier in the chapter.) So if a nucleotide is in place on one strand of DNA, and each type of nucleotide binds with only its complementary partner (A with T, C with G, and so on), what do you think is on the other DNA strand? Right! The other nucleotide of the complementary pair. Wherever there's a G on one strand, there's a C on the other, and the pair is attached in the middle. And if the two strands become separated (which they do), what do you think will happen? The nucleotides on each strand will attract and hold other molecules of their complementary partners, thus creating two new double-strands identical to the original double-strand.

Synthesizing protein

A gene that's active sends messages to its own cell or to other cells, ordering them to produce molecules of its particular enzyme, the only one it's capable of making. The message is sent from DNA through the intermediary of *mRNA* (messenger RNA), which places an order at the protein factory of the cell and stays around for a while to supervise production. The first part of the process, where DNA "writes the order" in the form of a sequence of nucleotides on mRNA, takes place in the nucleus and is called *transcription* ("writing across"). The next part of the process, where RNA places the order at the factory, is called *translation* ("carrying across"). The last part, where the amino acid monomers are sorted out and assembled into the polypeptide, starts in the ribosome. Figure 3-5 shows this process. The finishing touches are put on the enzyme molecule in the endoplasmic reticulum and the Golgi body. The process from transcription to the last finishing touch on the protein molecule is called *gene expression*.

Keep in mind the relationship between the *genome* and *gene expression*. Your entire genome is contained in identical DNA molecules in the nucleus of every one of your cells; it remains unchanged through your lifetime. Any given gene may be expressed in only a few cells, or only very occasionally in a few cells, or only under certain physiological conditions, or possibly never. The totality of gene expression changes every second throughout your lifetime, as quickly as nerves transmit impulses and cells react.

The terms *gene expression, protein synthesis,* and *transcription-and-translation* are essentially the same in meaning, but they're used in different contexts in cell biology and physiology.

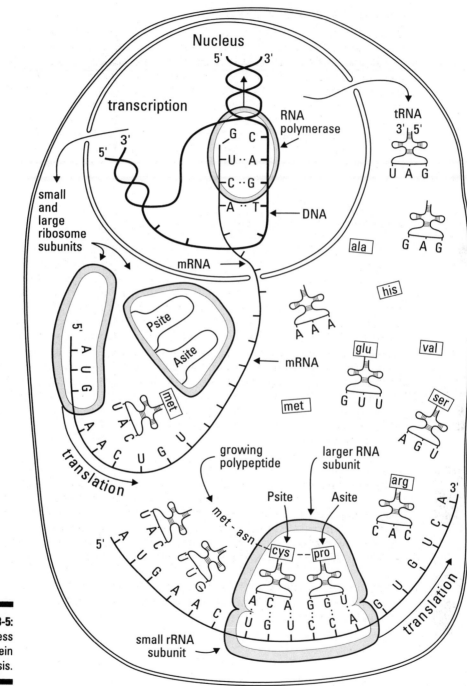

Figure 3-5:
The process
of protein
synthesis.

The Cell Cycle

The life cycle of an individual cell is called the *cell cycle.* The moment of *cell cleavage,* when a cell membrane grows across the "equator" of a dividing cell, is considered to be the end of the cycle for the mother cell and the beginning of the cycle for each of the daughter cells.

Typically, but by no means universally, *interphase* is the longest period of the cell cycle. Interphase comes to an end when the cell divides in the process of *meiosis.* We discuss both these periods of the cell cycle in the following sections.

Cells that divide, cells that don't

All cells arise from the division of another cell, but not all cells go on to divide again:

- **Zygote:** This is the diploid cell that comes into existence when the sex cells (ovum and sperm, both haploid) fuse at conception. Almost immediately, the zygote divides into two somatic cells.

- **Somatic cells:** These include all the cells of the body except the sex cells — in other words, all the diploid cells of the body. Somatic cells may be *relatively differentiated* (somewhat specialized), *terminally differentiated* (they never divide again), or stem cells.

- **Stem cells:** These are special kinds of rather "generic" somatic cells that divide to produce one new stem cell and one new somatic cell that goes on to differentiate into a particular type of cell in a particular type of tissue. The *embryo* (organism in the very early stages of development) has very special stem cells, called *pluripotent* ("many powers") stem cells, which have the ability to give rise to just about any kind of cell an organism needs, given the right chemical environment. When an organism has developed beyond the embryo stage, embryonic stem cells disappear, and other types of stem cells, called *adult* stem cells, arise in particular tissue types and specialize in producing new cells for that tissue. (Turn to Chapter 9 for a description of how stem cells in the bone marrow give rise to many types of blood cells.)

- **Sex cells (gametes):** These form when specialized somatic cells in the reproductive system divide by a process called *meiosis.* Meiosis is the only cellular process in the human life cycle that produces *haploid* cells. See Chapter 14 for more details on sex cells and the processes of meiosis.

Table 3-2 summarizes how different types of cells behave when it comes time to divide.

Table 3-2	Dividing Behavior of Different Cell Types		
Cell Type	**Arise From**	**Divide?**	**Give Rise To**
Zygote	Fusion of two sex cells	Y	Two somatic cells
Somatic cell	Somatic cell or stem cell	Y or N*	Somatic cells; sex cells**
Stem cell	Stem cell	Y	One specialized somatic cell and one stem cell
Sex cell	Somatic cell	N	NA

*Some somatic cells go on to terminal differentiation and never divide again.

**Sex cells arise from meiosis of certain somatic cells. They are haploid cells and never divide again.

Interphase

Interphase begins when the cell membrane fully encloses the new cell and lasts until the beginning of mitosis or meiosis. The duration of interphase may be anywhere from minutes to decades. Generally speaking (there are always exceptions in cell biology), cells do most of their differentiating and most of their routine metabolizing during interphase. Stem cells grow in size and reduplicate organelles during interphase, in preparation for mitosis (more on mitosis in a minute). Some other cells enter mitosis after an extended period of steady-state metabolism. Sometimes, a cell remains in interphase, carrying out its physiological function for years and years until it dies.

DNA replication

DNA replication is an early event in cell division, occurring during interphase, just prior to the beginning of mitosis or meiosis but within the protected space in the nuclear envelope. Maintaining the integrity of the DNA code is absolutely vital.

During DNA replication, the double helix must untwist and "unzip" so that the two strands of DNA are split apart. As shown in Figure 3-6, each strand becomes a template for building the new complementary strand. This process occurs a little at a time along a strand of DNA. The entire DNA strand doesn't unravel and split apart all at once. When the top part of the helix is open, the original DNA strand looks like a *Y*. This partly open/partly closed area where replication is happening is the *replication fork*.

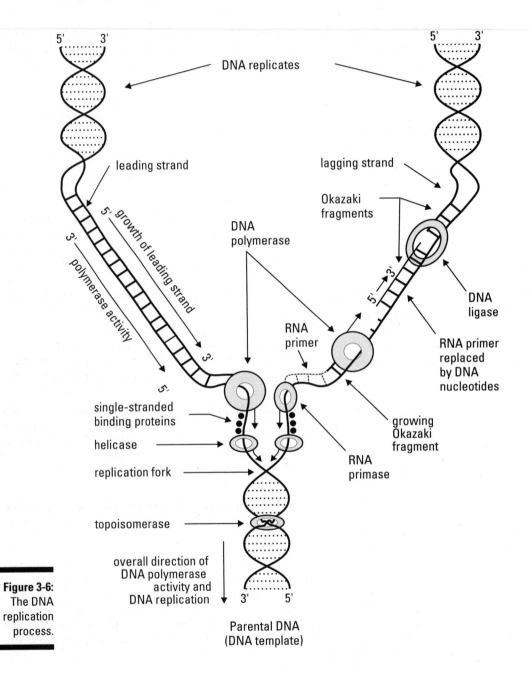

Figure 3-6:
The DNA replication process.

In Figure 3-6, the symbols *5'* and *3'* (read *five prime* and *three prime*) indicate the direction in which DNA replication is occurring. The template strand is read in the 3'-to-5' direction. The bases that are complementary to the template strand are added in the 5'-to-3' direction.

Mitosis

A cell enters a process of *mitosis* (division) in response to signals from the nucleus. As shown in Figure 3-7, mitosis is a multistage process, proceeding in the following stages:

1. **Prophase: The nuclear envelope is dismantled, and the duplicated DNA, in the form of *chromatin*, thickens and coils into *chromosomes.*** Each duplicated chromosome is composed of two identical strands of DNA referred to as *chromatids.* The chromatids are held together by a protein mass called the *centromere.* (**Note:** When the chromatids separate, each is considered a new chromosome.) Cellular structures called *centrioles* and *spindle fibers* move apart to the poles of the cells and move chromosomes to the middle of the dividing cell.

2. **Metaphase: The chromosomes are moved by the spindle fibers to form a perfect row at the center of the spindles, midway between the centrioles.** At this point, 92 chromatids are in a double set of 46 chromosomes.

3. **Anaphase: The centromere is split by enzymatic activity, and the chromatids are pulled by the spindle fibers toward one of the centrioles: 46 to one, 46 to the other.** Following this movement, the chromosomes are referred to as *daughter chromosomes,* and the set at one pole is identical to the set at the opposite pole. But the cell isn't quite ready to divide yet.

4. **Telophase: A fresh nuclear membrane is reassembled around each set of chromosomes.** The spindles dissolve, which frees the daughter chromosomes.

At this point, when each of the two identical nuclei is at one pole of the cell, mitosis is technically over. However, the cell's cytoplasm still has to actually split apart into two masses, a process called *cytokinesis.* The center of the mother cell indents and squeezes the cell membrane across the cytoplasm until two separate cells are formed. The two daughter cells are then in interphase and go on to differentiation or not, depending on the instructions to the cell from the genome.

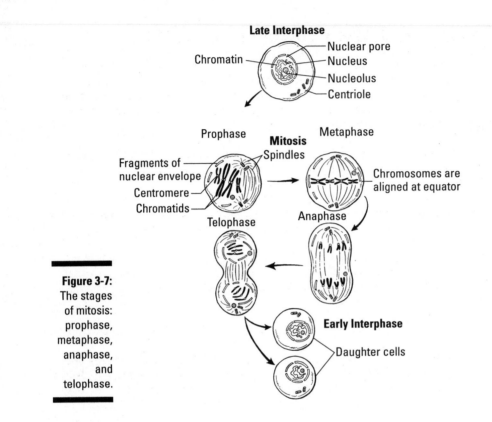

Figure 3-7:
The stages
of mitosis:
prophase,
metaphase,
anaphase,
and
telophase.

Almost 100 percent of the time, the DNA replication process results in two identical copies of the genome. Mutations that get passed on to the next generation through the sex cells are usually fatal to the organism, often as early as the zygote or embryo stage. But if the mutation permits the organism to survive and reproduce, it provides the raw material for "natural selection" in the process of evolution.

Organizing Cells into Tissues

A *tissue* is an assemblage of cells, not necessarily identical but from the same origin, that together carry out a specific function. As we discuss in Chapter 1, tissue is the second level of organization in organisms, above (larger than) the cell level and below (smaller than) the organ level.

Like just about everything else in anatomy, tissues are many and various, and they're grouped into a reasonable number of "types" to make talking about them and understanding them a little simpler. The tissues of the animal body are grouped into four types: *connective tissue, epithelial tissue, muscle tissue,* and *nervous tissue.* All body tissues are classified into one of these groups.

Connecting with connective tissue

Connective tissues connect, support, and bind body structures together. Generally, connective tissue is made up of cells that are spaced far apart within a gel-like, semisolid, solid, or fluid matrix. (A *matrix* is a material that surrounds and supports cells. In a chocolate chip cookie, the dough is the matrix for the chocolate chips.)

Connective tissue has many functions, and thus many forms. In some parts of the body, such as the bones, connective tissue supports the weight of other structures, which may or may not be directly connected to it. Other connective tissue, like adipose tissue (fat pads), cushions other structures from impact. You encounter lots of connective tissue in the chapters to come because every organ system has some kind of connective tissue.

We discuss the important connective tissues *bone* and *cartilage* in some detail in Chapter 4, and we discuss the important connective tissue *blood* in detail in Chapter 9. (What? Blood is a tissue? A connective tissue? Yes, and you'll see why.)

The following other types of connective tissue are found in the human body:

- ✔ **Areolar (a type of loose connective tissue):** This tissue surrounds and separates structures in every part of the body. Various types of cells are scattered through a gel-like matrix, called *amorphous ground substance,* along with wavy ribbons and cylindrical threads of protein fibers.

- ✔ **Dense regular connective tissue:** The secretory cells of this type of connective tissue produce dense bands and sheets of parallel protein fibers, like those in ligaments and tendons.

- ✔ **Dense irregular connective tissue:** The protein fibers in this tissue type are arranged in thick, tough, irregularly-oriented bundles. The dermis is a typical tissue of this type (see Chapter 6).

- ✔ **Adipose tissue (a type of loose connective tissue):** Composed of fat cells, adipose tissue provides fuel storage as well as support and protection to its underlying structures.

- ✔ **Reticular tissue (a type of loose connective tissue):** This type of tissue forms a web or net and functions as a filter in such organs as the spleen, the lymph nodes, and the bone marrow.

Continuing with epithelial tissue

Epithelial tissue forms the epidermis of the *integument* (the skin and the accessory structures such as sweat glands, oil glands, nails, and hair follicles; see Chapter 6), a continuous covering of the outside of the body, and the *endothelium,* a continuous lining of the internal surfaces of the blood vessels.

Epithelial tissue comes in ten types that are defined by the way epithelial cells are combined and shaped (see Figure 3-8).

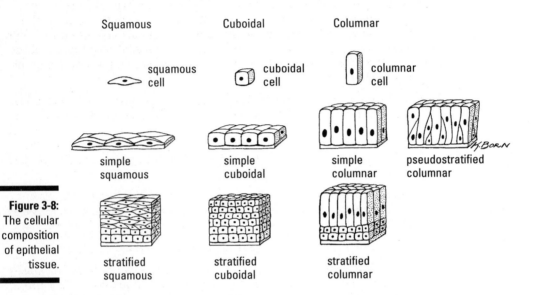

Figure 3-8: The cellular composition of epithelial tissue.

✔ **Simple squamous epithelium:** A single layer of flat cells, this tissue functions in rapid diffusion and filtration. The lining of the alveoli (small sacs) of the lung is a typical tissue of this type.

✔ **Simple cuboidal epithelium:** A single layer of cuboidal cells, this tissue functions in absorption and secretion. This tissue is typically found in glands. The cuboidal cells have the capacity to produce and modify the glandular product (for example, sweat, oil, or milk).

✔ **Simple columnar epithelium:** A single layer of cells that are elongated in one dimension (like a column), usually densely packed together. Like simple cuboidal epithelium, this tissue functions in secretion and absorption. This type of tissue is primarily found lining the portion of the digestive tract from the stomach to the anal canal.

✔ **Simple columnar ciliated epithelium:** This tissue possesses a type of organelle called *cilia* — hairlike structures that act to move substances along in waves. This tissue lines the small tubes of the respiratory system and moves mucus steadily along the tubes toward elimination through the nose and mouth.

✔ **Pseudostratified columnar epithelium:** A single layer of columnar cells. Note that the prefix *pseudo* means "false." The tissue appears stratified, or layered, because the cells' nuclei don't line up in a row, as they do in simple columnar epithelium. Other than that, they're the same, and have similar functions of absorption and secretion. This type of tissue forms a portion of the lining of the male urethra.

- ✔ **Pseudostratified columnar ciliated epithelium:** Try figuring out this term yourself. (If you figured out that these cells are the same as the pseudostratified columnar epithelium, but they also feature cilia, you're right!) This type of tissue is present in the linings of respiratory tubes, functioning in a more or less identical way to the simple columnar ciliated epithelium.

- ✔ **Stratified squamous epithelium:** This tissue consists of several layers of cells: squamous epithelial cells on the outside with deeper layers of cuboidal or columnar epithelial cells. It's found in areas where the outer layer is subject to wear and needs to be replaced continuously. The epidermis of the skin is an example of a specific type called *keratinized stratified squamous epithelial tissue.*

- ✔ **Stratified cubiodal epithelium:** Several layers of cuboidal cells that mainly act as protection in such structures as the conjunctiva of the eye.

- ✔ **Stratified columnar epithelium:** Multilayered absorptive and secretive tissue. The tissue lining the excretory ducts of glands is typical.

- ✔ **Transitional epithelium:** The cells of this tissue can change (or *transition*) from dome-shaped or scalloped-shaped to squamous (flat) and back again, as needed by the tissue. This tissue is found in the lining of the bladder, where having a little room to stretch is sometimes handy. See a diagram of transitional epithelium in Chapter 12.

Mixing it up with muscle tissue

Muscle tissue comes in three types: skeletal muscle, smooth muscle, and cardiac muscle. We discuss the similarities and differences in cellular composition in these three types of tissue in Chapter 5. We also discuss in Chapter 5 the anatomy and physiology of the large organ system called the *muscular system,* of which skeletal muscle tissue is a major component. In Chapter 9, we discuss the function of cardiac muscle in the context of the circulatory system as well as the role of smooth muscle in blood circulation. We also cover smooth muscle's role in the digestive system in Chapter 11.

Getting nervous about nervous tissue?

Don't be. Nerve tissue is relatively simple in one way: Your body has only one type of nerve tissue, and it's made mostly of only one type of cell, the *neuron.* You can get lots more information about the nervous system in Chapter 7, if you get the *impulse* to find out more.

Part II
Sizing Up the Structural Layers

The 5th Wave · By Rich Tennant

"Believe me, if there were an easy cure for baldness, our team would have found it by now."

In this part . . .

Part II begins the survey of the organ systems of human anatomy and physiology. The three chapters in this part cover the anatomical framework of the human body: the bones, muscles, and skin that comprise more than half the body's weight and give structure, support, and protection to the other organ systems. The musculoskeletal and integumentary systems build and continuously maintain their own specialized cells and tissues from the nutrients and oxygen furnished by other organ systems. They're important reservoirs for energy and minerals, and they contribute cells and metabolites to other systems.

Chapter 4

Scrutinizing the Skeletal System

In This Chapter

▶ Listing the functions of your skeleton

▶ Breaking down the skeleton's structure

▶ Joining everything together with joints

▶ Looking separately at the axial and appendicular skeletons

▶ Noting some skeletal pathologies

*I*f you have any skeletons in your closet, now is the time to pull them out. We're not talking about your deep, dark secrets. Seriously! Actually looking at a model of a skeleton is the best way to figure out what's connected to what. If you don't have any skeletons in your closet, try your refrigerator. Observantly cutting up a chicken can show you a lot about bones and joints. And you'll get a head start on dinner!

Starting the discussion of gross anatomy with the skeleton seems logical. The skeleton determines humans' general shape and size as a species, and also humans' very distinctive upright posture and bipedal gait. To get an overview of the skeleton, refer to the "Major Bones of the Skeleton" color plate in the center of this book.

At a much higher taxonomic level, humans are classified into a large group of animals that includes all mammals, birds, reptiles, amphibians, and fish, based on a component of the skeleton. That component is the spine, or more technically, the *vertebrae*. The large taxonomic group is the *Vertebrata*. The "other" group, the *Invertebrata*, includes such wildly successful (evolutionarily speaking) but otherwise very different groups as the insects and the mollusks.

In humans, as in all vertebrates, the skeleton is part of the *musculoskeletal system*. The other part, the muscular system, is the subject of Chapter 5.

The skeleton consists of all your bones, all the joints that connect your bones, and various kinds of fibrous tissue that cover, protect, and bind bones and joints together. In this chapter, we look at the special structures of these tissues and name some of the most important bones and joints. Other important functions of bone tissue, like mineral storage and blood cell production, are mentioned briefly or covered in detail in other chapters.

Reporting for Duty: The Jobs of Your Skeleton

The structural functions of the skeletal system are these:

✔ **Protection:** Bones and joints are strong and resilient. The rib cage provides a protected inner space for your more delicate internal organs. The vertebral column partially encases and protects the spinal cord, and the skull completely encases the brain.

✔ **Movement:** The musculoskeletal system is a motion machine: The bones anchor the skeletal muscles and act as levers, the joints act as fulcrums, and muscle contraction provides the force for movement. (See Chapter 5 for information about muscles and muscle contraction.)

✔ **Support:** The curved vertebral column supports most of your body's weight (see the section "Setting you straight on the curved vertebral column" later in the chapter). The arches of your feet support the weight of your body in a different way (see the sidebar "Strong feet, firm foundation").

Checking Out the Skeleton's Makeup

This section is about how your body builds the tissues of the bones and the joints and how they all fit together to protect, move, and support the entire body.

Caring about connective tissue

The skeleton is made up mainly of three types of connective tissue: *bone tissue, cartilage,* and *fibrous connective tissue.*

Bone tissue

Bone tissue is physiologically very active, constantly generating and repairing itself, and has a generous blood supply all through it. Not only that, but bone makes a huge amount of "product for export," notably the very cells of the blood. (Yes, new blood cells are made by bones — see Chapter 9.) Bone contains dozens of specialized cell types, very few of which we even mention.

But all the skeletal system's functions depend on the functioning of specialized cells in bone tissue.

Keep in mind the difference between *bone tissue* and a specific, named bone. Both the *femur* (thigh bone) and the *humerus* (arm bone) contain bone tissue, but each bone has its own specialized configuration of the components of bone tissue.

Cartilage

Cartilage is a firm but flexible tissue made up of mostly protein fibers. If you put your finger on the end of your nose and push gently, you can get a very good idea of the rubbery texture and flexibility of cartilage. Cartilage is the main component of joints.

Cartilage is less complicated in its structure than bone tissue, having fewer cells, fewer cell types, and little or no direct blood supply. However, among the functions of cartilage tissue is the building of new bone. The two types of cartilage in the skeletal system are hyaline cartilage and fibrocartilage.

- ✔ *Hyaline cartilage* is the type that forms the septum of your nose. It also forms a portion of the very first version of the fetal skeleton. It's the most abundant type of cartilage in several kinds of joints — it's a major component of the freely movable joints called synovial joints.

- ✔ *Fibrocartilage* is a fibrous, spongy tissue that acts as a shock absorber in the vertebral column (spine) and the pelvis.

Cartilage isn't generated and replaced as actively as bone, so cartilage gets by with fewer cells, and mature cartilage has no blood supply.

Fibrous connective tissue

Fibrous connective tissue (FCT) can be compared to the kind of packing tape that has fibers in it. FCT contains very few living cells and is composed mainly of protein fibers, complex sugars, and water.

FCT forms a structure called the *periosteum,* a protective sheet that covers bones. The sheet morphs into cordlike structures wherever these are needed. These cordlike structures are called *ligaments,* which connect a bone to another bone, and *tendons,* which connect a muscle to a bone.

The periosteum is said to be *continuous* with the ligaments and tendon, because there's no real separation between the "sheet" and the "cords."

The structure of a bone

The structures called *bones* (the femur, the vertebrae, the finger bones) are made of bone tissue. (No surprise there, you say, but note that structures called *joints* are made of the tissue called *cartilage*.) It's important to remember that different individual bones have different forms of bone tissue.

Long bones such as your thighbone (femur) or forearm bone (radius) are the type of bones people usually think of first. And in fact, they make a good illustration of general anatomy and physiology of bone tissue (see Figure 4-1).

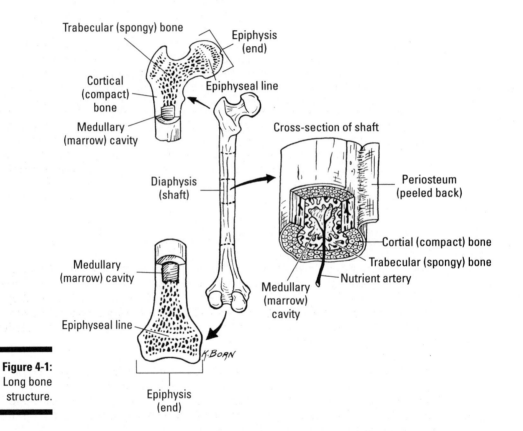

Figure 4-1:
Long bone
structure.

In cross section, bone is structured in concentric layers, that is, an outer layer surrounds a middle layer, which in turn surrounds an inner layer. In longitudinal section, a bone has two ends, mostly similar, and a long middle area, which has cells and tissues mostly different from the ends. The following list names and briefly describes cellular and material composition of the areas of a long bone:

✔ **Compact bone** (the outer layer) is a dense layer of cells in a hard matrix of protein fibers and compounds made of calcium and other minerals. This is the layer that gives bones their amazing strength.

✔ **Spongy bone** (the middle layer) is, like compact bone, a variety of cell types within a matrix of mineralized protein fibers. But spongy bone is more open in structure than compact bone, a physiological trade-off between strength and lightness. (Birds took this evolutionary trend much further.) Structures called *trabeculae,* which follow stress lines in the bone, act like braces, providing support. In adults, the spongy bone at the ends of the long bones contains the red marrow, described in the next subsection.

✔ **The medullary cavity and bone marrow** is the inner layer in the shaft of a long bone (the *diapysysis*) and the inner portion of other bones. There's *yellow marrow,* which is mostly fat (think butter), and *red marrow,* the site of *hematopoiesis,* the production of blood cells. In adults, red marrow is found in the skull, ribs, vertebrae, breastbone (sternum), and the spongy bone at the ends of the long bones. In infants, medullary cavities of the long bones are mostly filled with red marrow.

✔ **Epiphysis** is the enlarged, knobby end of a long bone. It consists of an outer layer of compact bone overlying spongy bone. This is the site of bone elongation. Within the epiphysis, bone and cartilage tissue are intimately connected: As cartilage cells divide, cartilage morphs into bone tissue. This process continues from before birth until the bones reach their full adult size.

How the skeleton develops

When long bones develop in a fetus, they're formed from hyaline cartilage. The softer cartilage allows the fetus to bend into the poses that would make a yoga instructor beam with pride. The shape of the bone is determined by the shape of the cartilage, so it serves as a template. Mineral salts, such as calcium, are deposited onto the template, and the cartilage becomes calcified in the *endochondral ossification* process.

The skeleton isn't fully ossified at birth. (Ever notice the "rubbery" feel of babies and toddlers?) As a child grows up, ossification and bone growth occur simultaneously — the cartilage cells divide, the bone lengthens, and the cartilage becomes calcified. Eventually, the only areas of growth are the epiphyses at the ends of certain bones.

When adult height is reached, the cartilage cells in the epiphyses no longer divide, and the bone stops growing in length.

That's not the end of the bone-development story. Bone development continues throughout life, in a constant process of minerals being deposited in and dissolved out, and the birth, growth, and death of cells. A similar process, but slower and less active, occurs in cartilage. Early in life, the processes of building are more active. As a person ages, building slows but never actually stops. The destructive processes continue or even accelerate (as happens in women after menopause), and the entire skeleton gradually shrinks and weakens.

Classifying bones

Bones come in different shapes and sizes. Appropriately, many bone type names match what they look like, such as flat bones, long bones, short bones, and irregular bones. Check out Table 4-1 for the differences among the four types of bones.

Table 4-1	Characteristics of Bone Types	
Bone Type	*Example Location in the Body*	*Characteristics*
Flat	Skull, shoulder blades, ribs, sternum, pelvic bones	Like plates of armor, flat bones protect soft tissues of the brain and organs in the thorax and pelvis.
Long	Arms and legs	Like steel beams, these weight-bearing bones provide structural support.
Short	Wrists (carpal bones) and ankles (tarsal bones)	Short bones look like blocks and allow a wider range of movement than larger bones.
Irregular	Vertebral column, kneecaps	Irregular bones have a variety of shapes and usually have projections that muscles, tendons, and ligaments can attach to.

Joints and the Movements They Allow

A *joint* is a connection between two bones. Some joints move freely, some move a little, and some never move. Joints, which vary greatly in their size and shape, are classified by the amount of movement they permit. This section tells you about the different joint structures and the movements they allow.

Categorizing the types of joints

Not all joints provide the same range of motion. In fact, one type of joint doesn't allow any movement at all. Check out the types of joints in the following sections.

Immovable joints

Synarthroses are joints that don't move, such as those between the bones of the skull. A thin layer of fibrous connective tissue, called a *suture,* joins them together. The sutures in the cranium are named as follows:

- ✔ **Coronal suture:** Joins the parietal bones and the frontal bone
- ✔ **Lambdoidal suture:** Joins the parietal bones and the occipital bone
- ✔ **Sagittal suture:** Between the parietal bones
- ✔ **Squamosal sutures:** Between the parietal and temporal bones

Slightly movable joints

Amphiarthroses are slightly movable joints connected by fibrocartilage or hyaline cartilage. Examples include the *intervertebral disks,* which join each vertebrae and allow slight movement of the vertebrae.

Freely movable joints

Diarthroses are joints that are freely movable. The numerous types of diarthroses are shown in Table 4-2.

Diarthroses are also called *synovial joints* because a cavity between the two connecting bones is lined with a synovial membrane and filled with *synovial fluid,* which helps to lubricate and cushion the joint. The ends of the bones are cushioned by hyaline cartilage. Diarthroses are joined together by ligaments (FCT).

Table 4-2	Types of Diarthroses (Synovial Joints)		
Type of Joint	*Description*	*Movement*	*Example*
Ball-and-socket joint	The ball-shaped head of one bone fits into a depression (socket) in another bone	Circular movements; joints can move in all planes, and rotation is possible	Shoulder, hip
Condyloid joint	Oval-shaped condyle of one bone fits into oval-shaped cavity of another bone	Can move in different planes but can't rotatet	Knuckles (joints between metacarpals and phalanges)
Gliding joint	Flat or slightly curved surfaces join	Sliding or twisting in different planes	Joints between carpal bones (wrist) and between tarsal bones (ankle)
Hinge joint	Convex surface joins with concave surface	Up and down motion in one plane, bend (flex) or straighten (extend)	Elbow, knee
Pivot joint	Cylinder-shaped projection on one bone is surrounded by a ring of another bone and ligament	Rotation is only movement possible	Joint between radius and ulna at elbow and joint atlas and axis at top of vertebral column
Saddle joint	Each bone is saddle shaped and fits into the saddle-shaped region of the opposite bone	Many movements are possible; can move in different planes but can't rotate	Joint between carpal and metacarpal bones of the thumb

Knowing what your joints can do

You know that certain types of joints can perform certain kinds of movements. The following list is a quick overview of those special movements. The two basic types of movements are angular and circular.

Angular movements make the angle formed by two bones larger or smaller. Examples of these include:

✔ **Abduction** moves a body part to the side, away from the body's middle. When you make a snow angel and you move your arms and legs out and up, that's abduction.

✔ **Adduction** moves a body part from the side toward the body's middle. When you're in snow angel position and you move your arms and legs back down, that's adduction.

✔ **Extension** makes the angle larger. Hyperextension occurs when the body part moves beyond a straight line (180 degrees).

✔ **Flexion** decreases the joint angle. When you flex your arm, you move your forearm to your arm.

Circular movements occur only at ball-and-socket joints like in the hip or shoulder. Examples include the following:

✔ **Circumduction** is the movement of a body part in circles.

✔ **Depression** is the downward movement of a body part.

✔ **Elevation** is the upward movement of a body part, such as shrugging your shoulders.

✔ **Eversion** only happens in the feet when the foot is turned so the sole is facing outward.

✔ **Inversion** also only happens in the feet when the foot is turned so that the sole is facing inward.

✔ **Rotation** is the movement of a body part around its own axis, such as shaking your head to answer, "No."

✔ **Supination** and **pronation** refer to the arm, and they stem from the terms *supine* and *prone.* Supination is the rotation of the forearm to make the palm face upward or forward. Pronation is the rotation of the forearm to make the palm face downward or backward.

The Axial Skeleton

The *axial skeleton* consists of the bones that lie along the midline (center) of your body, such as your vertebral column (backbone). An easy way to remember what bones make up the axial skeleton is to think of the vertebral column running down the middle of your body and then the bones that are directly attached to it — the thoracic cage (rib cage) and the skull. The hyoid bone, although it's not attached to another bone, is in line with the skull and vertebral column, so it's considered part of the axial skeleton.

The following sections give you a closer look at the main parts of the axial skeleton.

Keeping your head up: The skull

Rather than one big piece of bone, like a cap that fits over the brain, the skull comprises the cranium and the facial bones.

A human skull (see Figure 4-2) is made up of the *cranium,* which is formed of several bones, and the *facial bones.* The facial bones contain cavities called the *sinuses,* which have a purpose other than harboring upper respiratory infections.

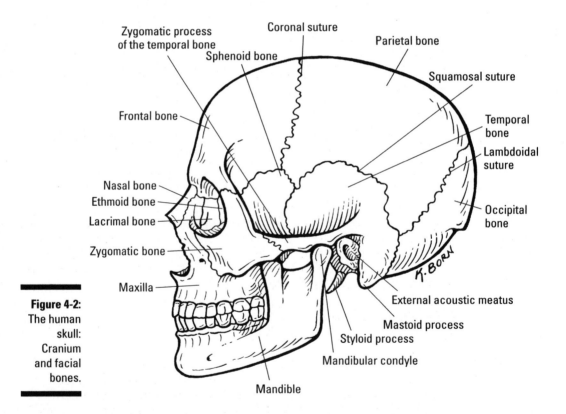

Figure 4-2:
The human skull: Cranium and facial bones.

Zygomatic process of the temporal bone

Coronal suture

Sphenoid bone

Parietal bone

Squamosal suture

Frontal bone

Temporal bone

Lambdoidal suture

Nasal bone

Ethmoid bone

Lacrimal bone

Occipital bone

Zygomatic bone

Maxilla

External acoustic meatus

Mastoid process

Styloid process

Mandibular condyle

Mandible

Cracking the cranium

The eight bones of your cranium protect your brain and have immovable joints between them called *sutures*. (These look a lot like the sutures, or stitches, that you may receive to close an incision or wound.) The bones of the cranium that are joined together by sutures include the following:

- **Frontal bone:** Gives shape to the forehead, the eye sockets, and part of the nose.

- **Parietal bones:** Two bones that form the roof and sides of the cranium.

- **Occipital bone:** Forms the back of the skull and the base of the cranium. The foramen magnum, an opening in the occipital bone, allows the spinal cord to pass into the skull and join the brain.

- **Temporal bones:** Form the sides of the cranium near the temples. The temporal bone on each side of your head contains the following structures:

  - **External auditory meatus:** The opening to your ear canal

  - **Mandibular fossa:** Articulates with the mandible (the lower jaw)

  - **Mastoid process:** Provides a place for neck muscles to join your head

  - **Styloid process:** Serves as an attachment site for muscles of the tongue and voice box (larynx)

- **Ethmoid bone:** Contains several sections, called *plates,* most of which form the nasal cavity.

- **Sphenoid bone:** Shaped like a butterfly or a saddle (depending on how you look at it), the sphenoid forms the floor of the cranium and the sides of the eye sockets (orbits). A central, sunken portion of the sphenoid bone called the *sella turcica* shelters the pituitary gland, which is very important in controlling major functions of the body. (See Chapter 8 for more on the pituitary gland.)

Facing the facial bones

The fairly small bones that form facial structures are

- **Lacrimal bones:** Two tiny bones on the inside walls of the orbits. A groove between the lacrimal bones in the eye sockets and the nose forms the *nasolacrimal canal.* Tears flow across the eyeball and through that canal into your nasal cavity, which explains why your nose "runs" when you cry.

- **Mandible:** The lower jaw and the only movable bone of the skull.

- **Maxillae:** Two of these bones form the upper jaw.

- **Nasal bones:** Two rectangular-shaped bones that form the bridge of your nose. The lower, movable portion of your nose is made of cartilage.

- **Palatine bones:** Form the hard palate (roof of your mouth). The top of the hard palate is the floor of the nasal cavity.

- **Vomer bone:** Joins the ethmoid bone to form the nasal septum — that part of your nose that can be deviated by a strong left hook.

- **Zygomatic bones:** Form the cheekbones and sides of the orbits.

Sniffing out the sinuses

The sinuses allow air into the skull, making it much lighter. The air in your sinuses also gives resonance to your voice, which means that when you talk, the sound waves reverberate in your sinuses.

Several types of sinuses are named for their location:

- **The frontal sinus** is a hollowed-out area in the frontal bone.

- **Mastoid sinuses** drain into the middle ear.

- **Maxillary sinuses** are large and within the bones of the upper jaw (the maxilla).

- **Paranasal sinuses** consist of the frontal, sphenoidal, and ethmoidal sinuses, which, along with the maxillary sinuses, drain into the nose (*para* means "near"; *nas-* means "nose").

Setting you straight on the curved spinal column

The spinal column (see Figure 4-3) begins within the skull and extends down to the pelvis. It's made up of 33 bones in all: 24 separate bones called *vertebrae* (singular, *vertebra*), plus the fused bones of the sacrum and the coccyx.

Your spinal column is the central support for the upper body, carrying most of the weight of your head, chest, and arms. Together with the muscles and ligaments of your back, your spinal column enables you to walk upright.

An important purpose of the vertebral column is to protect your spinal cord, the big data pipe between your body and your brain. Nearly all your nerves are connected, either directly or through networked branches, to the spinal cord, which runs directly into the brain through the opening in the skull called the *foramen magnum*. Turn to Chapter 7 for more information about the spinal cord.

The bone that floats

The *hyoid* bone, a tiny, U-shaped bone that resides just above your voice box (larynx), anchors the tongue and muscles used during swallowing. However, the hyoid bone itself isn't attached to anything. It's the only bone in the body that does not articulate (connect) with another bone. The hyoid bone hangs by ligaments attached to the styloid processes of the temporal bones.

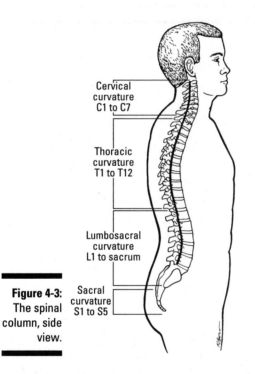

Cervical
curvature
C1 to C7

Thoracic
curvature
T1 to T12

Lumbosacral
curvature
L1 to sacrum

Sacral
curvature
S1 to S5

Figure 4-3:
The spinal
column, side
view.

If you look at the spine from the side, you notice that it curves five times: outward, inward, outward, inward, and outward. The curvature of the spine helps it absorb shock and pressure much better than if the spine were straight. A curved spine also affords more balance by better distributing the weight of the skull over the pelvic bones, which is needed to walk upright. A curved spine keeps you from being top-heavy. Each curvature spans a region of the spine: cervical, thoracic, lumbar, sacral, and coccygeal. The number of vertebrae in each region and some important vertebral features are given in Table 4-3.

Table 4-3	Regions of the Vertebral Column	
Region	*Number of Vertebrae*	*Features*
Cervical	7	The skull attaches at the top of this region to the vertebrae called the *atlas*.
Thoracic	12	The ribs attach to this region.
Lumbar	5	Commonly referred to as the small of the back, it takes the most stress.
Sacral	5 (fused into one; the sacrum)	The sacrum forms a joint with the hipbones and the last lumbar vertebra.
Coccygeal	4 (fused into one; the coccyx, also called the tailbone)	The coccyx is the remains of a tail, which was eliminated during evolution.

The vertebral column also provides places for other bones to attach. The skull is attached to the top of the cervical spine. The first cervical vertebra (abbreviated C-1; "C" for cervical, "1" for first) is the *atlas,* which supports the head and allows it to move forward and back (for example, the "yes" movement). The second cervical vertebra (C-2) is called the axis, and it allows the head to pivot and turn side to side (that is, the "no" movement).

You can differentiate these two important bones by recalling the Greek story about Atlas, who held the world on his shoulders. Your atlas holds your head on your shoulders.

Being caged can be a good thing

The rib cage (also called the *thoracic cage*) consists of the thoracic vertebrae, the ribs, and the sternum (see Figure 4-4). The rib cage is essential for protecting your heart and lungs and for providing a place for the pectoral girdle to attach.

Another kind of skeleton

Animals with backbones (*vertebrates* — yes, that's us) wear their skeleton on the inside, below layers of skin and muscle. This is called an *endoskeleton* ("inside skeleton"). Another very successful group of animals, the insects, wear their skeleton on the outside of their bodies. This is called an *exoskeleton*. Many *entomologists* (biologists who specialize in the study of insects) consider the exoskeleton to be a major factor in the evolutionary success of insects. At least 1 million known living species of insects exist.

The exoskeleton contains no bones or bone tissue. Its hardness and strength come from a substance called *chitin*. At the molecular level, chitin is a polysaccharide, similar to wood. At the tissue level, chitin fibers are embedded in a protein matrix, in much the same way that, in bone tissue, calcium compounds are embedded in a protein matrix. As wood is lighter than stone, chitin is much lighter than bone and amazingly strong, just what a tiny flying creature needs.

The insect's exoskeleton performs many of the some functions as the vertebrate's exoskeleton: It protects the internal organs and provides an attachment site for muscles, on the inside.

The exoskeleton also performs some of the functions performed in vertebrates by the skin: It has important roles in maintaining homeostasis (fluid balance, temperature regulation, excluding foreign matter, for example), and protects the insect from natural enemies. The exoskeleton has numerous types of sense organs for detecting light, pressure, sound, temperature, wind, and odor. These sense organs are distributed over the entire body, rather than being concentrated in the head, as in vertebrates. The exoskeleton also contains glands that produce a wide variety of substances, from a waxy covering for the exoskeleton itself to chemicals used in navigation, communication, and attracting mates.

You have 12 pairs of *bars* in your cage. Some of your ribs are *true,* some are *false,* and some *float.* All ribs are connected to the bones in your back (the thoracic vertebrae). In the front, true ribs are connected to the sternum (breastbone) by individual *costal cartilages* (*cost*- means "rib"); false ribs are connected to the *sternum* by fused *costal cartilage.* The last two pairs of ribs are called *floating ribs* because they remain unattached in the front. The floating ribs give protection to abdominal organs, such as your kidneys, without hampering the space in your abdomen for the intestines.

The sternum has three parts: the *manubrium,* the *body,* and the *xiphoid* (pronounced *zi*-foid) *process.* The notch that you can feel at the top center of your chest, in line with your collarbones (the *clavicles*), is the top of the manubrium. The middle part of the sternum is the body, and the lower part of the sternum is the xiphoid process.

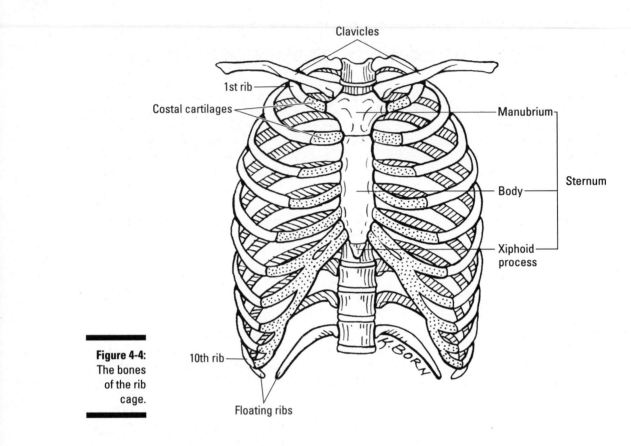

The Appendicular Skeleton

The *appendicular skeleton* is made up of the bones and joints of the *appendages* (upper and lower limbs) and the two *girdles* that join the appendages to the axial skeleton. We describe each of these categories in the following sections.

Wearing girdles: Everybody has two

The word *girdle* is a verb than means "to encircle." It has nothing to do with that funny undergarment all polite women wore in the early 20th century.

The body contains two girdles: the *pectoral girdle,* which encircles the vertebral column at the top, and the *pelvic girdle,* which encircles the vertebral column at the bottom. The girdles serve to attach the appendicular skeleton to the axial skeleton.

Why weight-bearing exercise is good for you

You probably know that aerobic exercise and lifting weights are good for your heart and your muscles, but did you realize it benefits your bones, too? Exercise — especially weight-bearing exercise (such as exercises that use the hips and legs, like walking, running, bicycling, and weight lifting) — increases the activity of osteoblasts, regardless of your age. Osteoblasts are the bone-building cells that mature into osteocytes. So if you're forming more bone cells, your bones are getting stronger. Exercise staves off osteoporosis, which is the depletion of osteocytes (bone cells) and the weakening of bones. And exercising the muscles also exercises the bones and joints, which maintains flexibility and strength. Nothing beats a firm foundation!

The pectoral girdle consists of the two clavicles (collarbones) and the two triangle-shaped *scapulae* (shoulder blades). The scapulae provide a broad surface to which arm and chest muscles attach. Refer to Figure 4-4 earlier in the chapter to see the individual parts of the pectoral girdle.

The clavicles are attached to the sternum's *manubrium* (breastbones). Significantly, this is the only point of attachment of the pectoral girdle and the axial skeleton. Because of this relatively weak attachment, the shoulders have a wide range of motion but are prone to dislocation.

The pelvic girdle (see the pelvis in Figure 4-4) is formed by the hipbones (called *coxal bones*), the *sacrum,* and the *coccyx* (tailbone). The hipbones bear the weight of the body, so they must be strong.

The hipbones (coxal bones) are formed by the *ilium,* the *ischium,* and the *pubis.* The ilia are what you probably think of as your hipbones; they're the large, flared parts that you can feel on your sides. The part that you can feel at the tip of the ilium is the *iliac crest.* In your lower back, the ilium connects with the vertebral column at the sacrum; the joint that's formed is appropriately called the *sacroiliac joint* — a point of woe for many people with lower back pain.

The *ischium* is the back part of your hip. You have an ischium on each side, within each buttock. You're most likely sitting on your *ischial tuberosity* right now. These parts of your hips are also called the *sitz bones* because they allow you to sit. The *ischial tuberosity* points outward, and the *ischial spine* — which is around the area where the ilium and ischium join — is directed inward into the pelvic cavity. The distance between a woman's ischial spines is key to her success in delivering an infant vaginally (see Chapters 14 and 15); the opening between the ischial spines must be large enough for a newborn's head to pass through.

Why women have bigger hips than men

Okay. You know it's true. Women aren't built like men. Most men tend to be straight up and down with few curves. Women, on the other hand, are hourglasslike — their hips tend to be wider than men's hips. In a woman, the iliac bones flare wider than they do in a man. And the *true pelvis* of a woman — the ring formed by the pubic bones, ischium, lower part of the ilium, and the sacrum — is wider and more rounded. The true pelvis of a man is shaped somewhat like a funnel.

These differences in anatomy have a physiological purpose: Women have babies, and when those babies are ready to be born, they need to pass through a woman's true pelvis without getting stuck. Other differences also relate to giving birth: The sacrum in women is wider and tilted back more so than in men; the coccyx of women moves easier than it does in men. These two features allow a little more "give" when a baby is passing through the pelvis. When a woman is pregnant, she creates a hormone called *relaxin* that allows the ligaments connecting the pelvic bones to relax a little. Thus, the bones can spread farther apart and flex a bit during delivery.

All these female characteristics are wonderful for giving birth, but as a woman who has done so several times, my experience leads me to believe that the bones don't return to their original positions as readily. The hips tend to stay a bit wider after giving birth. Maybe the bones staying loose and spreading out is a physiological "reward," making delivery of a subsequent infant even easier. Too bad female bodies don't know when that last infant has been delivered so the hips can go back to their pre-pregnant sizes!

Going out on a limb: Arms and legs

Your arms and legs are *limbs* or appendages. The word *append* means to attach something to a larger body. Your appendages are attached to the axial skeleton by the girdles (see the preceding section).

Giving you a hand with hands (and arms and elbows)

Your upper limb or arm is connected to your pectoral girdle. The bones of your upper limb include the *humerus* (arm), the *radius* and *ulna* (forearm), the *carpals* of the wrist, and the hand, which is made up of the *metacarpal bones* and the *phalanges* (refer to Figure 4-4).

Your humerus connects to your scapula at the *glenoid cavity*. Muscles that move the arm and shoulder attach to the *greater* and *lesser tubercles,* two points near the head (shoulder end) of the humerus. The greater tubercle is larger than the lesser tubercle. Between the greater and lesser tubercles is the *intertubercular groove,* which holds the tendon of the biceps muscle to the humerus bone. The humerus also attaches to the deltoid muscle of the shoulder at a point called the *deltoid tuberosity.* The muscle attached to the deltoid tuberosity allows you to raise and lower your arm.

The bones of the forearm attach at the elbow end of your humerus in four different spots:

- ✔ **Capitulum:** Condyle (knob) that allows the radius to articulate with the humerus.

- ✔ **Trochlea:** Condyle on the humerus that lies next to the capitulum and allows the *trochlear notch* of the ulna to articulate with the humerus.

- ✔ **Coronoid fossa:** Depression in the humerus that accepts a projection of the ulna bone (called the *coronoid process*) when the elbow is bent.

- ✔ **Olecranon fossa:** Depression in the humerus that accepts a projection of the ulna (called the *olecranon process*) when the arm is extended. Fitting, isn't it?

The radius is the bone on the thumb side of your forearm. When you turn your forearm so that your palm is facing backward, the radius crosses over the ulna so that the radius can stay on the thumb side of your arm. The radius is shorter but thicker than the ulna. The head of the radius looks like the head of a nail. The ulna is long and thin, and its head is at the opposite end of the bone compared with the head of the radius.

Both the radius and ulna connect with the bones of the wrist. The wrist contains eight small, irregularly shaped bones called the *carpal bones.* The ligaments binding the carpal bones are very tight, but the numerous bones allow the wrist to flex easily. The eight carpal bones are the *pisiform, triangular, lunate, scaphoid, trapezium, trapezoid, capitate,* and *hamate.* The palm of your hand contains five bones called the *metacarpals.* When you make a fist, you can see the ends of the metacarpals as your knuckles. Your fingers are made up of bones called the *phalanges;* each finger has three phalanges (phalanx is singular): the *proximal phalanx,* which joins your knuckle, the *middle phalanx,* and the *distal phalanx,* which is the bone in your fingertip. The thumb has two phalanges, so sometimes it's not considered a true finger. So you may have eight fingers and two thumbs or ten fingers depending on how you look at it.

Getting a leg up on your lower limbs

Your lower limb consists of the *femur* (thigh bone), the *tibia* and *fibula* of the leg, the bones of the ankle *(tarsals),* and the bones of the foot *(metatarsals* and *phalanges;* refer to Figure 4-4).

The term *phalanges* refers to the finger bones *and* the toe bones.

The femur is the strongest bone in the body; it's also the longest. The head of the femur fits into a hollowed out area of the hip bone called the *acetabulum.* In women, the acetabula (plural) are smaller but spread farther apart than in men. This anatomic feature allows women to have a greater range of movement of the thighs than men. The *greater* and *lesser trochanters* of the femur are surfaces to which the muscles of the legs and buttocks attach.

Trochanters are large processes found only on the femur. The *linea aspera* is a curve along the back of the femur to which several muscles attach.

The femur forms the knee along with bones of the leg. The *patella* (commonly known as the kneecap) articulates with the bottom of the femur. The femur also has knobs (*lateral* and *medial condyles*) that articulate with the top of the tibia. The ligaments of the patella attach to the tibial tuberosity. The bottom, inner end of the tibia has a bulge called the *medial malleolus,* which forms part of the inner ankle.

The tibia, also called the shinbone, is much thicker than the fibula and lies on the inside (medial) portion of the leg. Although the fibula is thinner, it's about the same length as the tibia. The bottom, outside end of the fibula is the *lateral malleolus,* and it is the bulge of the outside of your ankle.

Your foot is designed in much the same way as your hand. The ankle, which is akin to the wrist, consists of seven tarsal bones. Altogether, the bones of the ankle are called the *tarsus,* but only one of those seven bones is part of a joint with a great range of motion — the *talus.* The talus bone joins to the tibia and fibula and allows your ankle to rotate. The largest tarsal bone is the *calcaneus,* which is the heel bone. The calcaneus and talus help to support your body weight.

The instep of your foot is akin to the palm of your hand, and just as the hand has metacarpals and phalanges, the foot has metatarsals and phalanges. The ends of the metatarsals on the bottom of your foot form the ball of your foot. As such, the metatarsals also help to support your body weight. Together, the tarsals and metatarsals held together by ligaments and tendons form the arches of your feet. Your toes are also called phalanges, just like your fingers. And, just as your thumbs have only two phalanges, your big toes have only two phalanges. But, the rest of your toes have three: proximal, middle, and distal.

Strong feet, firm foundation

The arches of your feet help to absorb the shock created by the pounding of your feet when walking or running, and they help to distribute your weight evenly to the bones that bear the brunt of it: the *calcaneus* (heel) and *talus* (ankle) in each foot. The heel, ankle, and metatarsals bear a significant amount of your body weight and are thus called *weight-bearing bones.* If the ligaments and tendons that form those arches along with the tarsals and metatarsals weaken, the arches can collapse, resulting in flat feet. A person with flat feet is more prone to damage to the feet bones (such as the metatarsals) because of increased pressure placed on those bones. Examples of damage include bunions, which is a painful displacement of the first metatarsal (the big toe), and heel spurs, which are bony outgrowths on the calcaneus that cause pain when walking. Flat feet also can contribute to knee pain, hip pain, and lower back pain.

Pathophysiology of the Skeletal System

Bones and joints are very strong, but they're prone to injuries, the effects of aging, and disease, just like any other body part. This section gives you some information on a few of the most common problems that occur in bones and joints.

Abnormal curvature

Abnormal curvatures of the spine can cause plenty of pain and can lead to several problems. When the curve of the lumbar spine is exaggerated, the abnormal condition is *lordosis,* more commonly known as *swayback.* The lumbar spine of a pregnant woman becomes exaggerated because the woman needs to balance the pregnant belly on her frame. However, sometimes the curve remains after pregnancy, when weakened abdominal muscles fail to support the lumbar spine in its normal position. Developing the habit of holding the abdominal muscles in (rather than letting it all hang out, so to speak) helps to strengthen the body's center and prevent swayback. Losing the beer belly helps, too.

Older men and women sometimes develop a condition called *kyphosis* (commonly known as *hunchback*), an abnormally curved spine in the thoracic region. Osteoporosis or normal degeneration and compression of the vertebrae tends to straighten the cervical and lumbar regions of the spine and push out the thoracic vertebrae, thus causing kyphosis.

You may recall being checked for *scoliosis* during junior-high gym class. The reason for that inspection is because scoliosis (twisted spine) first becomes obvious during the late childhood/early teen years — just when people are most self-conscious. When you look at the spine from the back, it appears to be straight — the normal curvature is evident when you view the spine from the side. However, in people with scoliosis, the spine curves side to side and looks S-shaped when viewed from the back.

Osteoporosis

Osteoporosis is a disease in which bones become fragile, progressively and painlessly. To some extent, the process is nearly inevitable with age. The continuous process of bone resorption continues but the osteoblast (bone-building cells) are less and less active, so more bone is lost than replaced. Osteoporosis occurs most often in postmenopausal women because they lose the protective effect of estrogen on the bones.

Osteoporosis affects all bones, but of special concern are fractures of the hip and spine. A hip fracture almost always requires hospitalization and major surgery. It can impair a person's ability to walk unassisted and may cause prolonged or permanent disability or even death. Spinal or vertebral fractures also have serious consequences, including loss of height, severe back pain, and deformity.

Cleft palate

A *cleft palate* is a relatively common birth defect that occurs when the *palatine bones* (a pair of the facial bones) fail to fuse during fetal development. This defect creates a problem in which the nasal cavity and oral cavity are open to each other. This problem can affect the palatine bones only or can be part of a syndrome of development problems. Cleft palate, also called *harelip,* is treated with surgery, usually when the child is very young.

Arthritis

Arthritis is a name for any of numerous conditions characterized by inflammation of the joints. The inflammation is painful in itself, and it also makes movement difficult and painful. The chronic inflammation can eventually erode the joint's tissues (bone and cartilage). Treatment consists of controlling pain, reducing inflammation, and slowing the progress of joint damage.

Arthritis conditions are closely associated with immunity: Inflammation is a normal response of the immune system, but chronic inflammation of the joints is pathophysiological. Several arthritis conditions are autoimmune disorders. (See Chapter 13 for a discussion of autoimmunity.) Here are the common forms of arthritis:

- ✔ **Osteoarthritis (OA)** is the most common form, and not only in humans — it reportedly affects other species, too. As the joints age and the ravages of normal use accumulate, low-level inflammation sets in. Eventually, the inflammation causes the joint's cartilage to become thinner and lose elasticity. It can affect people at any age. Many people develop some osteoarthritis of the finger joints in late middle age.

- ✔ **Rheumatoid arthritis (RA)**, an autoimmune condition, starts with inflammation of the *synovium* (joint lining). Later, often years later, the inflamed cells of the synovium begin to produce enzymes that actively destroy both bone and cartilage, restricting movement and increasing pain further.

✔ **Juvenile arthritis (JA)**, the most common form affecting children under 16, is an autoimmune condition. Most likely, JA is several different autoimmune conditions that vary in the number of affected joints and the age of onset. As with other forms of arthritis, the symptoms are inflammation, joint pain, and stiffness. Sometimes, JA causes the limbs to grow to different lengths.

✔ **Ankylosing spondylitis (AS)** affects the spine and the sacroiliac joints. The severity of the pain and inflammation varies, but in its worst form, the chronic inflammation can cause the spine to fuse into a rigid, brittle column, prone to fracture. The eyes, heart, lung, and kidneys can also be affected.

✔ **Gout** is a form of arthritis caused by crystallized deposits of uric acid in the joints. Think "sand in the gears." As the uric acid crystals fill the joints, they damage cartilage, synovial membranes, tendons, and even the muscles adjoining the bone. Complications include kidney stones, nerve damage, and circulatory problems. Drugs are available that reduce the amount of uric acid in the blood, preventing its deposition in the joints.

Chapter 5

Muscles: Setting You in Motion

In This Chapter

▶ Seeing how your muscular system moves you

▶ Differentiating the three types of muscle tissue

▶ Understanding how muscles contract

▶ Taking a tour of the skeletal muscles

▶ Looking at some skeletal muscle afflictions

M uscle tissues always have work to do. They pull things up and push things down and around. They move things inside you and outside you. And they move you, of course. They work together with all the other systems in your body, but more than any other organ system, the muscular systems specialize in movement: of food and air into and out of your body; of the circulation of your blood within your body; of different parts of your body relative to one another, as when you change position; and of your body through space — what you normally think of as "movement."

All muscle tissue is strong. Most is enduring, some of it astoundingly so. Its cells are crowded with *mitochondria,* thousands of little factories constantly turning out molecules of *ATP,* a refined fuel. The muscle cells use the fuel to manufacture strong and flexible proteins, with which they build and repair themselves and do work.

All the work that muscle tissues do is done by the coordinated contraction and release of millions of *sarcomeres,* tiny structures within the muscle cells. Muscle activity accounts for most of the body's energy consumption.

In this chapter, we give you an overview of some of the things muscles do and how they do it, and then we name the muscles. At the end of the chapter, we list some common muscle ailments, one of which you've probably experienced.

Functions of the Muscular System

This section focuses mainly on the functions of the *skeletal muscles,* the muscles that move your bones. The skeletal muscles comprise a substantial portion of your body mass, and most of what you eat goes to fuel their metabolism. In this section, you'll find out something about what they do with all that energy.

Supporting your structure

Muscles are attached to bones on the inside of your body and skin on the outside, with various types of connective tissue between the layers. Thus, they hold your body together. Along with your skin and your skeleton, your muscles shield your internal organs from injury from impact or penetration.

Moreover, and don't take this personally, but your body is heavy. As it does with everything else, gravity pulls your weight downward (toward the planet's center). But gravity doesn't only pull on the soles of your feet — it pulls on all your weight. If gravity had its way, you'd be lying on the floor right now. Your muscles pull your weight up ("oppose" the pull of gravity) and hold you upright. Gravity is a relentless cosmic force, and eventually, gravity will win. But while you're still fighting, your muscles need fuel and rest.

Moving you

Contracting and releasing a muscle moves the bone it's attached to relative to the rest of the body. The movement of the bone, in turn, moves all the tissue attached to it through space, as when you raise your arm. Certain combinations of these types of movements move the entire body through space, as when you walk, run, swim, skate, or dance.

Muscle contraction is responsible for little movements, too, like blinking your eyes, dilating your pupils, and smiling.

Poised positioning

A very close interaction outside of your conscious control between some muscle cells and the nervous system keeps you not just upright, but in balance. Nerve impulses throughout the muscular system cause muscles to contract or relax to oppose gravity in a more subtle way when, say, you're shifting your weight from one side to the other as you step. This interaction is called *muscle tone.*

When you step down a steep incline in rough terrain, your muscle tone brings your abs and your back muscles into action in a different way than when you step across your living room rug. The mechanisms of muscle tone may move your arms up and away from your body to counterbalance the pull of gravity with an accuracy and precision you could never calculate cognitively. Below your conscious level, the mechanisms of muscle tone are active every minute of every day, even when you're asleep.

Muscle tone relies on *muscle spindles* — specialized muscle cells that are wrapped with nerve fibers. (See the "Skeletal muscle" section later in this chapter for information about spindle fibers.) The central nervous system stays in contact with the muscles through the muscle spindles. (Turn to Chapter 7 for the structures of the central nervous system.) Spindles send messages about your body position through the spinal cord to the brain; to initiate the fine adjustments, the brain sends signals through the spinal cord and nerves to the muscle spindles about which muscles to contract and which to release.

Maintaining body temperature

Muscle contributes to *homeostasis* (see Chapter 2) by generating heat to balance the loss of heat from the body surface. Muscle contraction uses energy from the breakdown of ATP and generates heat as a byproduct. The heat generated by the muscles interacts with other physiological processes that release heat from the body — sweating, for example — to maintain thermoregulation. Shivering is a series of muscle contractions that generate extra heat to keep your temperature up in cold situations.

Pushing things around inside

The other two types of muscle tissues, smooth muscle and cardiac muscle, have their own important functions, discussed in other chapters. Here, we give you an overview of some of the muscles that keep things moving within your body, all without any thought from you.

Cardiac muscle

The heart is a muscle that contracts rhythmically, pumping blood into the arteries. A muscular lining rhythmically dilates and contracts the arteries, pushing blood along with enough force (blood pressure) to drive this relatively viscous fluid out into the capillary beds. The rhythmic movements of the heart and arteries are detectable as your pulse. The ability of the arterial wall muscles to dilate and contract in response to physiological stimuli enables the subtle control of blood pressure. Damage to this muscular lining causes *arteriosclerosis* (hardening of the arteries), which inhibits the muscle lining's ability to move (dilate and contract) the vessel, an underlying factor in much cardiovascular disease and dysfunction. See Chapter 9 for more information about the circulatory system.

Diaphragm

The *diaphragm* is a skeletal muscle whose contraction and release forces air in and out of the lungs. Chapter 10 covers the respiratory system in all its breathtaking glory.

Digestive smooth muscle

Your digestive system is lined with a kind of muscle tissue called smooth muscle that contracts in pulsating waves, pushing ingested material along the digestive tract. Think of this muscular lining as a conveyor belt on a disassembly line. Refer to Chapter 11 for details.

Controlling release

Sphincter muscles are essentially valves: rings of smooth muscle that are fully in contraction in their resting state, holding some material in one place, and then relaxing only briefly to allow the material to move through. You find sphincters at various places in the digestive system (refer to Chapter 11), from the very beginning to the very end, and in other parts of the body as well.

Most sphincters aren't under conscious control. Two of them — the urinary sphincter, which holds urine in the bladder, and the anal sphincter, which holds feces in the colon — come under conscious control usually at around 2 years of age. This control allows the release of these bodily wastes under culturally appropriate circumstances. Its acquisition is considered a milestone in infant development.

By the way, human males have much stronger urinary sphincter muscles than do females, meaning that they can retain about twice as much urine in the bladder (up to 800mL) for twice as long. Kindly keep that in mind, fellas, when you're on a road trip with girls.

Talking about Tissue Types

A "muscle tissue type" is not the same as "a muscle." Your left bicep is a muscle; in all, you have hundreds of named muscles. (Refer to the "Naming the Skeletal Muscles" section later in the chapter and to the "Muscular System" color plate in the center of the book.) There are only three muscle tissue types: *skeletal muscle tissue, cardiac muscle tissue,* and *smooth muscle tissue.*

Defining unique features of muscle cells

Your muscle tissue is made up of cells that are different from the other cells of your body. These cells are so unique that they're even different from each other, based on the type of muscle tissue they belong to. The three muscle types are distinguishable anatomically by their characteristic cells and structures and physiologically as *voluntary* or *involuntary*.

Muscle cells feature these characteristics:

- ✔ **Single or multiple nuclei:** Cardiac muscle cells and smooth muscle cells have one nucleus apiece, like most other cells. Skeletal muscle cells (fibers) are *multinucleate,* meaning numerous nuclei are found within one cell membrane. Skeletal muscle cells don't grow extra nuclei; during the development of skeletal muscle tissue, numerous skeletal muscle cells merge into one large cell, and most of the nuclei are retained within one continuous cell membrane, along with most of the mitochondria.

- ✔ **Striation:** Skeletal muscle is *striated,* meaning that, under a microscope, alternating light and dark bands are visible in the fiber (muscle cell). Striation is the result of the subcellular structure of skeletal muscle cells (see the "Skeletal Muscle" section later in the chapter) and is integral to the mechanism of contraction called the *sliding filament model.* (See the "Getting a Grip on the Sliding Filament" section later in the chapter). Cardiac muscle cells are striated as well, and they also contract by a variation of the sliding filament model. Smooth muscle cells are not striated in appearance but follow a version of the sliding filament model as well.

See Figure 5-1 to get an idea of how muscle cells and tissues are similar and different.

Muscle cells can also be categorized by the type of contraction they perform. Smooth and cardiac muscle cells are *involuntary,* meaning their contraction is initiated and controlled by parts of the nervous system that are far from the conscious level of the brain. You have no practical way to consciously control, or even become aware of, the smooth muscle contractions in your stomach that are grinding up this morning's muffin. The involuntary contraction known as the *heartbeat* isn't even under the nervous system's control, as we discuss in Chapter 9.

Skeletal muscle is classified as *voluntary* because you make a decision at the conscious level to move the muscle. At least you do sometimes — you decide to reach for a doorknob and turn it, for example, and your muscles carry out the command from your brain to do so.

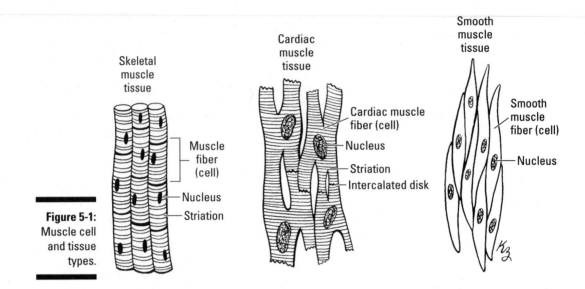

Figure 5-1:
Muscle cell
and tissue
types.

But note this: If the doorknob is charged with static electricity, your arm pulls your hand away before you're even consciously aware of being zapped. This *somatic reflex arc* is still classified as voluntary movement, however, because it involves skeletal muscle, which is voluntary muscle.

Not everything in anatomy and physiology makes sense at first. Just remember that skeletal muscle is classified as voluntary.

Table 5-1 sums up the characteristics and classifications of muscle cells.

Table 5-1	Cell Characteristics of Muscle Cells		
	Skeletal	*Cardiac*	*Smooth*
Multinucleate	Yes	No	No
Striated	Yes	Yes	No
Voluntary	Yes	No	No

Skeletal muscle

Skeletal muscle tissue is, essentially, bundles of fibers bundled together. Like fibrous material of every kind, skeletal muscle tissue gets its strength from assembling individual fibers together into strands, and then bundling and rebundling the strands. Two properties make this particular fibrous material very special: The strands are made of protein, and they renew and repair themselves constantly. Refer to Figure 5-1 for details of skeletal muscle tissue anatomy.

At the cellular level

Individual muscle cells, which physiologists call *fibers*, are slender cylinders that sometimes run the entire length of a muscle. Each fiber (cell) has many nuclei located along its length and close to the cell membrane, which is called the *sarcolemma* in skeletal muscle fibers. Outside the sarcolemma is a lining called the *endomysium*, a type of connective tissue.

Muscle *spindles* are specialized skeletal muscle fibers that are wrapped with nerve fibers. Figure 5-2 shows how skeletal muscle is connected to the nervous system. Spindles are distributed throughout the muscle tissue and provide sensory information to the central nervous system.

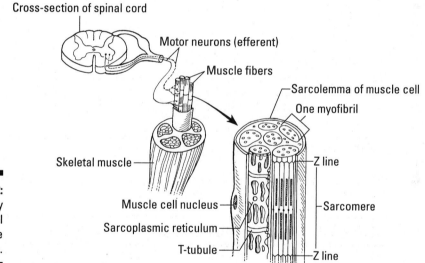

Figure 5-2:
Anatomy of skeletal muscle tissue.

Within the muscle fibers are *myofibrils.* The myofibrils are composed of *sarcomeres*, which are distinct units arranged linearly (end to end) along the length of the myofibril. Within the sarcomere is where muscle contraction actually happens. (Refer to the "Getting a Grip on the Sliding Filament Model" section later in the chapter for more on muscle contraction within sarcomeres.)

At the tissue level

Muscle fibers are bound together into bundles called *fascicles.* Each fascicle is bound by a connective-tissue lining called a *perimysium.* Spindle fibers are distributed throughout each fascicle. The fascicles are then bound together to form a muscle, a discrete assembly of skeletal muscle tissue, like the *biceps brachii* (your biceps), with a connective-tissue wrapper called an *epimysium* holding the whole package together.

Tendons — ropy extensions of the connective tissue covering the skeletal bones — grow into the epimysium, holding the muscle firmly to the bone. (See Chapter 4 for more on connectivity in the skeletal system.)

How many ways can you say "fiber"? Anatomists need them all when they're talking about the muscular system. Make sure you're thinking at the right level of organization *(subcellular, cellular,* or *tissue)* when you see these terms: *filament, myofibril, fiber,* and *fascicle.*

Working together: Synergists and antagonists

Groups of skeletal muscles that contract simultaneously to move a body part are said to be *synergistic.* The muscle that does most of the moving is the *prime mover.* The muscles that help the prime mover achieve a certain body movement are *synergists.* When you move your elbow joint, the bicep is the prime mover and the brachioradialis stabilizes the joint, thus aiding the motion.

Antagonistic muscles also act together to move a body part, but one group contracts while the other releases, a kind of push-pull. One example is flexing your arm. When you bend your forearm up toward your shoulder, your biceps muscle contracts, but the triceps muscle in the back of your arm relaxes. The actions of the biceps and triceps muscles are opposite, but you need both actions to allow you to flex your arm. Antagonistic actions lower your arm, too: The biceps relaxes, and the triceps contracts.

Cardiac muscle

The heart has its own very special type of muscle tissue, called *cardiac muscle.* The cells (fibers) in cardiac muscle contain one nucleus (they're *uninucleated)* and are cylindrical; they may be branched in shape. Unlike skeletal muscle, where the fibers lie alongside one another, cardiac muscle fibers interlock, which promotes the rapid transmission of the contraction impulse throughout the heart. Cardiac muscle cells are striated, like skeletal muscle cells, and cardiac muscle contraction is involuntary, like smooth muscle contraction. Cardiac muscle fibers contract in a way very similar to skeletal muscle fibers, by a sliding filament mechanism (more on that in a minute).

Cardiac muscle tissue is on the job, day and night, from before birth to the moment of death. The cardiac muscle cells contract regularly and simultaneously hundreds of millions of times throughout your lifetime. When cardiac muscle tissue gives up, the game is over.

Unlike skeletal muscle and smooth muscle, contraction of the heart muscle is *autonomous,* which means it occurs without stimulation by a nerve. In between contractions, the fibers relax completely.

Smooth muscle

Smooth muscle tissue lines the organs and structures of many organ systems, including the digestive system, the urinary system, the respiratory system, the circulatory system, and the reproductive system. Smooth muscle tissue is fundamentally different from skeletal muscle tissue and cardiac muscle tissue in terms of cell structure and physiological function. However, smooth muscle sarcomeres are similar, and contraction is effected by a variation of the sliding filament model.

Smooth muscle fibers (cells) are *fusiform* (thick in the middle and tapered at the ends) and arranged to form sheets of tissue. Smooth muscle cells aren't striated. However, smooth muscle contractions are affected by the same sliding filament mechanism as skeletal muscle cells (see the next section for more).

Smooth muscle contraction is typically slow, strong, and enduring. Smooth muscle can hold a contraction longer than skeletal muscle. In fact, some smooth muscles, notably the sphincters, are in a constant or nearly constant state of contraction. Childbirth is among the few occasions in life when humans (some humans, anyway) consciously experience smooth muscle contraction (although they don't consciously control it).

Getting a Grip on the Sliding Filament

A muscle contracts when all the sarcomeres in all the myofibrils in all the fibers (cells) contract all together. The *sliding filament model* describes the fine points of how this happens.

The key to the sliding filament model is the distinctive shapes of the protein molecules *myosin* and *actin,* and their partial overlap in the sarcomere. The special chemistry of ATP supplies the energy for the filaments' movement. The following sections explain how sarcomeres create muscle contraction.

The *sarcomere* is the functional unit within the *myofibril.* Sarcomeres line up end to end along the myofibril.

Assembling a sarcomere

The sarcomere is composed of *thick filaments* and *thin filaments.* The thick filaments are molecules of the protein *myosin,* which is dense and rubbery. The thin filaments are primarily made up of two strands of the lighter (less dense) protein *actin,* wrapped in a double helix which, as in DNA, is springy. The thin and thick filaments line up together in an orderly way to form a sarcomere. One end of a thin filament touches and adjoins the end of another thin filament.

Adjoined thin filaments adjust themselves so the joining points form a structure called a *Z line* — a straight line that runs perpendicular to the filament axis. The sarcomere begins at one Z line and ends at the next Z line. The thick filaments line up precisely between the thin filaments. Sarcomeres and Z lines are shown in Figure 5-1.

The two types of filaments overlap only partially when the sarcomere is at rest. The partial overlap gives skeletal muscle cells their *striations:* where thick and thin filaments overlap, the tissue appears dark (dark band); where only thin filaments are present, the tissue appears lighter (light band).

Contracting and releasing the sarcomere

Myosin molecules have binding sites with a high affinity for ATP. Actin molecules have binding sites with a high affinity for myosin. When a nerve impulse sends calcium ions into the cytoplasm of the muscle fiber, things start to happen.

The binding of calcium ions on actin molecules exposes the myosin binding sites of actin. Myosin, with its ATP cargo, binds to actin, forming cross-bridges between the thick and thin (actin) filaments of the sarcomere. The bond with actin distorts the myosin-ATP binding site, leading to the hydrolysis of the ATP and the release of its energy. This energy fuels the motion of the cross-bridges that pulls (slides) the thin filaments past the thick filaments toward the middle of the sarcomere. The distance between the Z lines becomes shorter because the length of the nonoverlapping portion of the two types of filaments becomes shorter. All sarcomeres in a fiber contract simultaneously, transmitting the force to the fiber ends.

The myosin then drops the products of the hydrolysis (ADP and P_i) from the binding site. Another molecule of ATP takes its place, reshaping the myosin molecule once again and pulling the actin-myosin bond apart. Now the cycle begins again.

Both the binding action and the release action require energy in the form of ATP: one molecule of ATP for each binding and each release of each filament pair within each sarcomere; thousands of molecules of ATP for every second of muscle contraction.

Muscle cells contain a large number of mitochondria, which produce ATP (see Chapter 2 for more information).

The last contraction

The time comes for every animal — humans included — to die. The fact that every animal gets cold and stiff tells others when that time has come. Do you know why? The cells no longer make ATP.

At the moment of death, the lungs stop filling with oxygen, the heart stops pumping blood through the body, and the brain stops sending signals. The cells — without incoming oxygen, nutrients, or stimulus from the brain — cease performing their metabolic reactions. So ATP can no longer be produced.

Without ATP flooding the myofibrils, contractions can't occur, but neither can the last step of muscle contraction. In order for a myofibril to relax, ATP must hook onto myosin and dissolve the actin-myosin cross-bridges. But when ATP is unavailable to generate a subsequent contraction, the last contraction becomes permanent, and the corpse stiffens. *Rigor mortis*, which means *rigidity of death*, occurs in every muscle throughout the body. And remember that movement of muscles generates heat, so when the muscles stop their physiological reactions and warm blood stops flowing through the blood vessels, the corpse gets cold.

Naming the Skeletal Muscles

Get ready, because we're about to tell you the muscle names from head to toe — literally. Check out the "Muscular System" color plate in the center of the book as you go through this section.

To name muscles, anatomists had to come up with a set of rules to follow so the names would make sense. They chose to focus on certain characteristics from which to derive each muscle's Latin name. As you go through the following sections, refer to Chapter 1 if necessary for information on anatomic position. Examples of characteristics in muscle names are given in Table 5-2.

Table 5-2	Characteristics in Muscle Names
Characteristics	*Examples*
Muscle size	The largest muscle in the buttocks is the *gluteus maximus* (*maximus* means *large* in Latin); a smaller muscle in the buttocks is the *gluteus minimus* (*minimus* means *small* in Latin).
Muscle location	The *frontalis muscle* lies on top of the skull's frontal bone.
Muscle shape	The *deltoid muscle,* shaped like a triangle, comes from *delta* — the Greek alphabet's fourth letter, which is also shaped like a triangle.

(continued)

Table 5-2 *(continued)*

Muscle action	The *extensor digitorum* is a muscle that extends the fingers or *digits.*
Number of muscle attachments	The *biceps brachii* attaches to bone in two locations, whereas the *triceps brachii* attaches to bone in three locations.
Muscle fiber direction	The *rectus abdominis muscle* runs vertically along your abdomen (*rectus* means *straight* in Latin).

Starting at the top

Your head contains muscles that perform three basic functions: chewing, making facial expressions, and moving your neck. Ear wiggling falls into this category, too.

To chew, you use the muscles of *mastication* (a big, fancy word that means "chewing"). The *masseter,* a muscle that runs from the *zygomatic arch* (your cheekbone) to the *mandible* (your lower jaw), is the prime mover for mastication, so its name is based on its action (*ma*sseter, *ma*stication). The fan-shaped *temporalis muscle* works with the masseter to allow you to close your jaw. It lies on top of the skull's temporal bone, so its name is based on its location. Figure 5-3 shows the muscles of the head and neck.

To smile, frown, or make a funny face, you use several muscles. The *frontalis muscle* along with a tiny muscle called the *corrugator supercilii* raises your eyebrows and gives you a worried or angry look when wrinkling your brow. (Think of the appearance of corrugated cardboard, and then feel the skin between your eyebrows when you wrinkle your brow.) The *orbicularis oculi muscle* surrounds the eye (the word *orbit,* as in *orbicularis,* means "to encircle"; *oculi* refers to the eye). This muscle allows you to blink your eyes and close your eyelids, but it also gives you those little crow's feet at the corners of your eyes. The *orbicularis oris* surrounds the mouth. (*Or* refers to mouth, as in "oral.") You use this muscle to pucker up for a kiss. Figure 5-4 shows the facial muscles.

If you play the trumpet or another instrument that requires you to blow out, you're well aware of what your *buccinator muscle* does. This muscle is in your cheek. (*Bucc* means "cheek," as in the word *buccal,* which refers to the cheek area.) It allows you to whistle and also helps keep food in contact with your teeth as you chew. Remember that your zygomatic arch is your cheekbone? Well, the *zygomaticus muscle* is a branched muscle that runs from your cheekbone to the corners of your mouth. This muscle pulls your mouth up into a smile when the mood strikes you.

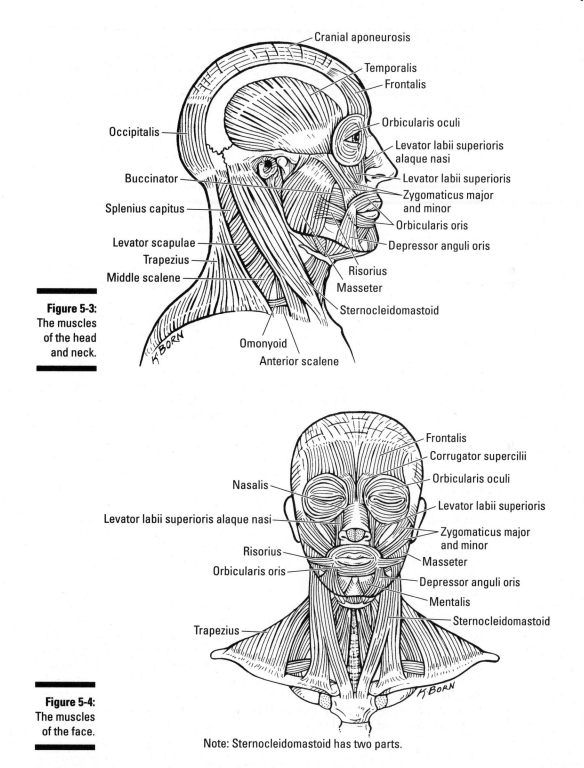

Figure 5-3:
The muscles
of the head
and neck.

Figure 5-4:
The muscles
of the face.

Note: Sternocleidomastoid has two parts.

When you want to nod *yes, no,* or tilt your head into a *maybe so,* your neck muscles come into play. You have two *sternocleidomastoid muscles,* one on each side of your neck. We know this is a long name, but the name reflects the locations of its attachments: the sternum, collarbone, and mastoid process of the skull's temporal bone. When both sternocleidomastoid muscles contract, you can bring your head down toward your chest and flex your neck. When you turn your head to the side, one sternocleidomastoid muscle contracts — the one on the opposite side of the direction your head is turned. So if you turn your head to the left, your right sternocleidomastoid muscle contracts, and vice versa. If you lean your head back to look up at the sky or to shrug your shoulder, your *trapezius muscle* allows you to do so.

The trapezius is an antagonist to the sternocleidomastoid muscle. If you remember basic geometry, a trapezoid is shaped like a diamond, and that's exactly what the trapezius looks like. It runs from the base of your skull to your *thoracic vertebrae* and connects to your shoulder blades. Therefore, the trapezius and sternocleidomastoid muscles connect your head to your torso and provide a nice segue to the next section. The trapezius is shown in both Figures 5-3 and 5-4.

Twisting the torso

The torso muscles have important functions. They not only give support to your body but also connect to your limbs to allow movement, allow you to inhale and exhale, and protect your internal organs. In this section, we cover the muscles that run along the front of you (called your anterior or ventral side) and then cover the muscles of your back (your posterior or dorsal side).

In your chest (see Figure 5-5), your *pectoralis major muscles* connect your torso at the sternum and collarbones to your upper limbs at the humerus bone in the upper arm. Your "pecs" also help to protect your ribs, heart, and lungs. You can feel your pectoralis major muscle working when you move your arm across your chest. Also in your chest are the muscles between and around the ribs. The *internal intercostal muscles* help to raise and lower your rib cage as you breathe. However, the torso's largest muscles are the abdominal muscles.

The *abdominal muscles* really form the center of your body. If the abdominal muscles are weak, the back is weak because the abdominal muscles help to flex the vertebral column. So if the vertebral column doesn't flex easily, the muscles attached to it can become strained and weak. And the muscles of the abdomen and back join to the upper and lower limbs. Therefore, if the abdomen and back are weak, the limbs can have problems.

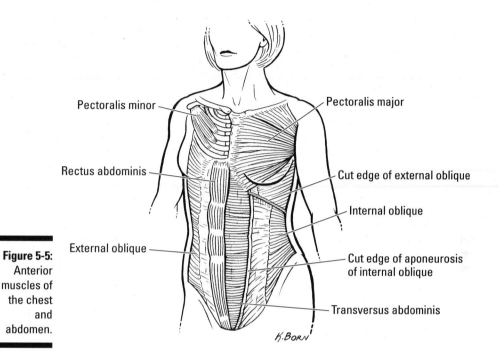

Figure 5-5:
Anterior
muscles of
the chest
and
abdomen.

Pectoralis minor

Pectoralis major

Rectus abdominis

Cut edge of external oblique

Internal oblique

External oblique

Cut edge of aponeurosis
of internal oblique

Transversus abdominis

K.BORN

The muscles of the abdomen are thin, but the fact that these muscle fibers run in different directions increases their strength. This woven effect makes the tissues much stronger than they would be if they all went in the same direction. Think about how a child connects building blocks. Laying a top layer of blocks perpendicular to the blocks underneath helps the structure stay together, which is similar to how the abdominal muscle tissues provide strength and stability.

The "six-pack" muscle of the abdomen, the *rectus abdominis,* forms the front layer of the abdominal muscles, and it runs from the pubic bones up to the ribs and sternum. The function of the rectus abdominis muscle is to hold in the organs of the abdomino-pelvic cavity and allow the vertebral column to flex.

Other layers of abdominal muscles also help to hold in your organs on the side of your abdomen and provide strength to your body's core. The *external oblique muscles* attach to the eight lower ribs and run downward toward the middle of your body (slanting toward the pelvis). The *internal oblique muscles* lie underneath the external oblique (makes sense, eh?) at right angles to the external oblique muscles. The internal oblique muscles extend from the top of the hip at the iliac crest to the lower ribs.

Together, the external and internal oblique muscles form an *X,* essentially strapping together the abdomen. The abdomen's deepest muscle, the *transversus abdominis,* runs horizontally across the abdomen; its function is to tighten the abdominal wall, push the diaphragm upward to help with breathing, and help the body bend forward. The transversus abdominis is connected to the lower ribs and lumbar vertebrae and wraps around to the pubic crest and *linea alba.* The linea alba ("white line") is a band of connective tissue that runs vertically down the front of the abdomen from the xiphoid process at the bottom of the sternum to the *pubic symphysis* (the strip of connective tissue that joins the hip bones).

The muscles in your back (refer to Figure 5-6) serve to provide strength, join your torso to your upper and lower limbs, and protect organs that lie toward the back of your trunk (such as your kidneys). The *deltoid muscle* joins the shoulder to the collarbone, scapula, and humerus. This muscle is shaped like a triangle (think of the Greek letter delta): △. The deltoid muscle helps you raise your arm up to the side (that is, laterally). The *latissimus dorsi muscle* is a wide muscle that's also shaped like a triangle. It originates at the lower part of the spine (thoracic and lumbar vertebrae) and runs upward on a slant to the humerus. Your "lats" allow you to move your arm down if you have it raised, and also to reach, such as when you're climbing or swimming.

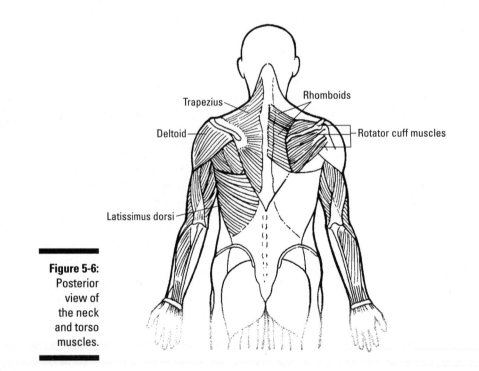

Figure 5-6:
Posterior view of the neck and torso muscles.

Spreading your wings

Your upper limbs have a wide range of motions. Obviously, your upper limbs are connected to your torso. One of the muscles that provides that connection, the *serratus anterior,* is below your armpit (the anatomic term for armpit is *axilla*) and on the side of your chest. The serratus anterior muscle connects to the scapula and the upper ribs. You use this muscle when you push something or raise your arm higher than horizontal. Its action pulls the scapula downward and forward.

Although the *biceps brachii* and *triceps brachii* are muscles located in the top (anterior) part of your upper arm, their actions allow your forearm (lower arm) to move. Figure 5-7a shows an anterior view of the upper limb. The name *biceps* refers to this muscle's two origins (points of attachment); it attaches to the scapula in two places. From there, it runs to the radius of the forearm (its point of insertion). The triceps brachii is the only muscle that runs along the back (posterior) side of the upper arm. Figure 5-7b shows a posterior view of the upper limb. The name *triceps* refers to the fact that it has three attachments: one on the scapula and two on the humerus. It runs to the ulna of the forearm. You can feel this muscle in motion when you push or punch. Other muscles of the arm include the *brachioradialis,* which helps you flex your arm at the elbow, and the *supinator,* which rotates your arm from a palm-down position to a palm-up position (remember the word supinator has the word "up" in it).

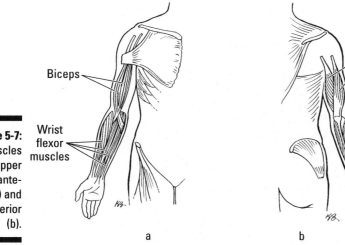

Figure 5-7:
The muscles of the upper limb: anterior (a) and posterior (b).

Biceps

Wrist flexor muscles

Triceps

Wrist extensor muscles

a

b

Thumbs up!

A key feature of all primates is a *prehensile thumb;* that is, a thumb that's adapted for grasping objects. Many animals have digitlike structures, but only primates can grasp things with their hands. And the only way to grasp things is to have a thumb.

Imagine having webbing between your four fingers so that you couldn't spread them apart; you wouldn't be able to pick things up. That's why animals such as dogs, cats, and birds hold things in their mouth (or beak). But primates — apes, monkeys, and humans — can easily grasp things between their thumb and fingers. However, of those primates, only humans have an *opposable thumb* (one that can touch each of the other fingers; the thumb can be "opposite" from each finger).

Because of the ability to oppose the thumb to each finger, the muscles in your digits are capable of performing minute movements. As you touch your thumb to your pinky finger, your palm becomes arched, which only happens in humans because of short bones in the pinky and an opposable thumb.

Your hand contains muscles that perform the fine movements of your fingers. When you type or play the piano, you're using your *extensor digitorum* and *flexor digitorum muscles* to raise and lower your fingers onto the keyboard and move them to the different rows of keys. As you lift your hands off the keyboard, the muscles of your wrist kick into gear. The *flexor carpi radialis* (attached to the radius bone) and *flexor carpi ulnaris* (attached to the ulna bone) allow your wrist to flex forward or downward. The *extensor carpi radialis longus* (which passes by the carpal bones), the *extensor carpi radialis brevis,* and the *extensor carpi ulnaris* allow your wrist to extend; that is, bend upward.

Getting a leg up

Your lower limbs are connected to your buttocks, and your buttocks are connected to your hips. The *iliopsoas muscle* connects your lower limb to your torso and consists of two smaller muscles: the *psoas major,* which joins the thigh to the vertebral column, and the *iliacus,* which joins the hipbone's ilius to the thigh's femur bone. Originating on the iliac spine of the hip and joining to the inside surface of the tibia (a bone in your shin), the *sartorius muscle* is a long, thin muscle that runs from the hip to the inside of the knee (see Figure 5-8a). These muscles stabilize your lower limbs and provide strength for them to support your body's weight and to balance your body against the pressure of gravity.

Some muscles in the lower limb allow the thigh to move in a variety of positions. The buttocks muscles allow you to straighten your lower limb at the hip and extend your thigh when you walk, climb, or jump. The *gluteus maximus* — the largest muscle in the buttocks — is the largest muscle in the body (see Figure 5-8b). The gluteus maximus is antagonistic to the stabilizing *iliopsoas muscle,* which flexes your thigh. The *gluteus medius muscle,* which lies behind the gluteus maximus, allows you to raise your leg to the side so that you can form a 90-degree angle with your two legs (this action is *abduction* of the thigh). Several muscles serve as *adductors;* that is, they move an abducted thigh back toward the midline. These muscles include the *pectineus* and *adductor longus,* which become injured when you "pull a groin muscle," as well as the *adductor magnus* and *gracilis,* which run along the inside of your thigh.

Other muscles in the thigh serve to move the lower leg. Along your thigh's front and lateral side, four muscles work together to allow you to kick. These four muscles — the *rectus femoris, vastus lateralis, vastus medialis,* and *vastus intermedius* — are better known as the *quadriceps (quadriceps femoris). Quad* means "four," as in *quadrilateral* or *quadrant.* Refer to Figure 5-8a.

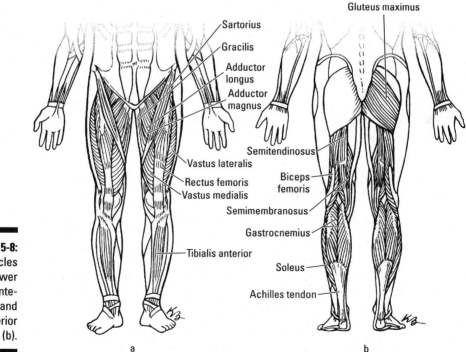

Figure 5-8:
The muscles of the lower limb: anterior (a) and posterior (b).

Where did these names come from?

Some muscle names have a pretty interesting history. Take the hamstring muscles and the sartorius muscle. First the hamstrings — ham may make you think of pigs, and yes, pigs have hamstrings in their legs. And the *biceps femoris, semimembranosus,* and *semitendinosus muscles* have the same strong tendons in a pig as they do in you. When butchers smoked hams (thigh meat from a pig), they hung the hams on hooks in the smokehouse by these ropelike tendons, which generated the name "hamstrings." Nobody said butchers were creative.

The sartorius muscle goes into action when you sit cross-legged, like tailors used to do when they pinned hems or cuffs (and maybe still do). So the sartorius muscle is sometimes referred to as the "tailor's muscle." And guess what means "tailor" in Latin? Yep, *sartor.*

The *hamstrings* are a group of muscles that are antagonistic to the quadriceps. The hamstrings — the *biceps femoris, semimembranosus,* and *semitendinosus* — run down the back of the thigh (refer to Figure 5-8b) and allow you to flex your lower leg and extend your hip. They originate on the ischium of the hipbone and join (insert at) to the tibia of the lower leg. You can feel the tendons of your hamstring muscles behind your knee.

Your lower leg's shin and calf muscles move your ankle and foot. The *gastrocnemius,* better known as the "calf muscle," begins (originates) at the femur (thigh bone) and joins (inserts at) the *Achilles tendon* that runs behind your heel. You can feel your gastrocnemius muscle contracting when you stand on your toes. The antagonist of the gastrocnemius, the *tibialis anterior,* starts on the surface of the tibia (shinbone), runs along the shin, and connects to your ankle's metatarsal bones. You can feel this muscle contract when you raise your toes and keep your heel on the floor. The *fibular longus* and *fibular brevis* (*brevis* meaning "short," as in *brevity*) run along the outside of the lower leg and join the fibula to the ankle bones. In doing so, the fibular muscles help to move the foot. The *extensor digitorum longus* and the *flexor digitorum longus* muscles join the tibia to the feet and allow you to extend and flex your toes, respectively, like your fingers.

Pathophysiology of the Muscular System

The body has so much skeletal muscle tissue, and it performs so many functions that wear and tear is normal and expected, not really "pathophysiological." Sore or strained muscles need only time to repair themselves and even become better than ever. However, many serious conditions can affect skeletal muscle and leave the sufferer disabled, in considerable pain, and even with a much-shortened life span. The following sections give you an overview of conditions that can affect muscles.

Muscular dystrophy

The muscular dystrophies (MD) are a group of more than 30 diseases characterized by progressive weakness and degeneration of the skeletal muscles. All are inherited, each in its own pattern; some of them are passed on only to sons — that is, they're *X-linked diseases.* The most common form, *Duchenne muscular dystrophy* (DMD), is one such.

The disorders differ in terms of the distribution and extent of muscle weakness, age of onset of symptoms, and rate of progression. The prognosis for people with MD varies according to the disorder's type and progression. Some cases may be mild and progress very slowly over a normal life span, while others produce severe muscle weakness and functional disability beginning at a young age. Some children with MD die in infancy, while others live into adulthood with only moderate disability.

The symptoms of DMD become evident usually before a boy is 3 years old. The muscles slowly weaken, shorten, and degenerate. Fat and connective tissue replace normal muscle tissue, thus causing problems in the heart and lungs. DMD patients are often in a wheelchair by about age 12 and usually die in their teens.

Myotonic muscular dystrophy affects males and females, and the onset of symptoms can occur at any age. Progressive muscle weakness and stiffness is usually evident first in the muscles of the face and neck. Turning the head becomes difficult. Eventually, those afflicted have problems with actions such as swallowing, because their muscles don't relax after contractions. Then, the arms and legs become affected. Eventually, the patient may require a wheelchair or even be confined to bed.

There's no specific treatment to stop or reverse any form of MD. Treatment to manage symptoms may include drug therapy, physical therapy, respiratory therapy, speech therapy, orthopedic appliances used for support, and corrective orthopedic surgery. Some patients may need assisted ventilation to treat respiratory muscle weakness and a pacemaker for cardiac abnormalities.

Muscle spasms

A *muscle spasm* is a sudden, violent contraction, sometimes causing severe pain. Any muscle can go into spasm, and the effects vary according to the location and nerves that are nearby. For example, the *piriformis* is a tiny, stringlike muscle that originates at the ilium (hip) and the sacrum at the base of the spine, and runs to a point at the top of the femur. It rotates the thigh out to the side (laterally). Spasm in this muscle can irritate the sacral nerves and cause pain in the lower back; afterward, the buttock and sometimes the hip (the piriformis muscle is also attached to the hip) feel bruised.

Not all spasms are painful. Hiccups, which are the result of spasm in the diaphragm, aren't usually painful — annoying, but not painful. The same with facial tics, such as when your eyelid twitches repeatedly. Muscle cramps are spasms, too. The *gastrocnemius* (calf muscle) is a common place for sudden cramps to occur.

Fibromyalgia

Fibromyalgia, a chronic pain condition, isn't exactly a muscle disease, but severe and widespread muscle pain is a major symptom of this mysterious condition. Strictly speaking, fibromyalgia isn't a disease at all but a *syndrome* — a group of symptoms that appear to be closely related, although the underlying cause may not be known. Recent research has suggested a genetic component. The disorder is often seen in families, among siblings or mothers and their children.

The muscle pain can be more or less severe, more or less chronic, and more or less debilitating. Fibromyalgia patients often have more of a neurotransmitter called *substance P* in their spinal fluid, which is believed by some clinicians to alter the affected person's perception of pain. A sensation that someone else may not even notice can be experienced by the fibromyalgia sufferer as excruciating. Some drug treatments are available, and some pain management techniques are helpful for some patients. Fibromyalgia is an active area of clinical research.

Chapter 6

Getting the Skinny on Skin, Hair, and Nails

In This Chapter

▶ Seeing what the integument does

▶ Explaining the integument's structure

▶ Taking a closer look at hair, nails, and glands

▶ Understanding what your skin does for you

▶ Checking out the integument's pathophysiology

The skin and its appendages (hair and nails), together called the *integument,* make up the body's largest organ system. In a human adult, the skin covers an area of about 20 square feet (almost 2 square meters), or about the size of a small blanket — a soft, pliable, strong, waterproof, and self-repairing blanket. It weighs around 8 pounds (3.6 kilograms), about 15 percent of the body's weight.

Your skin is one of the most beautiful parts of you. It's also a complex organ system with numerous types of tissue and many specialized structures. The average square inch of skin holds 650 sweat glands, 20 blood vessels, more than 1,000 hair follicles, half a million melanocytes (pigment cells), and more than 1,000 nerve endings. Refer to the "Skin (Cross Section)" color plate in the center of the book to see a detailed cross section of the skin.

Skin is made in layers, and layers within layers. New cells begin life in the lower levels and are gradually pushed to the surface to replace the old, dead ones. By the time the new cells reach the top, they've become hard and flat, like roof shingles. Eventually, they pop off like shingles blown from a roof in a strong wind.

Amazingly, every minute, 30,000 to 40,000 dead skin cells fall from your body. In approximately a month's time, all the cells on the top layer have blown off and been replaced.

In this chapter, we show you what your skin is made of, and we explain how your hair and nails form. We also look at the functions of your skin and what diseases and conditions can take a toll on your integument.

Functions of the Integument

The entity known as *you* is bounded by your integument. Everything inside the outermost layer of skin is *you.* Everything outside the outermost layer of skin is *not-you.* (Well, not quite everything — the stuff in your alimentary canal is not *you* yet. More about that in Chapter 11.)

Your skin mediates much of the interaction between *you* and *not-you,* which we henceforth call the *environment.* Your integument identifies *you* to other humans, a very important function for members of the hypersocial human species. Here's a look at the integument's other important functions:

- ✔ **Protection:** Skin and its appendages protect the rest of the body by keeping out many threats from the environment, such as infection and predation by other organisms, damaging solar radiation, and nasty substances everywhere.

- ✔ **Thermoregulation:** The skin supports *thermoregulation* (the maintenance of optimum body temperature) in several ways. See the "Controlling your internal temperature" section later in this chapter.

- ✔ **Water balance:** The skin's outer layers are more or less impermeable to water, keeping water and salts at an optimum level inside the body and preventing excess fluid loss. A small amount of excess water and some bodily waste (urea) are eliminated through the skin.

- ✔ **Incoming messages:** Many types of sensory organs are embedded in your skin, including receptors for heat and cold, pressure, vibration, and pain. See the "Your skin is sensational" section later in the chapter.

- ✔ **Outgoing messages:** The skin and hair are messengers to the outside environment, mainly to other humans. People get information about your state of health by looking at your skin and hair. Your emotional state is signaled by pallor, flushing, blushing, goose bumps, and sweating. The odors of sweat from certain sweat glands signal sexual arousal.

- ✔ **Substance production:** *Sebaceous glands* in the skin, usually associated with a hair follicle, produce a waxy substance called *sebum.* Sweat glands in the skin make sweat. In fact, your skin has several different types of glands, and each makes a specific type of sweat. See the "Nothing's bland about glands" section later in the chapter.

  Skin cells produce *keratin,* a fibrous protein that's an important structural and functional component of skin and is, essentially, the only component of hair and nails. See the "Caring about Keratins" sidebar later in the chapter.

The root of integument is *integr-,* meaning "whole" as in *integer* (a whole number), *integrate* (to bring different parts together into one), *integral* (inseparable from the whole), and *integrity* (always a positive term, but more difficult to define).

Structure of the Integument

The integument wraps around the musculoskeletal systems, taking the shape of your bones and muscles and adding its own shape-conferring structures. Although your skin feels tight, it's really loosely attached to the layer of muscles below. In spots where muscles don't exist, such as on your knuckles, the skin is attached right to the bone.

Without losing sight of the reality, it's sometimes helpful to imagine that you can unzip and remove your skin, spread the living skin out on a table, and look at it. What would you see, feel, and smell?

One of the most obvious features you'd notice is that the skin, itself a thin layer, is made up of more layers. The layering is visible to the unaided eye because each layer is different from the others and the transitions between layers appear to be relatively abrupt. We look at each of these layers in the following sections.

The epidermis is on top of the dermis. The prefix *epi-* means "on top of." Many anatomical structures are on top of other structures, so you see this prefix in many chapters throughout this book. Other related terms you may see are *endo* (within), *ecto* (outside), and *hypo* (beneath or lower).

In this discussion, *up* and *above* mean "toward the surface of the body," and *down* and *below* mean "toward the center of the body." These terms *don't* mean "toward the head" and "toward the feet," respectively. Sometimes, anatomists use the terms *superficial* and *deep* to mean the same thing.

Touching the epidermis

The most familiar aspect of the integument is the *epidermis,* the outermost surface that you see on yourself and other people. The epidermis feels soft, slightly oily, elastic, resilient, and strong. In some places, the surface has a dense stand of coarse hairs; in other spots, it has a lighter covering of light hairs; and in a few places, it has no hairs at all. The nails cover the tips of the fingers and toes.

The epidermis itself is made up of four (thin skin) to five (thick skin) layers (not visible to the unaided eye), from the *stratum corneum* at the top to the *stratum germinativum* (or *stratum basale*) at the bottom. All layers of the epidermis are composed of *stratified squamous epithelial tissue,* but the layers perform different functions. The epidermis has no blood supply; it's nourished by diffusion from the dermis layer below.

Cosmetics and the stratum corneum

The use of cosmetics is far older than history, along with other appearance-enhancers and subgroup identifiers like scarifications and tattoos. Two types of cosmetics in common use in culture today are moisturizers and exfoliants.

Moisturizers

What do *moisturizers* do? After the water in them evaporates into the air, the lipids they contain remain on the skin's surface, adding another waterproof barrier to prevent the escape of water from the skin (thereby indirectly increasing the very scant water in the stratum corneum) and to trap particles and chemicals of many types. That's it.

As far as their effects on the skin, all moisturizers are essentially the same, whether it's petroleum jelly, a luxury-brand "serum," a drugstore lotion, or a vitamin E cream from the health food store. Not that they don't contain the ingredients the manufacturers claim; it's just that none of them gets past the stratus corneum. All those age-defying peptides, metallic microparticles, organic extracts, vitamins, rare botanicals, and antioxidants remain within the lipid layer on the skin's surface and go down the drain with the lipid and the trapped dirt when the skin is washed.

Exfoliants

Speaking of washing, what about *exfoliants? Exfoliation* means to remove the topmost cells of the stratum corneum by mechanical or chemical means. Your skin sheds these dead cells all the time, but exfoliation hurries up the process for those cells it reaches. Generally, exfoliation neither hurts nor helps any physiological process in the skin: The stratum corneum remains an effective barrier. Some people think that having some slightly younger dead cells on the skin's surface enhances its appearance (a secondary sexual characteristic in humans).

Shaving is one common form of mechanical exfoliation. A mechanical exfoliant cosmetic usually contains an abrasive substance embedded in a soap or detergent matrix. Different people, and different skins, prefer different abrasives. Chemical exfoliation usually involves the application of an acidic substance that breaks up the fibers holding the keratinized cells of the stratum corneum together. But no worries — ranks of keratinized cells are always moving up from below to take their place. The chemicals may also produce a slight irritation that leads to a temporary inflammation reaction, pulling in lymph and, while the inflammation lasts, puffing out some shallow wrinkles in the skin (secondary sexual characteristic again).

Does that mean moisturizers and exfoliants are a waste of money? Not necessarily. Keeping the skin moist, lubricated, and clean feels good and is a good thing generally. You may choose one cosmetic over another for a number of reasons: nicer texture, nicer perfume, the absence of perfume or ingredients that irritate the skin. And for some people, the use of an expensive self-care product enhances self-esteem ("Because you're worth it," as the advertising tag line of one brand says). Self-esteem and self-confidence enhance psychological well-being, which may enhance reproductive success and survival.

Cells are shed continuously from the top of the epidermis and replaced continuously by cells that are pushed up from the deeper layers. The entire epidermis is replaced approximately every six to eight weeks throughout life.

The thin, impervious cover

Think of the *stratum corneum* as a sheet of self-repairing fiberglass over the other layers of the epidermis. It's only 25 to 30 cells thick, dense, and relatively hard. All the cells are *keratinocytes,* which produce the fibrous protein *keratin.*

The keratinocytes of the stratus corneum originate as *squamous epithelial cells* in the *stratum germinativum.* As these cells move up through the epidermal layers, they become *keratinized,* or *cornified,* by the production of keratin fibers that gradually fill the cells. Eventually, the nucleus and cytoplasmic organelles disappear and metabolism ceases. The cells undergo a programmed death as they become fully keratinized, at which point they're hard and waterproof. In the cells of the dermis, keratin filaments and other intermediate filaments function as part of the cytoskeleton to mechanically stabilize the cell against physical stress.

The uppermost surface of the stratum corneum is covered with a waxy, waterproof coating of *sebum.* The stratum corneum protects the entire body by making sure that some things stay in and everything else stays out. (See the nearby sidebar "Cosmetics and the stratum corneum.")

One important thing that the stratum corneum doesn't seal out, however, is ultraviolet radiation. This form of energy goes right through the skin's surface and down to the layers below, where it stimulates the production of vitamin D. In high doses, it burns the skin and damages DNA, which can cause cells to become cancerous. Evolution's response to the threat of UV-induced cell damage is the pigment *melanin,* which absorbs harmful UV-radiation and transforms the energy into harmless heat.

The transit zone

The *stratum lucidum,* found only on the palms of hands and soles of the feet (thick skin); the *stratum granulosum;* and the *stratum spinosum* lie in distinct layers below the stratum corneum. Old cells slough off above and new cells push up from below, finally getting up into the stratum corneum. The process takes about 14 to 30 days. Like other epidermal cells, these cells live in the stratum corneum for about a month.

The keratinocytes produce *lipids* (fatty substances) and undergo successive stages of *keratinization* and other kinds of differentiation in these layers. These layers also contain *Langerhans cells,* immune system cells that arrest microbial invaders and transport them to the lymph nodes for destruction.

Caring about keratins

Keratins are fibrous proteins produced by skin cells in mammals, birds, and reptiles. Keratin-containing anatomical structures are hard, tough, and waterproof. The mammalian types, the α-keratins, are the main components of hair, including wool, nails, claws, hooves, and horns. (But not antlers — they're derived from bone.) The baleen of filter-feeding whales is made mainly of α-keratins. The other type, the β-keratins, are even harder and are found in human nails, mammalian claws, and porcupine quills; in the shells, scales, and claws of reptiles; and in the feathers, beaks, and claws of birds. People who make educated guesses about the anatomy and physiology of dinosaurs believe that keratins were likely a major component of the claws, horns, and armor plates of these creatures.

The cell farm

The *stratum germinativum,* also called the *stratum basale* or *basal layer,* is like a cell farm, constantly producing new cells and pushing them up into the layer above. This stratum contains *melanocytes,* which produce the *melanin* pigment that gives color to your skin, hair, and eyes and protects the skin from the damaging effects of UV radiation in sunlight. Melanin absorbs UV radiation and dissipates more than 99.9 percent of it as heat.

Everybody's stratus germinativum has about the same number of melanocytes (half a million to a million or more per square inch of skin), but the amount of melanin they produce varies depending mainly on genetics (heredity) and secondly on the amount of sunlight they receive (environment). Human groups living close to the equator have evolved genes that stimulate melanocytes to produce more melanin as protection from UV radiation, which burns the skin, damages DNA, and can cause skin cancer. Human groups living farther from the equator have evolved to produce less melanin.

Exploring the dermis

Below the layers of the epidermis and several times thicker is the *dermis.* The dermis itself is made up of two layers: the papillary region and the reticular region.

Under the basement

The *papillary region* is made up of the *basement membrane,* which sits just below the epidermis, and *papillae* (finger-like projections) that push into the basement membrane, increasing the area of contact between the dermis and the epidermis. In your palms, fingers, soles, and toes, the papillae projecting into the epidermis form *friction ridges.* (They help your hand or foot to grasp by increasing friction.) The pattern of the friction ridges on a finger is called a *fingerprint.*

The papillary region is an example of a common anatomical "strategy" for increasing the interface (direct contact) area between two structures or between an anatomical structure and the environment. We discuss another prominent example, the intestines, in Chapter 11.

A manufacturing site

The *reticular region* is chock-full of protein fibers and is a complex and metabolically active layer. Cells (which migrate down from the epidermis during development) and structures of the reticular region manufacture many of the skin's characteristic products: hair and nails, sebum, watery sweat, apocrine sweat. (See the "Accessorizing Your Skin" section later in the chapter.) The region also contains structures that connect the integument to other organ systems: sensors of pressure (touch) and heat, lymph vessels, and a very rich blood supply.

The blood vessels in the dermis provide nourishment and waste removal from its own cells as well as from the stratum germinativum. The blood vessels in the dermis dilate when the body needs to lose heat and constrict to keep heat in. They also dilate and contract in response to your emotional state, brightening or darkening skin color, thereby functioning as social signaling.

The sensory receptors in the dermis transmit sensations, such as pressure, vibration, and light touch, to the nervous system. The receptors with names, such as *lamellated (Pacinian) corpuscles, tactile (Meissner's) corpuscles,* and *Ruffini's corpuscles,* are sprinkled throughout the dermis and are connected to the nerves that run through the dermis and subcutaneous layer. At one spot on your skin, you may sense light touch, while a few centimeters away, you may sense pressure. Not every inch of skin is covered with receptors for every sensation. To find out more about how the nerves transmit sensations received in the skin's receptors, flip to Chapter 7.

Getting under your skin: The hypodermis

The *subcutaneous layer* (or *hypodermis,* or *superficial fascia*) is the layer of tissue directly underneath the dermis. It is mainly composed of connective and *adipose* tissue (fatty tissue). Its physiological functions include insulation, the storage of energy, and help in the anchoring of the skin. It contains larger blood vessels, lymph vessels, and nerve fibers than those found in the dermis. Its loosely arranged elastin fibers anchor the hypodermis to the muscle below.

The thickness of the subcutaneous layer is determined in some places by the amount of fat deposited into the cells of the *adipose tissue,* which makes up the majority of the subcutaneous layer. Within the last decade or so, adipose tissue has been reclassified to the endocrine system (see Chapter 8).

The rise and fall of a tan

Melanogenesis, the production of melanin in response to UV radiation, is the physiology term for tanning. The more exposure skin has to UV, the greater the production of melanin. The melanin migrates from the melanocytes into the skin cells that journey to the stratus corneum and are sloughed off. So your tan falls off, cell by cell, but the DNA damage to the cells of the stratum germinativum remains for your lifetime.

Accessorizing Your Skin

This section has nothing to do with tattoos or body piercing. It's all about your skin's *accessory structures:* hair, nails, and glands — structures that work with the skin.

Now hair this

Your body has millions of hair follicles, about the same number as the chimpanzee, humanity's closest evolutionary relative. (Although we don't know who counted all of them!) Like chimps, humans have hairless palms, soles, lips, and nipples. Unlike a chimpanzee, most of your hair is lightweight, fine, and downy. The hair on your head is coarser and longer to help hold in body heat. Puberty brings about a surge of sex hormones that stimulate hair growth in the *axillary* (armpit) and pelvic regions and, in males, on the face and neck. Women with hormonal imbalances can develop facial hair, too. Turn to Chapter 8 for more about hormones.

A hair arises in a *hair follicle,* a small tube made up of epidermal cells that extend down into the dermis to take advantage of its rich blood supply. Cells at the bottom of the hair follicle continually divide to produce new cells that are added to the end of the hair and push the older cells up through the layers of the epidermis. On their way up and out, the hair cells become keratinized.

The hairs on your head live about three to four years before you shed them, and eyelashes live about three to four months before falling out. People don't go bald overnight. Baldness (called *alopecia*) occurs when follicles go dormant and hairs are no longer replaced.

Nailing nails

Your fingernails and toenails lie on a *nail bed* (not to be confused with a bed of nails). At the back of the nail bed is the *nail root.* Just like skin and hair, nails start growing near the blood supply that lies under the nail bed, and the cells move outward at the rate of about 1 millimeter per week. As they move out over the nail bed, they become keratinized. (See Figure 6-1.)

At the bottom of your nails is a white, half-moon-shaped area called the *lunula.* (*Lun-* is the Latin root for moon, as in *lunar.*) The lunula is white because this is the area of cell growth. In the nail body, the nail appears pink because the blood vessels lie underneath the nail bed. But many more cells fill in the area of growth. This layer is thicker, and you see white instead of pink.

Nothing's bland about glands

Glands in the skin make and secrete substances that are transported to your body's outer surface. The contraction of tiny muscles in the gland accomplishes this secretion. The two main types of skin glands are *sudoriferous glands* (sweat glands) and *sebaceous glands* (oil glands).

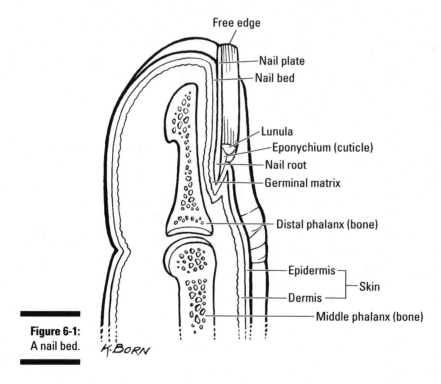

Free edge

Nail plate

Nail bed

Lunula

Eponychium (cuticle)

Nail root

Germinal matrix

Distal phalanx (bone)

Epidermis

Dermis

Skin

Middle phalanx (bone)

Figure 6-1:
A nail bed.

K. BORN

It will make your hair stand on end

Each hair follicle has a tiny *arrector pili* muscle. When these smooth muscles contract, the previously bent hair shafts become erect. This movement causes the epidermis to be pushed to the side, and a goose bump appears. When you're cold or frightened, the sudden tightening of the arrector pili muscle causes the hair to rise up, and as a result, air is trapped between the hair and the skin. When you're cold, the trapped air acts as insulation. When you're frightened, the hairs standing on end serve to make you look scary to whatever is scaring you, a form of signaling used by many mammals.

Sudoriferous glands

Your body contains two types of sudoriferous glands. *Eccrine sweat glands* are distributed all over the skin. These glands open to the skin's surface, and when you're hot, they let heat escape in the form of sweat to reduce body temperature by a process of evaporative cooling.

Apocrine sweat glands start to develop during puberty deep in the hair follicles of the armpits and groin. Apocrine sweat contains a milky white substance and may also contain *pheromones,* chemicals that communicate information to other individuals by altering their hormonal balance. (Some research has indicated that the apocrine secretions of one woman can influence the menstrual cycle of other women who live with her.) Apocrine glands become active when you're anxious and stressed, as well as when you're sexually stimulated. Bacteria on the skin that digest the milky white substance produce unpleasantly odiferous byproducts.

The milk-secreting glands in mammary tissue are thought to have evolved from apocrine sweat glands.

Sebaceous glands

Sebaceous glands secrete an oily substance called *sebum* into hair roots. Besides wreaking havoc with teenage facial pores, sebum has physiological functions. It helps maintain your hair in a healthy state, important in regulating body temperature. It flows out along the hair shaft, coating the hair and the epidermis, forming a protective, waterproof layer. Sebum prevents water loss to the outside. Sebum also helps to protect you from infection by making the skin surface an inhospitable place for some bacteria.

In the watery environment of the amniotic sac, the human fetus produces a thick layer of sebum, called the *vernix caseosa.* Ear wax *(cerumen)* is a type of sebum produced by specialized cells in the ear canal.

Your Skin Saving You

Your integumentary system participates in thousands of metabolic and homeostatic reactions, among them the thermoregulatory processes and the interaction of the skin and the nervous system.

Controlling your internal temperature

Your skin plays an important role in homeostasis, specifically thermoregulation (see Chapter 2). It has mechanisms that increase body temperature when you're cold and decrease it when you're hot.

Your body is continuously converting energy from food into energy in the form of ATP (see Chapter 3). About 60 percent of food energy is converted to heat in the metabolic reactions that produce ATP, and more heat is given off in the reactions that use ATP. This heat replaces the heat that your body loses continuously to the environment. (See Chapter 16 for an overview of ten basic chemistry and physics concepts.)

Human physiology employs specially adapted integumentary structures for thermoregulation. When your internal temperature rises above a physiologically optimum level, the blood vessels in the dermis dilate, dissipating heat from the blood to the environment through the epidermis. The sweat glands are activated, and heat escapes the body as the sweat evaporates.

When your skin is cold, the sweat glands aren't activated, thus slowing down evaporative cooling and raising your internal temperature by retaining the heat produced in metabolism. Blood vessels in the dermis constrict, limiting heat escape from the blood. If body temperature falls to about 97 degrees, shivering begins automatically, and these muscle contractions generate heat.

Normal body temperature for a human is 97 to 100 degrees Fahrenheit (F). This is the optimal range for the very temperature-sensitive reactions of metabolism. A few degrees higher, and the body goes into convulsions. A few degrees lower, and metabolism gradually shuts down, sending the body into what may be its final sleep.

Other mammals, including other primates, rely on a more or less dense covering of hair on the epidermis to conserve body heat. During the evolution of the human species, natural selection favored the development of a very light coat of hair over most of the body. Turn to Chapter 17 for some physiology information related to that evolution.

Vitamin D

Sunlight has a nasty way of damaging skin, but you need a regular dose of sunlight to keep your bones healthy. Bones need vitamin D to develop healthy new bone cells. Insufficient vitamin D can lead to a condition called *rickets,* which results in soft, curved bones that can't support the body's weight.

Skin cells contain a molecule that's converted to vitamin D when struck by ultraviolet rays (UV) in sunlight. The vitamin D leaves the skin and goes through the bloodstream to the liver and kidneys, where vitamin D is converted to the hormone *calcitriol.* Calcitriol then circulates through the entire body and regulates the physiology of calcium and phosphorus, important

minerals for the development and maintenance of healthy bones. (Scientists are discovering that vitamin D plays important roles in keeping your body healthy in other ways, too. To find out more, pick up *Vitamin D For Dummies* by Dr. Alan Rubin [Wiley].)

The use of sunscreen and sunblock products can prevent the UV radiation from reaching the skin cells and initiating vitamin D synthesis.

Just a few minutes of sunshine a day is all that's needed for your skin to make adequate amounts of vitamin D to keep your bones healthy. Anything beyond that amount of exposure is risky.

Your skin is sensational

So how does your body know when it's cold or hot? How do you know when you get a cut or splinter? How can you tell the difference between being tickled with a feather and punched with a fist? Because the dermis contains nerve endings that serve as specialized receptors for hot and cold, touch, pressure, and pain. See Chapter 7 for more about these receptors.

Not every square inch of skin contains all types of nerve endings. So in some spots, you can sense the differences when you touch various objects. You can feel the difference between your skin being lightly pressed on or heavily pushed on; some spots sense cold, and some spots sense heat. All the receptors connect to nerves that run through the subcutaneous layer and connect with the network of nerves up to your brain, which does a pretty good job of making sense of the different kinds of information.

Pathophysiology of the Integument

Being in contact with the external environment, the integument inevitably encounters some nasty stuff: *pathogens* (disease-causing organisms like some bacteria, fungi, protists, and viruses), UV radiation, and harmful forces like fire, chemicals, and sharp objects. There are also hereditary pathophysiologies. Chances are that you've experienced — or known someone who has experienced — one of the conditions described in the following sections.

Skin cancer

Many cases of skin cancer are related to UV exposure. Skin cancers are classified as a *melanoma* (a malignant or spreading type) or a *nonmelanoma* (a limited-to-one-portion-of-the-skin type).

- ✔ **Basal cell carcinoma** is the most common form of skin cancer. UV radiation can cause a cancerous tumor to develop in the stratum germinativum. The immune system becomes increasingly unable to detect the tumor as it grows. This type of tumor is relatively easily removed and is usually easily cured.

- ✔ **Squamous cell carcinoma** is a melanoma type that starts in the epidermis. Squamous cell carcinoma is more likely to spread to a nearby organ than basal cell carcinoma, and for every 100 people diagnosed with squamous cell carcinoma, one is likely to die from it.

- ✔ **Malignant melanoma** starts in the melanocytes, the cells that produce melanin. The lesions of a malignant melanoma are almost black with irregular borders. These cancerous spots look like a spot on your garage floor where you spilled oil. Malignant melanoma occurs mostly in light-skinned people who have a history of severe sunburns, especially when they were kids. One in five people diagnosed with malignant melanoma die from it within five years.

Dermatitis

Dermatitis is inflammation of the skin that takes the form of broken skin that itches and burns. The inflammation can have many causes: infection, insect bites, irritation from chemicals, allergy, skin abrasion from shaving, or sunburn. A genetic predisposition is frequently involved, as in seborrheic dermatitis, or eczema, which often affects the scalp and hair as well as the skin of the hands, feet, face, or just about anywhere else.

Many people have an episode of dermatitis at some time in their lives, and some people suffer chronic dermatitis. Although the cause can vary, the treatment is often relatively simple: some combination of avoiding the disturbing influence or eradicating the infectious organism and applying a hydrocortisone cream to calm the inflammation and allow the skin to heal.

Alopecia

Alopecia is the medical term for abnormal hair loss — that is, over and above the normal rate of hair loss caused by the follicles taking a break once in a while. Like dermatitis, the causes are various; unlike dermatitis, alopecia doesn't usually break the skin and allow invasion by infectious organisms, nor does it usually hurt, itch, or sting. However, cure or even treatment can be elusive.

The most prevalent type is *androgenetic alopecia,* or male pattern baldness. Androgenetic alopecia is an inherited condition that affects about 25 percent of men before the age of 30 and two-thirds of all men before the age of 60. The condition is less common and less extreme in women. It can develop in older adults, resulting in an overall thinning of all the scalp hair rather than complete baldness.

Alopecia areata is a type of hair loss in which the immune system attacks hair follicles. The root cause (sorry!) is unknown, but damage to the follicle is usually temporary. Alopecia areata is most common in people younger than 20, but children and adults of any age may be affected. Alopecia areata can be treated but not cured.

Temporary alopecia can result from a long list of causes, including stress, severe illness or surgery, nutritional deficiencies, treatment with certain drugs (notably chemotherapy agents for cancer, which attack all actively dividing cells), and certain medicines for arthritis, depression, heart problems, and high blood pressure. Use of some hair cosmetics or hair abuse (wearing the hair pulled back tightly over a long period of time) can harm the hair shaft or follicle. *Trichotillomania,* or compulsive hair pulling, can also eventually lead to alopecia.

Nail problems as signs of possible medical conditions

Unhealthy nails can be a symptom of an underlying disease and can be of help in diagnosis.

- ✔ Brittle, concave (spoon-shaped) nails with ridges may indicate iron-deficiency anemia.

- ✔ Nails separating from the nail bed can result from a thyroid disorder in which too much of a thyroid hormone is produced, such as Graves disease.

- ✔ Black marks that look like tiny splinters under the nails can help diagnose respiratory disease or heart disease.

- ✔ Hard, curved, yellow nails may indicate *bronchiectasis* (chronic dilation of bronchial tubes) and *lymphedema* (fluid retention in lymph glands).

Part III
Exploring the Inner Workings

The 5th Wave   By Rich Tennant

"Can anyone tell me, am I eating from the endocrine system or the nervous system? I always get those two mixed up."

In this part . . .

I n Part III, we continue our survey of the human organ systems, focusing on the internal systems — the organs that you don't see. Some of them you don't even feel, unless pain impulses signal injury or disease. We start this part with the electrical and chemical signaling systems that control and regulate an organism's physiology moment to moment, and then we discuss how the other organ systems respond.

Chapter 7

The Nervous System: Your Body's Circuit Board

In This Chapter

▶ Summarizing the functions of the nervous system

▶ Taking a detailed look at the CNS, the brain, and the PNS

▶ Getting down to a cellular level — neurons and neuroglial cells

▶ Relaying impulses through a cell and across a synapse

▶ Receiving impulses: Your five senses

▶ Noting a few nervous system disorders

*A*n organism's awareness of itself and of its environment depends on communication between one part of its body and another. In biology, such internal messaging is accomplished by several different mechanisms. Humans, like all mammals, use mechanisms involving chemistry and mechanisms involving electricity. We discuss the chemical messaging system in Chapter 8. This chapter is devoted to the body's electrical messaging system.

The nervous system is the body's electrical communications network. It generates and transmits information throughout the body in the form of electrical impulses. An electrical charge creates electrical energy, which has two important characteristics: It moves in distinct "packets," called *impulses*, and it moves very quickly.

The nervous system's structures reach into every organ and participate one way or another in nearly every physiological reaction. Perceiving the beauty of a flying bird and digesting your breakfast can happen simultaneously, and each action is dependent on the nervous system.

The human nervous system is, in fact, the single most distinctive feature of the human species. In particular, the human brain, an organ of the nervous system, functions differently from the brain of any other species.

Understanding the anatomical and physiological adaptations that accompanied the astoundingly rapid evolution of the human brain just a few million years ago (yesterday in evolutionary time) is an active area of research. The discussion in this chapter, however, focuses on the more basic aspects of the human nervous system — aspects that humans share with all other mammals. All have similar nerve cells, nervous tissues, and organs distributed in parallel networks throughout the body, and these are all specialized for the initiation and controlled transmission of electrical impulses.

Integrating the Input with the Output

The nervous system has just three jobs to do, and these jobs overlap.

- ✔ **Sensory input:** Specialized neurons called *sensory receptors* collect information from the entire body, create an impulse, and transmit the impulse to either the spinal cord or brain stem, and then to the brain. (See the "Making Sense of Your Senses" section later in this chapter.)

- ✔ **Integration:** The *central nervous system* (CNS) makes sense of the input it receives from all around the body. (See the "Central nervous system" section later in the chapter.)

- ✔ **Motor output:** In response to the integration of the sensory input, the peripheral nervous system (PNS) initiates and sends out impulses through nerves to muscles, glands, and other organs capable of the appropriate response. (See the "Peripheral nervous system" section later in the chapter.)

Neural Tissues

The nervous system is made up primarily of two categories of cells — neurons and neuroglial cells — that associate very closely in the brain and in tissues called *nerves*.

Neurons

A *neuron* is an individual cell and is the basic unit of the nervous system. Neurons are highly specialized for the initiation and transmission of electrical signals (impulses). The neuron is able to, in an instant, receive the *outputs* (pulses of electrical energy) of many other cells, process this incoming information, and "decide" whether to generate its own signal to be passed on to other neurons, muscles, or gland cells.

Here are the three types of neurons:

- **Sensory neurons:** Also called *afferent neurons* (*afferent* means "moving toward"), these neurons respond to sensory stimuli (touch, sound, light, and so on), passing the impulses ultimately to the spinal cord and brain.

- **Motor neurons:** Also called *efferent neurons* (*efferent* means "moving away"), these transmit impulses from the brain and spinal cord to effector organs (muscles and glands), triggering responses from these organs (muscle contraction or release of the gland's product).

- **Interneurons:** Also called *association neurons,* these connect neurons to other neurons within the same region of the brain or spinal cord.

Neurons in different parts of the nervous system perform diverse functions and therefore vary in shape, size, and electrochemical properties. However, neurons have a special cellular anatomy adapted to the quick transmission of an electrical charge. See Figure 7-1. All neurons have the same three parts, all enclosed within their cell membrane:

- **Cell body:** The body of a neuron is similar to a generic cell. It contains the nucleus, mitochondria, and other organelles.

- **Dendrites:** The dendrites are extensions that branch from one end of the cell body. They receive information from other neurons and send impulses in the direction of the cell body.

- **Axon:** The axon is a cable-like projection located on the opposite end of the cell body from the dendrite. Extending many times (tens, hundreds, or even tens of thousands of times) the cell body's diameter in length, the axon carries the impulses away from the cell body and toward the next neuron in the chain. (Think of electrical transmission wires.)

Fully differentiated neurons don't typically divide and may live for years, or even the whole lifetime of an organism.

Neuroglial cells

Various types of cells, collectively called *neuroglial cells* (or just *glial cells;* *glia* means "glue"), support neurons in various ways, including physically by holding them in place and supplying them with nutrients. They protect neurons from pathogens and remove dead neurons. Certain types of glial cells generate *myelin,* a fatty substance that wraps around axons and provides electrical insulation that allows the axons to transmit action potentials much more rapidly and efficiently. Scientists estimate that in the human brain, the total number of glia roughly equals the number of neurons, although the proportions vary in different brain areas.

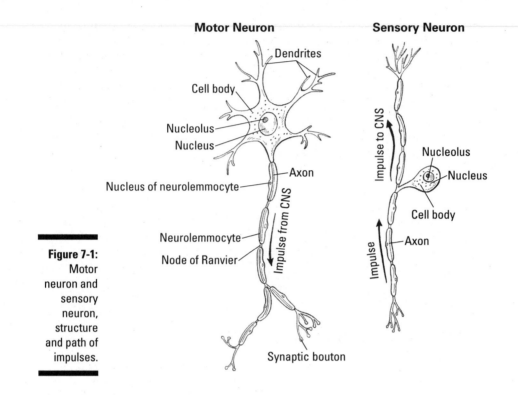

Figure 7-1:
Motor
neuron and
sensory
neuron,
structure
and path of
impulses.

Nerves

A *nerve* is a bundle of peripheral axons. An individual axon plus its myelin sheath is called a *nerve fiber*. Nerves provide a common pathway for the electrochemical nerve impulses that are transmitted along each of the axons. Nerves are found only in the peripheral nervous system. Nerve fibers can be of two types: *motor,* which send impulses away from the CNS, or *sensory,* which send impulses toward the CNS.

Ganglia and plexuses

A *ganglion* (plural, *ganglia*) is literally a bundle of nerves. Well, almost literally. A ganglion is an aggregation of neuron cell bodies. Ganglia provide relay points and intermediary connections among the body's neurological structures, especially between the CNS and the PNS.

Ganglia may be connected to form a *chain*. For example, the sympathetic nervous system contains a chain of ganglia referred to as the *paravertebral ganglia* or the *sympathetic chain of ganglia.*

Plexus is a general term for a network of anatomical structures, such as lymphatic vessels, nerves, or veins. (The term comes from the Latin "plectere" meaning "to braid.") A neural plexus is a network of intersecting nerves. The *solar plexus* serves the internal organs. The *cervical plexus* serves the head, neck, and shoulders. The *brachial plexus* serves the chest, shoulders, arms, and hands. The *lumbar*, *sacral*, and *coccygeal plexuses* serve the lower body.

Integrated Networks

The nervous system comprises two physically separate but functionally integrated networks of nervous tissue. Working together, these networks perceive and respond to internal and external stimuli to maintain homeostasis and simultaneously move the genetic development program forward. The following sections take a closer look at the central and peripheral nervous systems. Check out the "Nervous System" color plate in the middle of the book for a detailed look at the anatomy.

Central nervous system

The *central nervous system* (CNS), which consists of the brain and spinal cord, is the largest part of the nervous system. It integrates the information it receives from the sensory receptors and coordinates the activity of all parts of the body.

Both the brain and spinal cord are masses of neural tissue protected within bony structures (the skull and the vertebral column, respectively) and layers of membranes and specialized fluids, reflecting their prime importance to the continuation of the organism's life.

The brain and spinal cord are made up mainly of two types of tissue, called *gray matter* and *white matter*. Gray matter consists of unmyelinated neurons, neuron cell bodies, and neuroglial cells. White matter is made up of neuroglial cells and the myelinated axons extending from the neuron cell bodies in the gray matter. (See the "Neural Tissues" section earlier in the chapter for more on neurons and neuroglial cells.) The myelin has a high fat content, which results in white matter's white color.

In the brain, the gray matter forms a thin layer on the outside (the *cortex*). The white matter is beneath and makes up the brain's big data lines, carrying information around the brain. In the spinal cord, the tissue is arranged in a long cylinder; the gray matter forms the inner layer, and the white matter forms the outer layer.

The spinal cord extends from the bottom of the brain stem down the vertebral column within a cylindrical tubular opening created by the vertebrae and three tough membranes with cushioning fluid between them (see Figure 7-2).

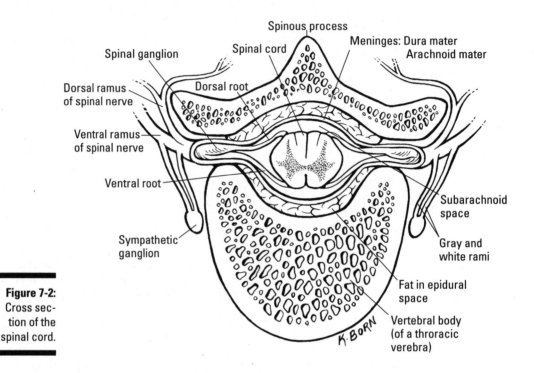

Figure 7-2: Cross section of the spinal cord.

Peripheral nervous system

The *peripheral nervous system* (PNS) consists of the nerves and ganglia outside of the brain and spinal cord. Unlike the CNS, the PNS isn't protected by bone or by the blood-brain barrier, leaving it exposed to toxins and mechanical injuries. Structures of the PNS include:

✔ **Cranial nerves:** Twelve pairs of nerves that emerge directly from the brain and brain stem. Each pair is dedicated to particular functions — some bring information from the sense organs to the brain, others control muscles, but most have both sensory and motor functions. Some are connected to glands or internal organs such as the heart and lungs. For example, the longest of the cranial nerves, called the *vagus nerves,* pass through the neck and chest into the abdomen. They relay sensory impulses from part of the ear, tongue, larynx, and pharynx; relay motor impulses to the vocal cords; and relay motor and secretory impulses to some abdominal and thoracic organs.

- ✔ **Spinal nerves:** Thirty-one pairs of nerves that emerge from the spinal cord. Each contains thousands of *afferent* (sensory) and *efferent* (motor) fibers.

- ✔ **Sensory nerve fibers:** Nerve fibers all over the body that send impulses to the CNS via the cranial nerves and spinal nerves.

- ✔ **Motor nerve fibers:** Nerve fibers that connect to muscles and glands and send impulses from the CNS via the cranial and spinal nerves.

The PNS is further divided into the somatic system and the autonomic system.

Somatic system

The *somatic nervous system* regulates activities that are under conscious control. Its sensory fibers receive impulses from receptors. Its motor fibers transmit impulses from the CNS to the (voluntary) skeletal muscles to coordinate body movements.

Autonomic system

The motor fibers of the *autonomic system* transmit impulses from the CNS to the glands, heart, and smooth (involuntary) organ muscles. The autonomic system controls internal organ functions that are involuntary and that happen subconsciously, such as breathing, heartbeat, and digestion.

The autonomic system is made up of the following:

- ✔ **Sympathetic nervous system:** Nerves originate in the thoracic and lumbar regions of the spinal cord (see Figure 7-3). The sympathetic nervous system responds to stress and is responsible for the increase of your heartbeat and blood pressure, among other physiological changes, along with the sense of excitement you may feel due to the increase of adrenaline in your system.

- ✔ **Parasympathetic nervous system:** Nerves originate in the brain stem and sacral portion of the spinal cord. The parasympathetic nervous system is evident when you rest or feel relaxed and is responsible for such things as the constriction of the pupil, the slowing of the heart, the dilation of the blood vessels, and the stimulation of the digestive and urinary systems. The parasympathetic nervous system is known as the "housekeeping" system because it maintains your normal functioning when you're not under stress.

- ✔ **Enteric nervous system:** This system manages every aspect of digestion, from the esophagus to the stomach to the small intestine to the colon.

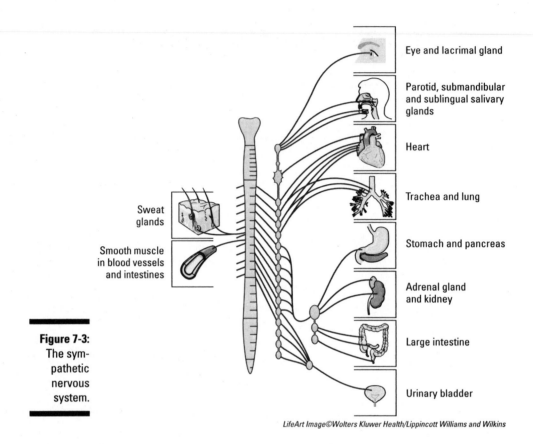

Eye and lacrimal gland

Parotid, submandibular and sublingual salivary glands

Heart

Trachea and lung

Stomach and pancreas

Adrenal gland and kidney

Large intestine

Urinary bladder

Sweat glands

Smooth muscle in blood vessels and intestines

Figure 7-3:
The sympathetic nervous system.

LifeArt Image©Wolters Kluwer Health/Lippincott Williams and Wilkins

Thinking about Your Brain

The brain is one of the largest organs in the human body, and different parts of it are responsible for different functions.

The brain's major parts are the *cerebrum, cerebellum, brain stem,* and *diencephalon.* The brain's four connecting, fluid-filled cavities are called *ventricles.* In this section, you find out some details about the parts of your brain and its ventricles. Take a look at Figure 7-4, and refer to it as necessary.

Keeping conscious: Your cerebrum

If you're conscious, you're using your cerebrum. The *cerebrum,* the brain's largest part, controls consciousness.

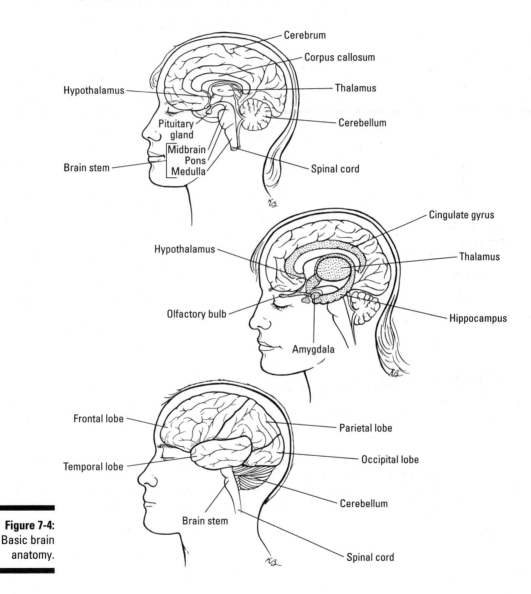

Figure 7-4:
Basic brain anatomy.

The cerebrum is divided into left and right halves, called the *left* and *right cerebral hemispheres,* and each half has four lobes: *frontal, parietal, temporal,* and *occipital.* The names of the lobes come from the skull bones that overlay them (see Chapter 4). Table 7-1 shows you what each lobe controls.

Table 7-1	Functions of Lobes within Cerebral Hemispheres
Lobe	*Functions*
Frontal lobe	Speech production, concentration, problem solving, planning, and voluntary muscle control
Parietal lobe	General interpretation area, understanding speech, ability to use words, and sensations including heat/cold, pressure, touch, and pain
Temporal lobe	Interpretation of sensations, remembering visually, remembering through sounds, hearing, and learning
Occipital lobe	Vision, recognizing objects visually, and combining images received visually

The *cerebral cortex* is the brain's outer layer of gray matter. It covers the entire surface of the cerebrum and overlays the deeper white matter. The brain's elevations are called *gyri* (singular *gyrus*). The shallow grooves that separate the elevations are called *sulci* (singular *sulcus*). Deep grooves in the brain are called *fissures.*

When you look at the top of a brain, you notice a deep groove running down the middle of the cerebrum. This groove is the *longitudinal fissure,* and it incompletely divides the cerebrum into the left and right hemispheres. The *corpus callosum,* located within the brain at the bottom of the longitudinal fissure, contains myelinated fibers that connect the left and right hemispheres.

Making your moves smooth: The cerebellum

The *cerebellum* lies just below the bottom of the cerebrum. A narrow stalk *(vermis)* connects the cerebellum's left and right hemispheres. The outside is gray matter, and the inside is white matter.

The cerebellum coordinates your skeletal muscle movements, making them smooth and graceful instead of stiff and jerky. The cerebellum also maintains normal muscle tone and posture (see Chapter 2 for an overview of the body's functions).

Coming up roses: Your brain stem

The *brain stem* consists of the *midbrain, pons,* and *medulla oblongata.* The medulla oblongata is continuous with the spinal cord after it passes through the hole called the *foramen magnum* at the bottom of the skull (see Chapter 4 for more on the skeletal system).

Inside your brain, just in front of (anterior to) the cerebellum, lie the midbrain and pons. The midbrain serves as a "station" for information that passes between the spinal cord and the cerebrum or between the cerebrum and the cerebellum. Impulses pass through the midbrain, which has centers for reflexes based on vision, hearing, and touch. If you see, hear, or feel something that scares you, alarms you, or hurts you, your midbrain immediately responds by sending out impulses to generate the appropriate type of scream, jump, or exclamation.

Reflex arcs sometimes create immediate, unconscious responses. Reflex arcs (see Figure 7-5) happen unconsciously whenever you touch something really hot or sharp. Sensory neurons detect pain, temperature, pressure, and the like. If sensory neurons detect something that could harm your body, such as heat that may cause a burn or a sharp object that may puncture the skin, an impulse passes from the receptor in the skin through the sensory neuron to the spinal cord, and then to motor neurons that cause a muscle to contract and pull the body part at risk of injury away from the heat or sharp object.

Reflexes occur so fast that you don't have time to think about how to react. By the time the impulse gets to your brain, the spinal cord has already taken care of the problem! In normal processes of the CNS, impulses travel to the brain for interpretation and production of the proper response. However, using the spinal cord rather than the brain to produce a response, reflex arcs save time and possibly damaging consequences.

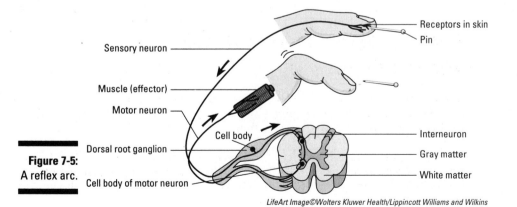

Figure 7-5:
A reflex arc.

LifeArt Image©Wolters Kluwer Health/Lippincott Williams and Wilkins

If the midbrain is a station for impulses, the pons is the bridge that joins the cerebellum with the cerebrum's left and right hemispheres. Axon bundles fill the pons and respond quickly to information it receives through the eyes and ears. (See the "Neurons" section earlier in the chapter for more on axons.)

The medulla oblongata, which transitions into the spinal cord, is responsible for several important functions, such as your breathing, the beating of your heart, and the regulation of your blood pressure. The medulla oblongata also contains the axons that send out the signals for coughing, vomiting, sneezing, and swallowing, based on information it receives from the respiratory or digestive systems. And whenever you get those annoying hiccups, blame your medulla oblongata.

Following fluid through the ventricles

Each cerebral hemisphere contains a *lateral ventricle* (the first and second ventricles). The other two ventricles are, believe it or not, the third and fourth ventricles. (Recall that a ventricle is a connecting, fluid-filled cavity.) The *third ventricle* lies just about in the center of your brain; the *fourth ventricle* lies at the top of the brain stem. The *cerebral aqueduct* (also called the *mesencephalic aqueduct*) connects the third and fourth ventricles. From the inferior portion of the fourth ventricle, a narrow channel called the *central canal* continues down into the spinal cord.

 When you think of an aqueduct, you may think of Rome. The Romans built aqueducts as a system for distributing water. Well, in your central nervous system, the ventricles and aqueducts serve as a system to circulate *cerebrospinal fluid* (CSF).

A clear fluid made in the brain, CSF is contained in the brain's four ventricles, the *subarachnoid space* (the space between the arachnoid and the pia mater), and the central canal of the spinal cord. The CSF picks up waste products from the CNS cells and delivers them to the bloodstream for disposal. The CSF also cushions the CNS. Along with your skull and vertebrae, CSF adds a protective layer around your brain and spinal cord.

Perhaps the most important function of CSF is to keep the ions in balance and thus stabilize membrane potentials. CSF circulates from the lateral ventricles to the third ventricle, through the cerebral aqueduct into the fourth ventricle, and then down through the central canal of the spinal cord. From the fourth ventricle, CSF oozes into the subarachnoid space just under the arachnoid membrane, which continuously covers the spinal cord and brain. In the subarachnoid space, CSF can seep through tiny spaces to get to the bloodstream.

The most interesting system of all

If you've ever fallen in love, enjoyed sex, stashed away some great memories, or felt enraged (seems like the full cycle of a relationship, doesn't it?), you were using your limbic system. The *limbic system* isn't an anatomical structure but rather a collection of areas in the brain — certain parts of the cerebrum and diencephalon — that are involved in some emotional issues. These areas control your libido, memory, pleasure or pain, and feelings such as happiness, sorrow, fear, affection, and rage. Although these reactions and emotions may not be key to your survival, they do make life interesting.

In a procedure known as a *spinal tap,* CSF is drawn through a needle from the subarachnoid space. Doctors can test the CSF for the presence of bacteria that cause meningitis or for the presence of proteins that can indicate other diseases, such as Alzheimer's.

Regulating systems: The diencephalon

Right smack in the middle of the brain, the *hypothalamus* and the *thalamus* that lie in the walls and floor of the third ventricle form the *diencephalon.* The hypothalamus regulates sleep, hunger, thirst, body temperature, blood pressure, and fluid level to maintain homeostasis. (Flip to Chapter 2 for an overview of homeostasis.)

The thalamus is the gateway to the cerebrum. Whenever a sensory impulse travels from somewhere in your body (except from the nose; sensations of smell are sent directly to the brain by the olfactory nerve), it passes through the thalamus. The thalamus then relays the impulse to the proper location in the cerebral cortex, which then interprets the message. Think of the thalamus as an e-mail server, routing your message through the correct lines.

Blood-brain barrier

Blood entering the CNS must pass through the *blood-brain barrier,* which restricts the entry of certain blood molecules into the CNS. The barrier consists of intercellular connections, called *tight junctions,* between the endothelial cells of the capillaries and the processes of the astrocytes that surround them. *Endothelial cells* restrict the diffusion of bacteria, including many common pathogens, thereby protecting the brain from infection. They also block large or hydrophilic molecules from entering the CSF, including some toxins and some drugs. The blood-brain barrier permits the diffusion of small hydrophobic molecules (oxygen, hormones, and carbon dioxide). Other molecules, such as glucose, are carried across the barrier by facilitated diffusion.

Transmitting the Impulse

To move a message from one part of your environment (whether internal or external) to your brain or spinal cord, an impulse must pass through each neuron and continue on its path. Through a chain of chemical events, the dendrites pick up the impulse, which travels through the cell body to the end of an axon, where a neurotransmitter is released, generating an impulse in the next neuron. The entire impulse passes through a neuron in about seven milliseconds, faster than a lightning strike. We take a look at this lickety-split transmission in detail in the following sections.

Across the neuron

When a neuron isn't stimulated (when it's "resting," that is — just sitting with no impulse to transmit), its membrane is *polarized:* The electrical charge on the outside of the membrane is positive while the electrical charge on the inside is negative. The fluid outside the cell contains excess Na^+ (sodium ions); the cytoplasm contains excess K^+ (potassium ions). This gradient is maintained by *Na^+/K^+ pumps* on the membrane. When the neuron is inactive and polarized, it's said to be at its *resting potential.*

In addition to the K^+, negatively charged protein and nucleic acid molecules also inhabit the cell; therefore, the inside is negative relative to the outside.

When a stimulus reaches the dendrite of a resting neuron, a *graded potential* is generated: Gated ion channels in the membrane open, allowing Na^+ to go rushing into the cell. As this happens, the neuron goes from being polarized to being *depolarized.* If the graded potentials are summed and the stimulus goes above the *threshold level,* more gated ion channels open and allow more Na^+ into the cell. This phenomenon is all or nothing — if the stimulus exceeds the threshold level, all the ion channels open. This influx of sodium results in an *action potential* (equivalent to approximately 100 millivolts). As the cytoplasm becomes flooded with Na^+, the Na^+ gates close, and gated ion channels in the membrane open to allow K^+ to flow out of the cell. The membrane is now *repolarized,* with the distribution of charge returned to that of a neuron in its resting state.

The K^+ gates stay open longer than they actually need to, so more K^+ leaves than is actually needed to return the neuron to its resting membrane potential. Thus, the membrane potential drops slightly lower than the resting potential, and the membrane is said to be *hyperpolarized.*

This is followed by the Na$^+$/K$^+$ pumps returning the ion concentration to its original state on either side of the membrane. The neuron doesn't respond to any stimulus as it returns to its normal polarized state; it stays in the resting potential until another impulse comes along.

Figure 7-6 shows the transmission of an impulse. The intensity of the electrical potential, measured in millivolts, a measure of potential energy, is shown on the Y-axis. Values above 0 show positive charge; values below 0 are negative charge. Time goes from left to right on the X-axis.

Across the synapse

Most neurons don't touch. A gap called a *synapse* or *synaptic cleft* separates the axon of one neuron from the dendrite of the next. To traverse this gap, a neuron releases a chemical called a *neurotransmitter* that may or may not cause the next neuron to generate an electrical impulse. In electrical synapses (where the neurons contact each other and are connected by "pipes" known as *gap junctions*), the ions — and an electrical impulse — flow from one neuron to the next.

At the end of the axon, the membrane depolarizes, voltage-gated ion channels open, calcium ions (Ca^{2+}) enter the cell, and as a result a neurotransmitter is released into the synapse. The neurotransmitter moves across the synapse and binds to receptor proteins on the membrane of the neuron (dendrite) that's about to receive the impulse.

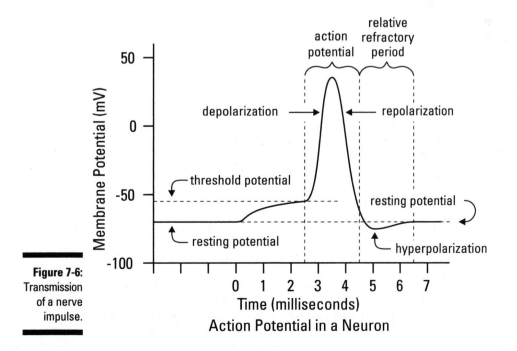

Figure 7-6:
Transmission
of a nerve
impulse.

Neurotransmitters: Facilitating the signal

Neurotransmitters are chemicals that, when released from the end of a neuron's axon, either excite or inhibit an adjacent cell. For example, neurotransmitters are released from the axon endings of motor neurons, where they stimulate or inhibit muscle fibers or glandular cells. Here are some details about some of the better-known neurotransmitters.

- *Acetylcholine* has many functions, notably the stimulation of muscles, including the digestive system's smooth muscles and all skeletal muscles. It's also found in sensory neurons and in the autonomic nervous system, and it plays a part in scheduling REM (dream) sleep.

- *Norepinephrine* (or *noradrenalin*) is released by the adrenal glands, along with its close relative *epinephrine (adrenalin).* It works to enhance the sympathetic nervous system, bringing the nervous system into high alert and increasing the heart rate and blood pressure. It's also important for forming memories.

- *Dopamine,* related to norepinephrine and epinephrine, is an *inhibitory neurotransmitter,* meaning that when it's bound to its receptor site, it blocks the firing of that neuron. Dopamine is strongly associated with reward mechanisms in the brain. If it feels good, dopamine neurons are probably involved. Too little dopamine in the brain's motor areas causes Parkinson's disease, which involves uncontrollable muscle tremors.

- *GABA* (gamma-aminobutyric acid), usually an inhibitory neurotransmitter, acts like a brake to the *excitatory neurotransmitters* that lead to anxiety. People with too little GABA tend to suffer from anxiety disorders, and drugs like Valium work by enhancing the effects of GABA. If GABA is lacking in certain parts of the brain, epilepsy results.

- *Glutamate* is an excitatory relative of GABA. It's the most common neurotransmitter in the central nervous system and is especially important in memory formation. Curiously, glutamate is actually toxic to neurons, and an excess will kill them. ALS, more commonly known as Lou Gehrig's disease, results from excessive glutamate production. Many researchers believe that excess glutamate may also be responsible for quite a variety of nervous system diseases and are looking for ways to minimize its effects.

- *Serotonin* is an inhibitory neurotransmitter that's intimately involved in emotion and mood. Serotonin also plays a role in perception. Serotonin depletion is linked with clinical depression, problems with anger control, obsessive-compulsive disorder, and suicide. It has also been tied to migraines, irritable bowel syndrome, and fibromyalgia.

- *Endorphin* is short for "endogenous morphine." It's structurally similar to the *opioids* and has similar functions — it's inhibitory and is involved in pain reduction and pleasure. Opioid drugs work by attaching to endorphin's receptor sites.

Each type of neurotransmitter has its own type of receptor. Whether the postsynaptic neuron is excited or inhibited depends on the neurotransmitter and its effect. For example, if the neurotransmitter is an excitor, the Na$^+$ channels open, the neuron membrane becomes depolarized, and the impulse is carried through that neuron. If the neurotransmitter is an inhibitor, the K⁺ channels open, the neuron membrane becomes hyperpolarized, and the impulse is stopped.

After the neurotransmitter produces its effect (excitation or inhibition), the receptor releases it, and the neurotransmitter goes back into the synapse. In the synapse, one of three things happens to the neurotransmitter. It may

✔ Be broken down by an enzyme in the postsynaptic neuron.

✔ Be taken back up by the cell that released it (a form of recycling).

✔ Simply float away into the surrounding fluid.

Making Sense of Your Senses

Some anatomists consider the sensory system to be part of the peripheral nervous system. Others regard it as a separate system. The five senses are touch, hearing, sight, smell, and taste. Each of the sense organs contains specialized types of receptors (see Table 7-2).

Table 7-2	Receptors Found in Sense Organs	
Sense Organ	*Function*	*Receptor*
Skin (touch, pressure, temperature, pain)	Mechanoreceptors (touch and pressure), thermoreceptors (hot and cold), and nociceptors (pain)	Specialized nerve endings detect different sensations (see Chapter 6 for more on skin).
Ears	Mechanoreceptors	The stereocilia of the "hair cells" (neurons) in the ears detect the fluid movement in the cochlea of the inner ear, which is generated by movement of the eardrum and ossicles (ear bones) that allow you to hear.
Eyes	Photoreceptors	The rods and cones of the retina of the eye detect light to allow vision.
Nose	Chemoreceptors	The olfactory cells (neurons) in the nasal cavity detect chemical molecules present in the air.
Tongue	Chemoreceptors	The gustatory cells (neurons) of the taste buds detect various chemical molecules present in foods.

Touch

The sensory receptors all over the skin perceive at least five different types of sensation: pain, heat, cold, touch, and pressure. The five are usually grouped together as the single sense of touch.

Receptors vary in terms of their overall abundance (pain receptors are far more numerous than cold receptors) and their distribution over the body's surface (the fingertips have far more touch receptors than the skin of the back). Areas where the pressure sensors are packed in are especially sensitive.

The structure of the sensory receptors varies with their function. Some examples: Pain receptors are simply unencapsulated terminal branches of neurons; the root hair plexus forms a neuronal fiber net around the base of a hair to determine whether the hair has been moved; tactile disks (unencapsulated), encapsulated Krause bulbs, Ruffini corpuscles, and tactile corpuscles are touch or light pressure receptors; and deep pressure receptors (lamellated corpuscles) consist of nerve endings encapsulated by specialized connective tissues.

Nerve fibers that are attached to different types of skin receptors either continue to discharge during a stimulus or respond only when the stimulus starts and, sometimes, when it ends. That's why you're aware of your shoes on your feet when you first put them on, but the stimulus fades within a minute or two. In other words, *slowly adapting nerve fibers* send information about ongoing stimulation; *rapidly adapting nerve fibers* send information related to changing stimuli.

Hearing and balance

The ear changes sound pressure waves into nerve impulses sent to the brain (see Figure 7-7). Sound moves through air in waves of pressure. Your outer ear acts as a funnel to channel sound waves to the eardrum, causing the eardrum to vibrate. (The outer ear isn't necessary for hearing, but it helps.) The ear bones, called *ossicles,* receive and amplify the vibration and transmit it to the inner ear. The vibrations create tiny ripples in the inner ear's fluid. The hollow channels of the inner ear's *cochlea* are filled with liquid, and one of the channels contains a sensory epithelium studded with *hair cells* — mechanoreceptors that release more or less of a neurotransmitter when stimulated.

The nerve impulses travel from the left and right ears through the eighth cranial nerve to both sides of the brain stem and up to the temporal lobe — the portion of the cerebral cortex dedicated to sound.

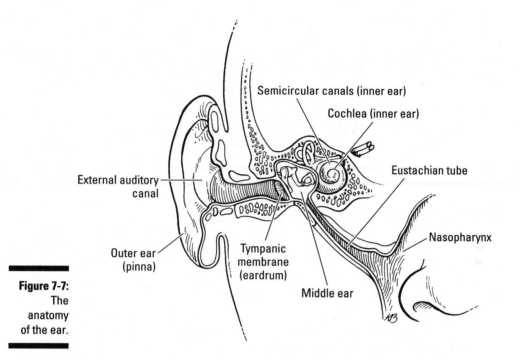

Figure 7-7:
The
anatomy
of the ear.

Semicircular canals (inner ear)

Cochlea (inner ear)

Eustachian tube

External auditory
canal

Nasopharynx

Outer ear
(pinna)

Tympanic
membrane
(eardrum)

Middle ear

Your inner ears also transmit information to your brain regarding what position your head is in; that is, whether you're horizontal or vertical, spinning or still, moving forward or backward. Therefore, your ears are the key organ of *balance*. The process of transmitting information to the brain about your body position is basically the same as that of hearing. When you're moving, the inner ear's fluid moves and causes hair cells to bend, sending impulses to the brain at varying speeds through the eighth cranial nerves. Your brain then processes the information about where you are spatially and initiates movements to help you keep your balance.

Sight

Vision is probably the most complex of the senses. Your eye's *pupil,* the dot in the center of your eye that's usually black in color, allows light in. The *iris,* the pretty, colored part of your eye, contains the muscle that controls the pupil's size and thus the amount of light that enters the eye. The iris muscle contracts to dilate the pupil and allow more light in, such as when you're in a dark room or outside at night. Likewise, another muscle in the iris contracts to make the pupil smaller and let less light in, such as when you're in the sun or a brightly lit place.

The *cornea* lies anterior to the iris and pupil. (You can kind of see a clear area if you look at an eyeball from the side.) The eye *lens* is behind the iris and pupil (see Figure 7-8).

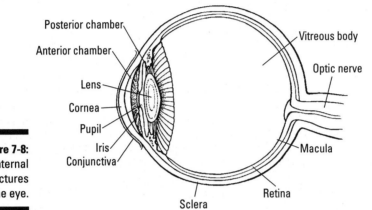

Figure 7-8:
The internal
structures
of the eye.

A clear, gelatinous material fills the *vitreous body,* which lies behind the lens. And we're not joking when we tell you that the gelatinous material is *vitreous humor.* The transparent vitreous humor gives the eyeball its rounded shape, and it also lets light pass through it to the back of the eyeball.

The *retina* is the innermost layer of the eyeball. The retina contains two types of photoreceptors — *rods,* which detect dim light and are sensitive to motion, and *cones,* which detect color and fine details. Three types of cones detect color — one each for detecting red, blue, and green. A missing or damaged cone (regardless of the type) results in color blindness.

When light strikes the rods and cones, nerve impulses are generated and sent to cells that form the *optic nerve.* The optic nerve joins your eyeball directly to your brain and sends impulses to the brain for interpretation in the occipital lobe.

Olfaction

The nose knows, but the *olfactory cells* know better. The nose is the sense organ for smell (*olfaction* is the proper term): The olfactory cells that line the top of your nasal cavity detect volatile, water-soluble molecules in the air as you breathe. As you take in air through your nostrils, those molecules waft right up to your olfactory cells, where the chemicals bind to the cilia-like "hairs" that line your nasal cavity. That action initiates a nerve impulse that's sent through the olfactory cell, into the *olfactory nerve fiber,* up to the *olfactory bulb,* and right to your brain. (The olfactory bulb is the expanded area at the end of the olfactory tract where the olfactory nerve fibers enter the brain.) The brain then "knows" what the chemical odors are from, and you know what you're smelling.

Taste

Your sense of taste *(gustation)* has a simple goal: to help you decide whether to swallow or spit out whatever is in your mouth. This extremely important decision can be made based on a few taste qualities. The tongue, the sense organ for taste, has chemoreceptors to detect certain aspects of the chemistry of food, certain minerals, and some toxins, especially poisons made by plants to deter predation by animals.

The human tongue has about 10,000 taste buds, each with between 50 to 150 chemoreceptor cells.

A taste bud's olfactory cells generate nerve impulses that are transmitted through the sensory nerve fiber to the gustatory areas of the brain via the seventh, ninth, and tenth cranial nerves. The brain then interprets the impulse and causes the release of the digestive enzymes needed to break down that food. So the sense of taste is tied to the endocrine system as well as to the digestive system.

Taste buds have receptors for sweet, sour, bitter, and salty sensations, as well as a fifth sensation called *umami* ("savoriness" — the flavor associated with meat, mushrooms, and many other protein-rich foods). Salty and sour detection is needed to control salt and acid balance. Bitter detection warns of foods that may contain poisons — many of the poisonous compounds that plants produce for defense are bitter. Sweet detection provides a guide to calorie-rich foods, and umami detection to protein-rich foods.

Each receptor cell in a taste bud responds best to one of the five basic tastes. A receptor can respond to the other tastes, but it responds strongest to a particular taste. And the taste buds detect only the rather unsubtle aspects of flavor. The nose detects more complex and subtler flavors.

The structure and function of taste buds is an area of active research and controversy, especially in the food industry. Even the exact number of basic tastes and their characterization is controversial, with new "basic tastes" being proposed from time to time.

Pathophysiology of the Nervous System

The discussion in this section is limited to anatomical disorders of nervous tissue affecting the "lower level" functions. Major brain disorders and physiology-based psychopathologies such as schizophrenia are beyond the scope of this book.

Chronic pain syndrome

Chronic pain is a disorder of the nervous system as much as it is a disorder of the organ causing the pain. Chronic pain has many sources (infection, trauma, and so on), but to be perceptible, the pain signal must be transmitted to the spinal cord and up to pain receptors in the brain. *Analgesics* (pain meds) work by blocking these receptors. Another approach that has proved effective in relieving chronic pain involves using a device to generate electrical impulses directly to the spinal cord to interrupt or cancel out the incoming pain impulses at the point where the spinal nerve transmitting the pain arrives at the spinal cord.

Multiple sclerosis

Multiple sclerosis (MS) affects the myelin sheath that covers myelinated axons. Lesions in the myelin sheath become inflamed and irritated. When the lesions heal, scar tissue (sclerosis) on the sheath interferes with the transmission of impulses through the axon, thus blocking movement or response in the innervated muscle. As the disease progresses, movement becomes increasingly difficult.

Macular degeneration

Macular degeneration is a vision disorder that's now a leading cause of blindness among older individuals. In macular degeneration, the *macula lutea* — a small area of the retina with a large concentration of cone photoreceptors that detect color and fine details — weakens and degenerates. Objects look smaller or larger than they really are, and colors appear faded.

One cause of macular degeneration is overgrowth of new blood vessels around the macula lutea. New growth sounds healthy, but often it's not. The new vessels leak, and as they ooze blood into the macula lutea, its delicate photoreceptors are destroyed. Macular degeneration can also result from excessive sun exposure, especially among people with blue or green eyes.

Chapter 8

The Endocrine System: Releasing Chemical Messages

In This Chapter

▶ Understanding what hormones do and how they work

▶ Reviewing the main glands and their functions

▶ Looking at the effects of hormone disorders

The glands of the endocrine system and the hormones they release influence almost every cell, organ, and function of your body. The endocrine system is instrumental in regulating mood, tissue function, metabolism, and growth and development, as well as sexual function and reproductive processes.

In general, the endocrine system is in charge of body processes that happen slowly, such as cell growth. The nervous system controls faster processes like breathing and body movement. Throughout this chapter, you can see that the nervous system (which we cover in Chapter 7) and the endocrine system work closely together. The nervous system controls when the endocrine system should release or withhold hormones, and the hormones control the metabolic activities within the body. This chapter explains the functions of hormones, the glands that they come from, and common disorders of the endocrine system.

Homing In on Hormones

A *hormone* is an *endogenous substance* (one that's produced within the body) that has its effects in specific target cells. Hormones are many and varied in their source, chemical nature, target tissues, and effects. However, they're characterized by the fact that they're synthesized in one place (gland or cell) and they travel via the blood until they reach their target cell. Hormones are bound by specific receptors in their target cells. The binding of the hormone on the receptor induces a response within the cell.

Among glands, the pituitary, thyroid, and adrenal glands are most well-known. These organs have no significant function other than to produce hormones. However, a number of other endocrine tissues and hormones, though less well-known, are just as important in controlling vital bodily functions. In fact, all the tissue in your body is, in some way, an endocrine tissue.

Hormones play a big part in *homeostasis* (see Chapter 2). When the blood passes certain checkpoints in the nervous system (such as the hypothalamus inside the brain), hormone levels are "measured." If the level of a certain hormone is too low, the gland that produces that hormone is stimulated to produce more of it. If a hormone level is too high, the gland that produces the hormone either doesn't receive any further hormonal stimulation or is "instructed" to stop or slow production. The hormone comes from the endocrine system, but the "instruction" comes from the nervous system.

Your body is always keeping tabs on the metabolic processes going on inside it. If your body temperature, glucose level, or pH level leaves the normal range, the checkpoints involved in homeostasis work with the endocrine system to bring your systems back into balance.

The following sections guide you through how hormones are structured chemically, where hormones come from, and how hormones work.

Hormone chemistry

Chemically, hormones fall into three types: those derived from lipids, from peptides, and from amines.

- ✔ **Lipid hormones:** Lipid and phospholipid hormones are derived from fatty acids. The best-known lipid hormones are the steroids, such as estrogen, progesterone, testosterone, aldosterone, and cortisol, which is synthesized from cholesterol. Another group of lipid hormones is called *prostaglandins*.

- ✔ **Peptide hormones:** Peptides are relatively short chains of amino acids. Peptide hormones include antidiuretic hormone (ADH), thyrotropin-releasing hormone (TRH), and oxytocin.

  Other hormones are *proteins* (chains of peptides), such as insulin, growth hormone, and prolactin.

- ✔ **Glycoprotein hormones:** More complex protein hormones bear carbohydrate side-chains and are called *glycoprotein hormones*. These include follicle-stimulating hormone (FSH), luteinizing hormone (LH), and thyroid-stimulating hormone (TSH).

✔ **Amine hormones:** Amine hormones are derivatives of the amino acids tyrosine and tryptophan. Examples are thyroxine, epinephrine, and norepinephrine.

Hormone sources

At one time, and not so long ago, by definition a hormone was produced in an endocrine gland (and an endocrine gland was a structure that produced one or more hormones). But as biologists have discovered and described more and more hormone substances and forms, they've expanded the definition to include similar, sometimes identical, substances that have a similar mechanism of action, wherever they're produced. Check out all the sources of hormones:

✔ **Endocrine glands:** An *endocrine gland* is an organ that synthesizes a hormone. It does so within a specialized cell type — the anterior pituitary gland, for instance, has cells that specialize in the production of such hormones as adrenocorticotropic hormone (ACTH), growth hormone, and TSH. Specialized cells within the thymus synthesize hormones that control the maturation of immune cells.

✔ **Various organs:** A number of organs not usually included within the endocrine system by anatomists and physiologists have cells and tissues specialized for the production of hormones. For example:

- While part of the pancreas is busy secreting enzymes for the digestion of food, other specialized cells of the pancreas produce insulin, and others produce glucagon.

- The stomach and intestines, too, synthesize and release hormones that control both physical and chemical aspects of digestion.

- Specialized cells in the ovaries and testes transform cholesterol molecules into molecules of estrogen and testosterone, respectively.

- Even the heart produces hormones, the secretion of which has an immediate strong effect on blood volume (fluid balance).

✔ **Neurons:** Neurons make hormones that are neurotransmitters. Or, to look at it another way, hormones are substances that transmit physiologically significant messages with considerable subtlety; therefore, not surprisingly, they're synthesized and released in different physiological messaging contexts — the transmission of nerve impulses across a synapse, for example (flip to Chapter 7 for more on the nervous system). The only difference between epinephrine synthesized in the adrenal glands and epinephrine synthesized in nerve cells is the distance the molecules travel to their target site.

Table 8-1 lists the body's major hormones, their sources, and their functions. We provide more information about some of the individual hormones in the "Grouping the Glands" section later in this chapter.

Table 8-1   Important Hormones: Sources and Primary Functions

Hormone	Source	Function(s)
Adrenocorticotropic hormone (ACTH)	Pituitary gland (anterior part)	Stimulates secretion of corticosteroids by the cortex of the adrenal gland.
Antidiuretic hormone (ADH)	Pituitary gland (posterior part)	Stimulates the kidneys to reabsorb water, preventing dehydration.
Calcitonin	Thyroid gland	Targets the bones, kidneys, and intestines to reduce the level of calcium in the blood.
Epinephrine Norepinephrine	Medulla of the adrenal gland	Stimulates the heart and other muscles during the fight-or-flight response; increases the amount of glucose in the blood.
Estrogen	Ovaries	Stimulates the maturation and release of ova; targets muscles, bones, and skin to develop female secondary sex characteristics.
Glucagon	Pancreas	Causes liver, muscles, and adipose tissue to release glucose into the bloodstream.
Glucocorticoids	Cortex of the adrenal glands	Stimulate the formation of glucose from fats and proteins.
Gonadocorticoids	Cortex of the adrenal glands	Stimulate the libido.
Growth hormone (GH)	Pituitary gland (anterior part)	Targets the bones and soft tissues to promote cell division, synthesis of proteins, and growth of bone tissue.
Insulin	Pancreas	Causes liver, muscles, and adipose tissue to store glucose as a way of lowering blood glucose level.
Melatonin	Pineal gland	Targets a variety of tissues to mediate control of biorhythms, the body's daily routine.

Hormone	Source	Function(s)
Mineralocorticoids	Cortex of the adrenal glands	Targets the kidney cells to reabsorb sodium and excrete potassium to keep electrolytes (ions) within normal level.
Oxytocin	Pituitary gland (posterior part)	Stimulates uterine contractions during childbirth and mammary glands to release milk. In males, it has been demonstrated to cause contraction of the spermatic ducts during ejaculation.
Parathyroid hormone	Parathyroid glands	Stimulates the cells in bones, kidneys, and intestines to release calcium so that blood calcium level increases.
Progesterone	Ovaries	Prepares the uterus for implantation of an embryo and maintains the pregnancy.
Prolactin	Pituitary gland (anterior part)	Targets the mammary gland to stimulate production of milk.
Testosterone	Testes	Stimulates the production of sperm in testes; in skin, muscles, and bones, causes development of male sex characteristics.
Thyroid-stimulating hormone (TSH)	Pituitary gland (anterior part)	Stimulates the thyroid gland to produce and release its important hormones, calcitonin and thyroxin.
Thyroxin	Thyroid gland	Distributed to all tissues to increase metabolic rate; involved in regulation of development and growth.

Hormone receptors

Hormones exit their cell of origin via *exocytosis*, which involves a sac or vesicle enveloping the substance and moving it across the cell membrane, or another means of membrane transport. A secreted hormone molecule goes directly into the blood and circulates until it enters a cell or binds with its specific receptor on the cell membrane, where through second messengers its effects may be profound.

The presence of a specific hormone receptor makes that cell a "target" for the hormone. Without the target receptor, the hormone has no effect.

The receptor may be on or embedded in the cell membrane, as is typically the case for peptide hormones. The hormone molecule, called the *first messenger,* is taken into the cell via active transport, stimulating the production of a compound, *cyclic AMP (cyclic adenosine monophosphate),* called the *second messenger,* thus causing the target cell to produce the necessary enzymes (that is, to induce the expression of a certain gene).

A steroid hormone molecule doesn't require a cell-membrane receptor or active transport. As a lipid, it enters a cell by diffusing through the membrane. After it's inside the cell, it binds with target receptor molecules in the cytoplasm. The receptor-hormone complex moves into the nucleus, where it activates the expression of the gene for a needed enzyme.

Grouping the Glands

In general, a *gland* is a structure that synthesizes a product that's exported from the cells. *Endocrine* (ductless) glands export their products (hormones) via the bloodstream to their target cells in anatomically distant organs. The following sections give you the lowdown on the endocrine glands. To see where these glands are located in the body, check out the "Glands of the Endocrine System" color plate in the center of the book.

The endocrine system, like any good communications system, functions in a very integrated way. That integration complicates the discussion of which anatomical structure belongs in what category. Remember that any grouping is somewhat arbitrary.

Please exit through the ducts: The exocrine glands

Exocrine glands (glands with ducts) produce substances, but not hormones, that are transported through a duct to the cavity of a body organ or to the surface of the body. For example, *sebaceous glands* are exocrine glands. So are sweat glands. Refer to the "Skin (Cross Section)" color plate in the center of the book for the anatomical position and structure of these glands. The sebaceous glands produce protective oils and secrete them directly onto their nearby tissues, the skin and hair. The oils don't travel around the body and don't have an effect anywhere else. The same with sweat: It's just deposited on your skin. Its only important effect, evaporative cooling, is limited to the skin immediately adjacent to the sweat gland.

Some physiologists use the term *diffuse endocrine system* to reflect the concept that many organs house clusters of cells that secrete hormones. The kidney, for example, contains scattered cells that secrete *erythropoietin,* a hormone essential for the production of red blood cells. The heart contains cells that produce *atrial naturetic hormone,* which is important in sodium and water balance.

The taskmasters: The hypothalamus and pituitary

The *hypothalamus* can be considered the location where the nervous system and the endocrine system meet. It's sometimes called the "master gland" because it ultimately controls the functioning of other glands, acting through the *pituitary.* Why so many "levels of management"? Evolution appears to have favored those organisms that had their hormones under tight control.

The hypothalamus and the pituitary gland adjoin in the central portion of the brain called the *diencephalon*.

Hypothalamus

The hypothalamus contains special cells that act as sensors that "analyze" the composition of the blood as it circulates through. It also contains other specialized cells that generate messengers (hormones) in response to the analysis. Tight pairing between these two types of cells is essential for homeostasis.

The hormones that the hypothalamus produces don't have target cells in the body. The hypothalamus synthesizes the hormones *hypothalamic releasing hormone* and *hypothalamic release-inhibiting hormone* and releases them into small blood vessels that connect to the anterior pituitary. The pituitary responds by synthesizing and releasing its hormones, which have target cells in distant organs. Figure 8-1 shows the relationship between the hypothalamus and pituitary glands.

Pituitary

The pituitary gland has two parts, called the *anterior pituitary* and *posterior pituitary,* that have different relationships with the hypothalamus.

The anterior pituitary gland secretes many hormones, including melanocyte-stimulating hormone (MSH). This hormone directly stimulates melanocytes to produce melanin pigment, which protects the skin from sunlight damage. This gland also secretes prolactin, which is responsible for the increase in size of the lactiferous glands in the breast and the production of milk.

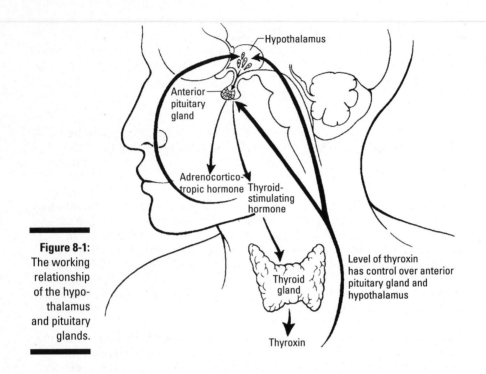

Figure 8-1:
The working relationship of the hypothalamus and pituitary glands.

Labels in figure:
Hypothalamus
Anterior pituitary gland
Adrenocortico-tropic hormone
Thyroid-stimulating hormone
Thyroid gland
Thyroxin
Level of thyroxin has control over anterior pituitary gland and hypothalamus

The anterior pituitary secretes the gonadotropic hormones FSH and LH, which target the ovaries and testicles, and ACTH, which targets the cortex of the adrenal glands. The function of these pituitary hormones is to stimulate the release of other hormones from their target glands. The same is true of growth hormone and TSH. They're messengers that stimulate the action of other endocrine glands. (This information is summarized in Table 8-1.)

The *posterior pituitary gland* is directly connected to the hypothalamus (refer to Figure 8-1). The hormones that the posterior pituitary gland releases are actually synthesized in the nerve cell bodies of the hypothalamus. The hormones travel down the axons that end in the posterior pituitary.

One such hormone is ADH. When the blood's fluid volume falls below the ideal range, the hypothalamus produces ADH, which travels down the axons into the posterior pituitary gland. Released by the pituitary into the blood, ADH reaches its target kidney cells. Via active transport, ADH enters the cells of the tubules and alters the cells' metabolism so that more water is removed from the urine that the kidney produces and is added to the blood.

Controlling metabolism

Two relatively small organs exert a major effect on the availability of energy for physiological processes.

Thy thyroid and thou

The *thyroid hormones* affect almost every physiological process in the body. Your *thyroid gland* looks somewhat like a butterfly that straddles your *trachea* (windpipe). (It's pictured in Figure 8-1 and in the "Glands of the Endocrine System" color plate in the center of the book.) Each lobe of the thyroid gland — the butterfly's wings — is to the side of the trachea; a stretch of tissue called the *isthmus* connects the lobes. Columnar epithelial cells lining the follicles within the lobes secrete a jellylike substance called *thyroglobulin.* Thyroglobulin "traps" iodine ions (consumed in food) in the *colloid* of the follicle and promotes a reaction with the amino acid tyrosine to form the amine hormones *thyroxin* and *triiodothyronine.* When TSH from the anterior pituitary binds the target receptors in the thyroid, the thyroid hormones adsorb (stick to) protein molecules in the thyroid cells and are released slowly into the bloodstream.

The thyroid hormones regulate the following physiological reactions, among others. We discuss some of the problems that arise from the malfunction of these hormones in the "Pathophysiology of the Endocrine System" section later in the chapter.

The thyroid hormones

- ✔ Regulate the rate at which cells metabolize and respire (use oxygen and release carbon dioxide).

- ✔ Increase the rate at which cells use glucose and stimulate the breakdown of glycogen (the storage form of glucose) into individual glucose molecules so that the blood level of glucose increases.

- ✔ Help maintain body temperature by increasing or decreasing metabolic rate.

- ✔ Regulate growth and differentiation of tissues in children and teens.

- ✔ Increase the amount of certain enzymes in the mitochondria that are involved in oxidative reactions.

- ✔ Influence the metabolic rate of proteins, fats, carbohydrates, vitamins, minerals, and water.

- ✔ Stimulate mental processes.

Topping the kidneys

The root *renal* means "kidney"; the prefix *ad* means "near." So the adrenal glands are near the kidneys. More specifically, they sit on top of the kidneys (see Figure 8-2 and refer to the "Glands of the Endocrine System" color plate in the center of the book). Like the skin on a kidney bean, a thin *capsule* covers the entire adrenal gland. Inside, each adrenal gland has two parts: the *cortex* and the *medulla,* which have different functions.

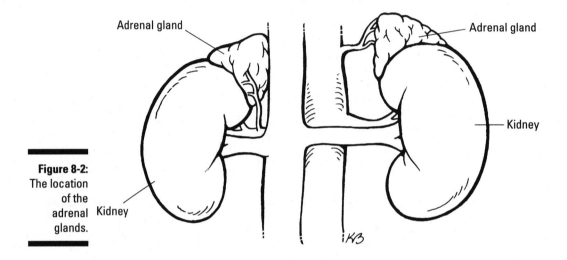

Adrenal gland

Adrenal gland

Kidney

Figure 8-2:
The location of the adrenal glands.

Kidney

Adrenal cortex

The adrenal cortex secretes *corticosteroids,* which include *mineralocorticoids, glucocorticoids,* and *gonadocorticoids* (refer to Table 8-1). One of the most important mineralocorticoids is *aldosterone,* which is responsible for regulating the concentration of *electrolytes,* such as potassium (K^+), sodium (Na^+), and chloride (Cl^-) ions. This regulation keeps the blood's salt and mineral content within the ranges required for homeostasis.

Electrolytes are substances that split apart into *ions* (atoms or molecules with a positive or negative charge) when in solutions, such as the watery tissue fluid around your cells or the cytoplasm within your cells. As their name suggests, electrolytes are capable of conducting electricity.

Aldosterone targets the kidneys' tubules and stimulates the resorption of sodium ions. When the sodium ions are reabsorbed into the bloodstream, chloride ions quickly follow. Na^+ and Cl^- love to be together as NaCl, commonly known as "salt." And where salt is, water follows. If salt ions move into the bloodstream, water does, too, increasing the blood's fluid volume. Ultimately, the fluid and electrolyte balance affects blood pressure (see Chapter 9 for everything you wanted to know about the circulatory system).

The *gonadocorticoids* were named such because they are identical to the steroid hormones made by the gonads, the testicles and ovaries. The gonadocorticoids consist of *testosterone, estrogen,* and *progesterone.* If you thought that as a woman you didn't have any testosterone or as a man you didn't have any estrogen or progesterone, you're wrong. Admittedly, these are secreted in small amounts and have little influence on the development of the reproductive system. Apparently, their primary responsibility is to heighten sex drive. Disruption in the production of their low amounts can lead to *feminization* in the male and *masculinization* in the female.

Cortisol, the main glucocorticoid hormone, regulates the metabolism of proteins, fats, and carbohydrates. Your body releases cortisol when you're stressed emotionally, physically, or environmentally. Cortisol affects metabolism in the following ways:

- ✔ It breaks down protein, decreases protein synthesis, and moves amino acids from tissues to liver cells to promote gluconeogenesis.
- ✔ It moves fat from adipose tissue to the blood.
- ✔ It reduces the rate at which cells use glucose, which causes the liver to store more glucose as glycogen.

Cortisol and other corticosteroids affect the immune system by decreasing the number of circulating immune cells and decreasing the size of the lymphoid tissue. Under severe stress and with large amounts of glucocorticoids circulating in the blood, the lymphoid tissue is unable to produce antibodies (see Chapter 13 for more on the immune system). The role of corticosteroids in susceptibility to infectious disease is an active area of medical research.

Adrenal medulla

The adrenal medulla developed from the same tissues as the sympathetic nervous system (we cover the nervous system in Chapter 7). Some of the adrenal medulla's functions involve regulating actions of structures of the sympathetic nervous system, including a class of hormones called the *catecholamines,* of which *epinephrine* and *norepinephrine* are the best known.

Epinephrine, also called *adrenaline,* initiates the "adrenaline rush" of the *fightor-flight response.* It stimulates the release of free fatty acid molecules from your adipose tissue. Your muscles — including your heart muscle and respiratory system muscles — use these fatty acid molecules for energy, saving glucose for use by your brain. After all, if you're fighting or fleeing, you need to think.

Unlike most hormones, epinephrine produces its effect almost instantaneously.

Like epinephrine, norepinephrine is a catecholamine that's tied to the nervous system. Norepinephrine causes *vasoconstriction* — that is, tightening of the blood vessels. Norepinephrine is released to increase blood pressure when the hypothalamus senses hypotension, as well as when you're stressed. (Your body still prepares to run or fight, even when these responses aren't quite appropriate.)

Getting the gonads going

Your *gonads* — *ovaries* if you're female or *testes* if you're male — produce and secrete the steroid sex hormones — *estrogen* and *progesterone* in females and *testosterone* in males. Your body secretes sex hormones throughout your lifetime at different levels. Their production increases at puberty and normally decreases as you age.

You may think that estrogen is limited to female animals, but you'd be wrong. Estrogen can be detected in the urine of male animals, and even, surprisingly, in growing plants! Just as men also have some estrogen, women also have some testosterone.

Estrogen

In women, the increased production of estrogen at puberty is responsible for initiating the development of the secondary sex characteristics, such as the enlargement of the breasts. Bone tissue grows rapidly, and height increases. Estrogen helps this process, causing calcium and phosphate to be transported in the bloodstream so they can be used for bone growth and for stimulating the activity of the osteoblasts.

Estrogen also allows the pelvic bones to widen to allow passage of an infant during childbirth. In addition, estrogen increases the deposition of fat around the body, thus giving women a more rounded appearance than men.

Progesterone

Progesterone works to prepare the uterus for implantation of a pre-embryo by causing changes in uterine secretions and in storing nutrients in the uterus's lining. Progesterone also contributes to breast development.

Testosterone

Testosterone causes the development of secondary sex characteristics in males. As a boy hits puberty, his muscle tissues start to grow, his sex organs enlarge, hair develops on his chest and face, and the hair on his arms and legs becomes darker and coarser. For more on development during puberty, turn to Chapter 15.

The short but ongoing history of hormone replacement therapy

As they age, men and women produce fewer sex hormones. In a man, the decline typically begins in the 30s and proceeds slowly. Many men retain their sexual and reproductive functions late in life. For women, the decline begins in the 30s, as with men, but becomes precipitous in the late 40s or 50s. This is the period of *menopause.*

The effects of sex hormone depletion are similar in both sexes, but they happen more suddenly in women. An early effect in women is the cessation of reproductive function — ovulation ceases and pregnancy is no longer possible. Acutely, faulty signaling in the parasympathetic nervous system disrupts the body's temperature-monitoring ability, causing hot flashes and sweat baths. Chronically, cellular metabolism slows down. *Bone remodeling* (the constant buildup and breakdown of bone tissue) tilts in the direction of breakdown. Replacement of the structural proteins in the skin slows to a crawl.

Because all of these undesirable effects could be shown to be related to decreased hormone levels, researchers questioned whether they could develop hormone therapy to slow down or stop some of this creeping decrepitude. Between the 1960s and the 1990s, various types of female hormone-replacement drugs became available at low costs. In the 1980s and 1990s, primary care physicians offered estrogen-replacement therapy routinely to women in menopause. Some 20 percent of women in menopause took hormone replacement therapy (HRT), a huge patient population. Many women, it seemed, hoped to enjoy their post-reproductive years in health and vitality.

Their expectations were based on some early research that had indicated beneficial effects on the cardiovascular and nervous systems of women who chose HRT. As is routine in research, larger studies were designed to test the results of these early studies. Exemplary data for this study even existed, from the Women's Health Initiative (WHI), a large, randomized, clinical trial, sponsored by the National Institutes of Health, that included more than 16,000 healthy women.

The medical "outcomes" in women who underwent HRT surprised nearly everyone when they were published in 2002. Although researchers expected some additional cancers from increased estrogen, the numbers were disturbing. Researchers had expected to see a "cardioprotective effect," based on the observation that heart disease remains low in women until menopause. But cardiovascular diseases were actually more prevalent and more severe in those who had taken HRT. The therapy had minimal effect on bone retention (exercise was better), which ceased immediately on discontinuation of the therapy. It didn't have a measurable effect on cognition or memory. Its only reliable effect, it seemed, was the diminution of hot flashes in the first months or years after menopause (not dismissed lightly by anyone who has suffered them). Follow-up analyses of these same data have supported the original conclusions.

Women's attitudes and doctors' prescribing habits changed abruptly. Though the current prescribing guidelines are considered safe, as supported by the data in the WHI study, some women are choosing to endure hot flashes.

Interest in the clinical uses of hormones to control the symptoms of aging remains strong, but the experience so far suggests that the endocrine function is subtler and more entwined than researchers imagined.

Enteric endocrine

Much of the endocrine function is *enteric* (related to digestive processes). Precise control of nutrient intake and storage and the excretion of toxins and digestive byproducts is essential for homeostasis and metabolism.

Flat as a pancreas

Your *pancreas* is a fibrous, elongated, flat (as a pancake) organ that lies nestled in your abdomen near your kidneys, stomach, and small intestine. (See Figure 8-3 and the "Glands of the Endocrine System" color plate in the center of the book.) Its two tissue types have different functions:

✔ **Digestive function:** It produces digestive enzymes that it secretes into the small intestine.

✔ **Endocrine function:** It produces the hormones *insulin* and *glucagon,* which it secretes into the blood.

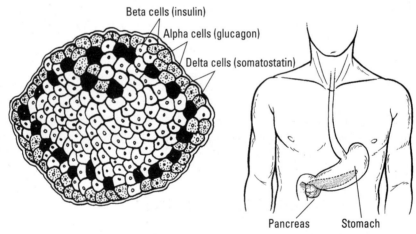

Beta cells (insulin)
Alpha cells (glucagon)
Delta cells (somatostatin)

Pancreas   Stomach

Figure 8-3:
Anatomy
of the
pancreas.

Glucose travels in the blood, making it available as an energy molecule for all cells. But too much glucose in the blood is harmful to the small vessels, especially those in the extremities, in the kidneys, and in the retina of the eye. The body must keep the concentration of glucose in the blood within limits. The pancreas's endocrine tissues produce insulin and glucagon, which work together to balance the blood's glucose level — insulin acts to lower it, and glucagon acts to raise it. The regulation of blood glucose is a classic example of a hormone mechanism of homeostasis. Refer to the "Pathophysiology of the Endocrine System" section later in this chapter for a discussion of disorders in this system.

Insulin is released when the blood's glucose level rises. It acts by stimulating glucose uptake in cells. With plenty of glucose available, the cell's metabolic rate increases and it produces more of its specialized anabolic product. Glucose is burned (calories) to fuel physiologically productive reactions. Insulin also stimulates glucose uptake activity in the energy storage cells in the liver, muscle cells, and fatty tissue.

A lower blood glucose level stimulates the pancreas to secrete insulin's partner glucagon, which pulls glucose from cells where it's stored and releases it into the blood. Glucagon keeps the metabolic fires burning at a steady level.

Insulin secretion is an example of a *negative feedback mechanism.* As the blood's glucose level declines, the body slows the secretion of insulin until the next surge of blood glucose following a meal or snack.

Stomaching another gland

Yes, the stomach is a key organ in digestion (Chapter 11 covers the digestive system in detail), but it also secretes hormones that are used during digestion, making it a gland. The stomach secretes a group of hormones called *gastrins.* Many types of gastrin molecules — small, medium, and large — are responsible for stimulating the secretion of gastric acid. Gastrins also control the sphincter muscle at the bottom of the esophagus, thereby controlling when food can be passed into the stomach.

Testing the intestines

The intestines secrete powerful digestive enzymes, but they require a pH much higher (less acidic) than the extremely acidic stomach acids. The small intestine produces hormones that control the release of neutralizing substances, such as bile from the gallbladder and pancreatic bicarbonate.

Other endocrine glands

Some endocrine glands aren't easily classifiable.

✔ **Parathyroid:** The *parathyroid glands* are four small glands that secrete *parathyroid hormone,* which regulates the concentration of calcium in the blood, making it available to the muscle fibers and neurons. The parathyroid essentially helps the nervous and muscular systems function properly. Calcium is the primary element that causes muscles to contract, and calcium levels are very important to the normal conduction of electrical currents along nerves.

The four parathyroids are typically found on the back side of the thyroid. They're about the size and shape of a grain of rice. They are related to the thyroid in their function of maintaining calcium levels.

✔ **Pineal gland:** The *pineal gland* is a small oval gland in the brain between the cerebral hemispheres that's considered part of the epithalamus of the diencephalon. The pineal gland secretes the hormone *melatonin,* thought to play a role in regulating the body's *circadian rhythm* — the normal fluctuation of physiology over the day-night cycle. The secretion of melatonin is influenced by the perception of light within the gland.

✔ **Thymus:** The *thymus* is a lobed gland situated in the thoracic cavity, just below the collarbones and just above the heart. The thymus's main function is to stimulate the maturation of T lymphocytes from the bone marrow into T cells (see Chapter 13). The thymus produces a group of hormones called *thymosins* that are involved in differentiating and stimulating the immune system's cells.

Pathophysiology of the Endocrine System

The body depends on its chemical messaging system to maintain control over its physiological processes. Malfunctions in the messaging system can disrupt target organ systems and the circulatory system that transports the hormones between the gland and the target organs.

Abnormalities in insulin metabolism

The following abnormalities in insulin metabolism cause blood glucose to rise and remain at high levels. This extra glucose damages the smallest blood vessels, like those in the retina of the eye and the glomeruli of the kidney.

Metabolic syndrome

Insulin moves glucose into cells. However, sometimes the cells develop a "resistance" to insulin's effects, and larger and larger concentrations of insulin are required to produce the same results. As long as the pancreas is producing enough insulin to overcome this resistance, blood glucose levels remain within the homeostatic range. Eventually, the pancreas can no longer produce enough insulin to overcome the resistance, initially after meals, when blood glucose levels are highest. Insulin resistance is thought to contribute to the accumulation of abdominal fat and other developments that increase risk for diabetes and cardiovascular disease. This set of risk factors is called the *metabolic syndrome.*

Diabetes mellitus type 1

Type 1 diabetes, previously called *juvenile diabetes,* is the result of the destruction of the pancreatic cells that produce insulin. The ultimate cause of the cell destruction has been an active area of medical research for a century. The effects of untreated type 1 are ultimately fatal. The development of intravenous insulin for treatment of this condition was one of the medical breakthroughs of the last century.

Diabetes mellitus type 2

Type 2 diabetes is caused by insufficient blood insulin concentration, either because insulin production is impaired or insulin resistance has developed. The high levels of blood glucose slowly damage small blood vessels, leading to impairment or failure of numerous organs and systems.

Diabetes insipidus

Diabetes insipidus is caused by the inability of the hypothalamus to produce the proper amount of antidiuretic hormone (ADH), which is responsible for stimulating the kidney to return water to the bloodstream. Without ADH, very little water is removed, and the concentration of glucose in the blood rises (along with that of other dissolved substances). Large amounts of watery urine lead to dehydration and thirst and carry electrolytes right out of the body. This disorder can be treated by administration of ADH therapy.

Gestational diabetes

Gestational diabetes is high blood glucose that develops at any time during pregnancy in a woman who doesn't have diabetes. Women who have gestational diabetes are at high risk for type 2 diabetes and cardiovascular disease later in life.

Thyroid disorders

The pervasive role of the thyroid hormones in metabolism can be seen by the widespread physiological effects of thyroid disorders.

Hypothyroid disorders

The prefix *hypo* means "below." The result of low levels of the thyroid hormones is *hypothyroidism.* This low level can result from a defect in the thyroid gland *(primary hypothyroidism),* or the hypothalamus or pituitary glands may not be sending the proper messaging hormones to the thyroid *(secondary hypothyroidism).* People with primary hypothyroidism may have inflammatory conditions similar to arthritis, or chronic conditions, such as *Hashimoto's thyroiditis* (a disease in which the body's immune system attacks the thyroid gland's cells). Dietary deficiency of iodine and medications that negatively affect the thyroid gland also may cause secondary hypothyroidism.

Hypothyroidism has many signs and symptoms because the thyroid hormones have such a widespread effect. Nearly every cell in the body is stimulated by the hormone *thyroxin,* which regulates the rate of metabolism. Symptoms of hypothyroidism are shown in Table 8-2.

Table 8-2	Symptoms of Hypothyroidism	
Initially	*As Disease Progresses*	*Severe*
Fatigue	Decreased sex drive	Psychiatric problems; changes in behavior
Cold sensitivity	Stiff joints	Carpal tunnel syndrome
Weight gain without increase in food intake or decrease in exercise	Muscle cramps	High cholesterol, poor circulation, heart problems
Constipation	Weight loss	Dry skin and hair, some hair loss; brittle, grooved nails
Memory problems	Numbness or tingling sensations	Impaired fertility
		Weak colon, intestinal obstruction, anemia

Myxedema is a life-threatening complication of hypothyroidism. As metabolism slows, the exchange of carbon dioxide and oxygen slows. As the amount of carbon dioxide in the blood rises, the patient is at risk of slipping into a coma, which may be fatal.

Whether primary or secondary, hypothyroidism has profound effects on metabolism in several organ systems (refer to Table 8-2). Treatment for people with hypothyroidism involves lifelong administration of a synthetic thyroid hormone. However, therapy must begin gradually so the heart isn't negatively affected.

Hyperthyroid disorders

Abnormally high levels of the thyroid hormones is the condition of *hyperthyroidism,* also called *Graves' disease.* Hyperthyroidism makes a person irritable, nervous, and unable to sleep. The thyroid gland may enlarge into a goiter, and swelling of the eye muscles may make the eyeballs protrude somewhat. Treatment options for patients with hyperthyroidism include oral medications, a single dose of radioactive iodine, or surgery to reduce the size and activity of the thyroid gland.

Androgen insensitivity

Androgen insensitivity syndrome (AIS) is a disorder caused by mutation of the gene for the receptor that binds testosterone, which regulates the expression of genes that stimulate male sexual development. Affected individuals are chromosomally XY but have a feminine phenotype and are sterile. AIS completely or partially prevents development of male sexual characteristics in the fetus, despite the presence of the Y chromosome. The extent of the syndrome ranges from complete androgen insensitivity and development of normal external (but not internal) female sexual anatomy, to partial insensitivity with altered or ambiguous male or female genitals, to mild insensitivity with normal male genitals, enlarged breasts, and possibly impotence.

Major Bones of the Skeleton

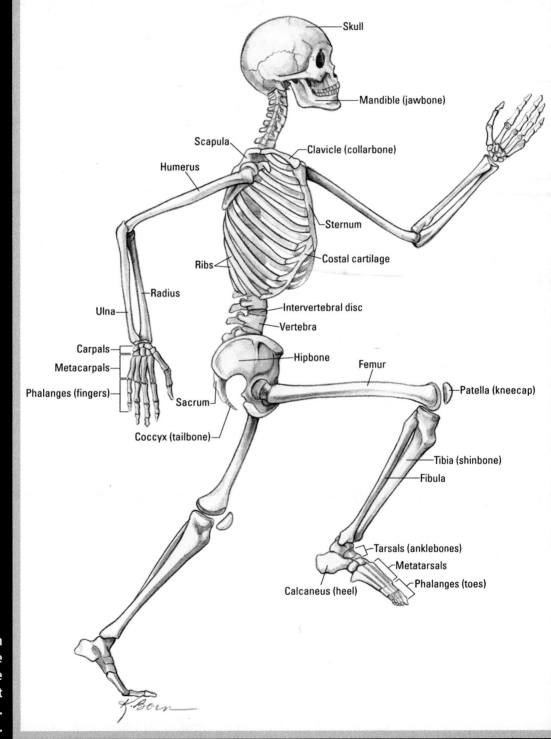

The skeleton comprises the bones and the joints that connect them. See Chapter 4.

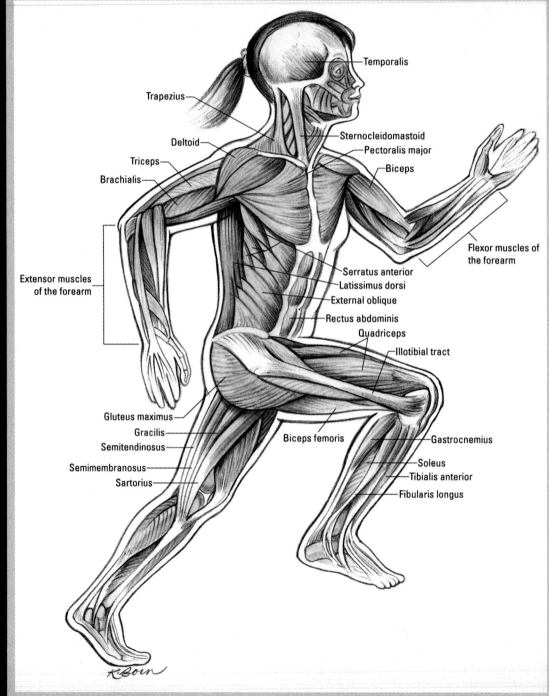

Temporalis

Trapezius

Deltoid

Sternocleidomastoid

Pectoralis major

Triceps

Biceps

Brachialis

Extensor muscles
of the forearm

Flexor muscles of
the forearm

Serratus anterior

Latissimus dorsi

External oblique

Rectus abdominis

Quadriceps

Illotibial tract

Gluteus maximus

Gracilis

Semitendinosus

Biceps femoris

Gastrocnemius

Semimembranosus

Soleus

Sartorius

Tibialis anterior

Fibularis longus

K Boin

The muscular
system
works with
the skeletal
and nervous
systems to
move the body
both spatially
and internally.
See Chapter 5.

Skin (Cross Section)

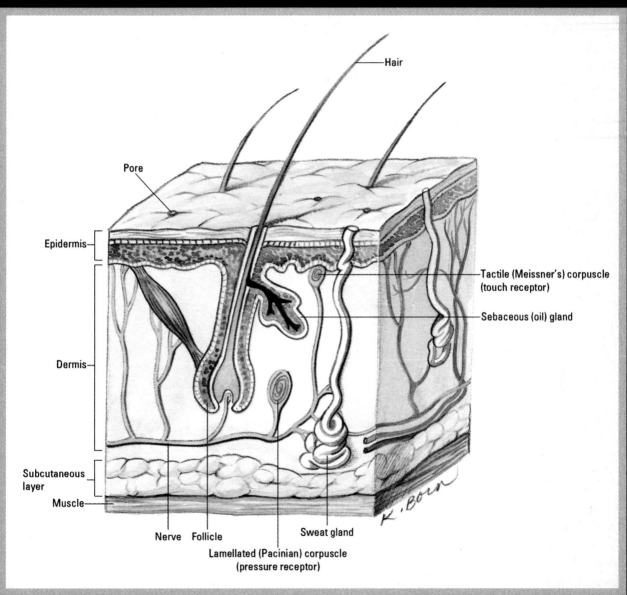

Hair

Pore

Epidermis

Tactile (Meissner's) corpuscle (touch receptor)

Sebaceous (oil) gland

Dermis

Subcutaneous layer

Muscle

Nerve Follicle

Sweat gland

Lamellated (Pacinian) corpuscle (pressure receptor)

K. Born

The skin's many layers protect the body from the environment. See Chapter 6.

Nervous System

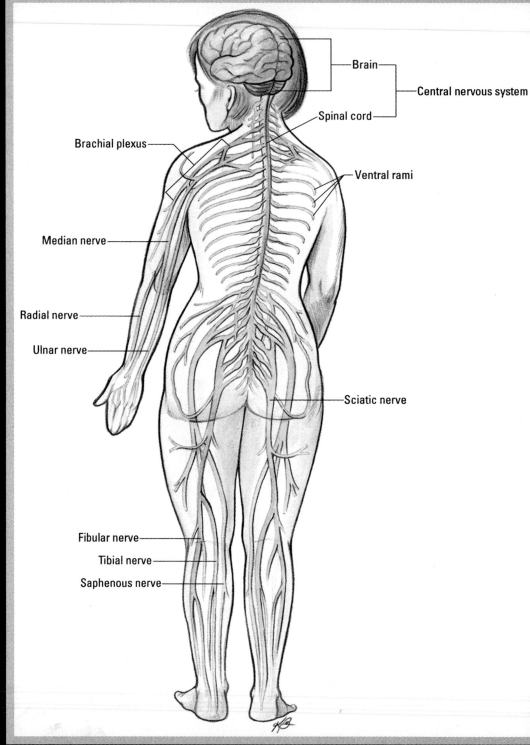

Brain

Central nervous system

Spinal cord

Brachial plexus

Ventral rami

Median nerve

Radial nerve

Ulnar nerve

Sciatic nerve

Fibular nerve

Tibial nerve

Saphenous nerve

The nervous system comprises the central nervous system and the peripheral nervous system. See Chapter 7.

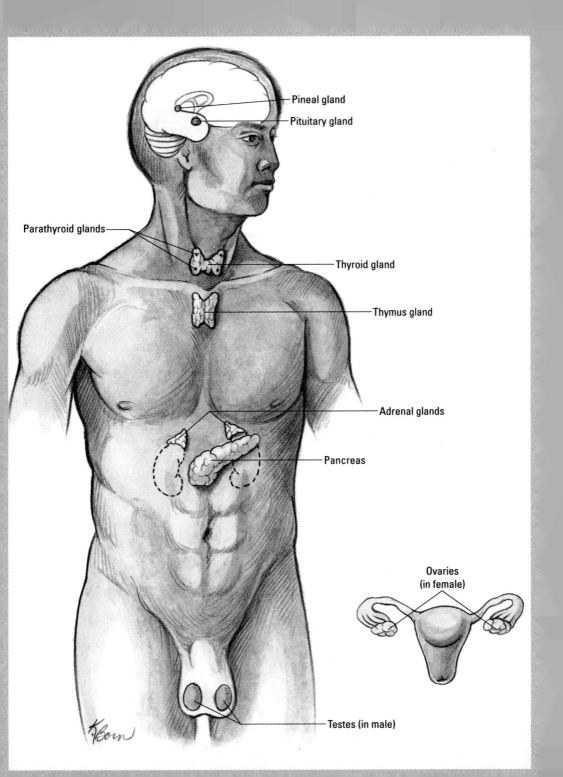

Pineal gland

Pituitary gland

Parathyroid glands

Thyroid gland

Thymus gland

Adrenal glands

Pancreas

Ovaries
(in female)

Testes (in male)

Heart

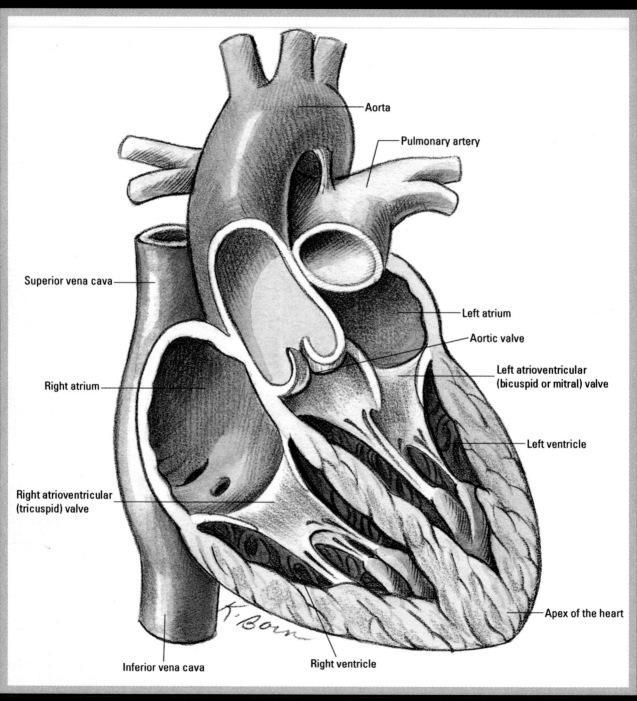

Aorta

Pulmonary artery

Superior vena cava

Left atrium

Aortic valve

Left atrioventricular (bicuspid or mitral) valve

Right atrium

Left ventricle

Right atrioventricular (tricuspid) valve

Apex of the heart

Inferior vena cava

Right ventricle

Effective blood circulation depends on the functioning of the heart's interior structure and its outer muscular layers. See Chapter 9.

Arterial Components of the Circulatory System

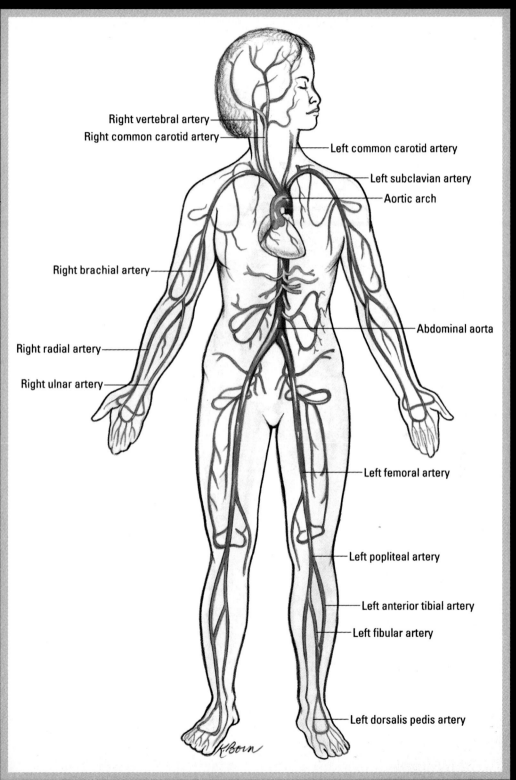

Right vertebral artery

Right common carotid artery

Left common carotid artery

Left subclavian artery

Aortic arch

Right brachial artery

Abdominal aorta

Right radial artery

Right ulnar artery

Left femoral artery

Left popliteal artery

Left anterior tibial artery

Left fibular artery

Left dorsalis pedis artery

K.Born

The arteries carry oxygenated blood from the heart to all parts of the body. Deoxygenated blood returns to the heart via the veins (not shown). See Chapter 9.

Respiratory System

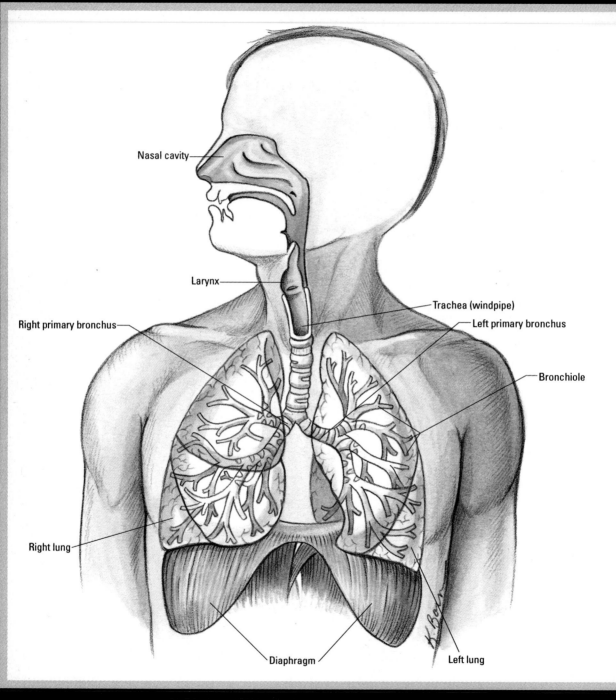

Nasal cavity

Larynx

Trachea (windpipe)

Right primary bronchus

Left primary bronchus

Bronchiole

Right lung

Left lung

Diaphragm

Contraction and release of the diaphragm alternately decreases and increases air pressure in the lungs. Air is drawn in and expelled through the airway. See Chapter 10.

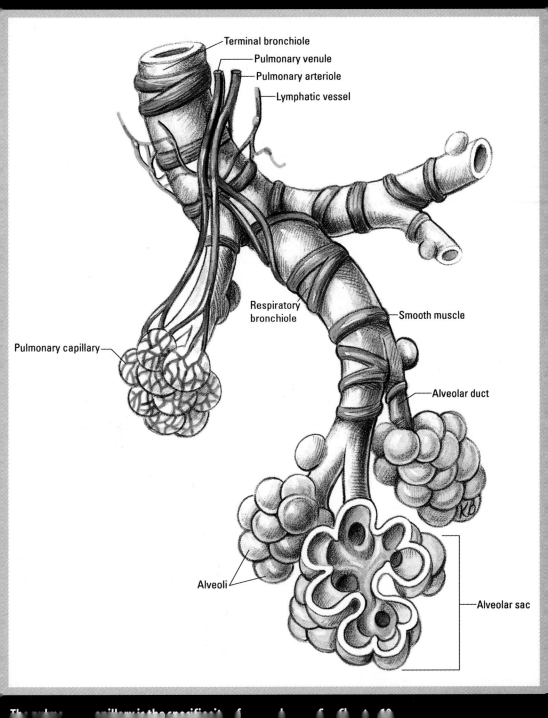

Digestive System

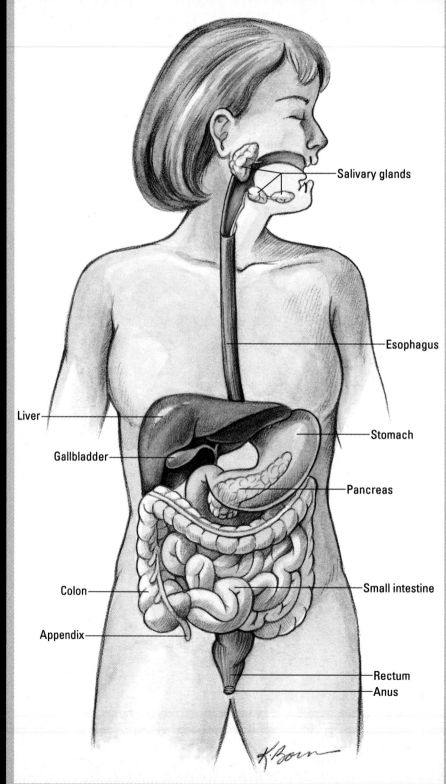

Salivary glands

Esophagus

Liver

Gallbladder

Stomach

Pancreas

Colon

Small intestine

Appendix

Rectum

Anus

The transformation of food into physiologically available nutrients involves the participation of many organs. See Chapter 11.

K. Born

Stomach

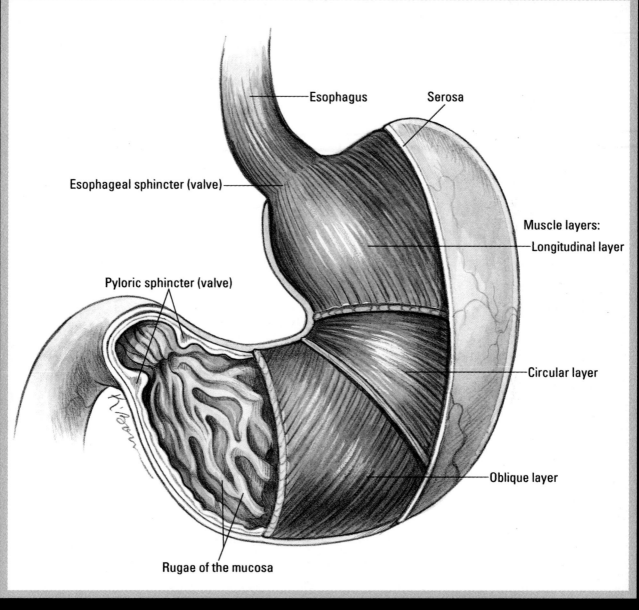

Esophagus

Serosa

Esophageal sphincter (valve)

Muscle layers:

Longitudinal layer

Pyloric sphincter (valve)

Circular layer

Oblique layer

Rugae of the mucosa

The stomach's tissue layers are a variation on a pattern that repeats all along the digestive tract: connective tissue, smooth muscle layers, and mucosa. See Chapter 11.

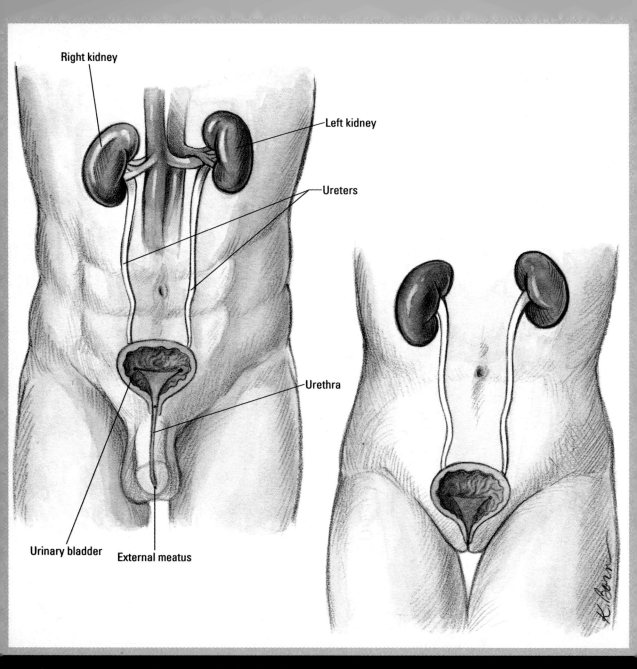

Right kidney

Left kidney

Ureters

Urethra

Urinary bladder   External meatus

The urinary system is specialized for the elimination of wastes and toxins. Its most complex organ, the kidney, also

Kidney and Nephron

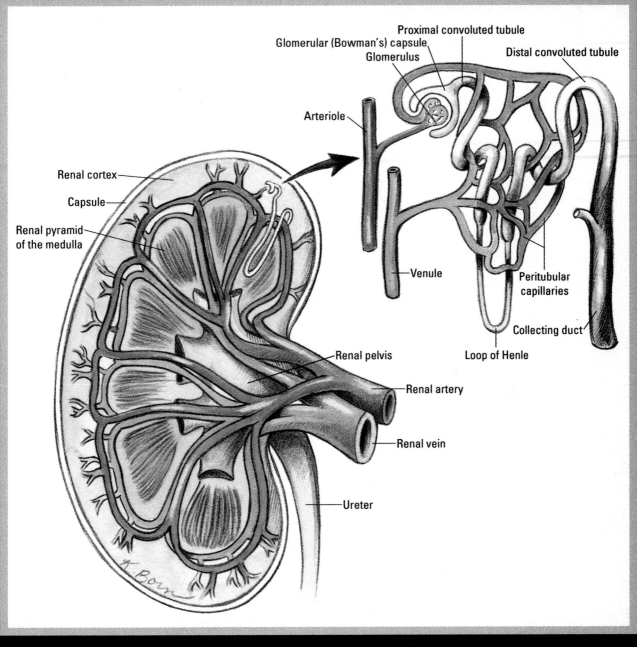

Proximal convoluted tubule

Glomerular (Bowman's) capsule

Glomerulus

Distal convoluted tubule

Arteriole

Renal cortex

Capsule

Renal pyramid of the medulla

Venule

Peritubular capillaries

Collecting duct

Loop of Henle

Renal pelvis

Renal artery

Renal vein

Ureter

The kidney is specialized for chemical processing, distribution, and disposal. The nephron is the kidney's filtering unit. See Chapter 12.

Lymphatic System

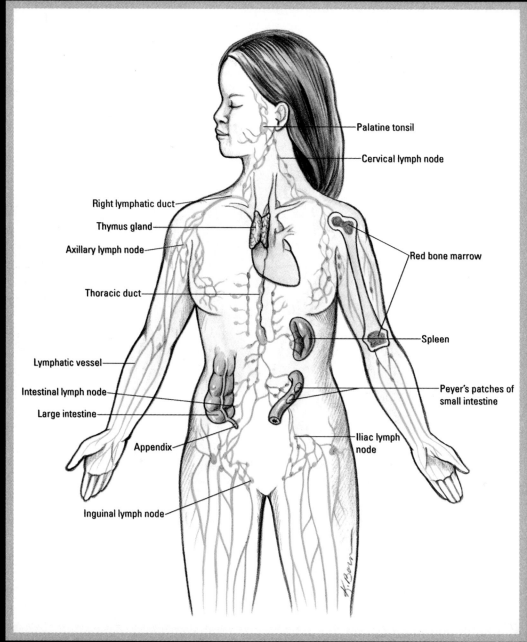

- Palatine tonsil
- Cervical lymph node
- Right lymphatic duct
- Thymus gland
- Axillary lymph node
- Red bone marrow
- Thoracic duct
- Spleen
- Lymphatic vessel
- Intestinal lymph node
- Peyer's patches of small intestine
- Large intestine
- Appendix
- Iliac lymph node
- Inguinal lymph node

The lymphatic system forms the infrastructure of the body's immune surveillance system. See Chapter 13.

Reproductive System (Female and Male)

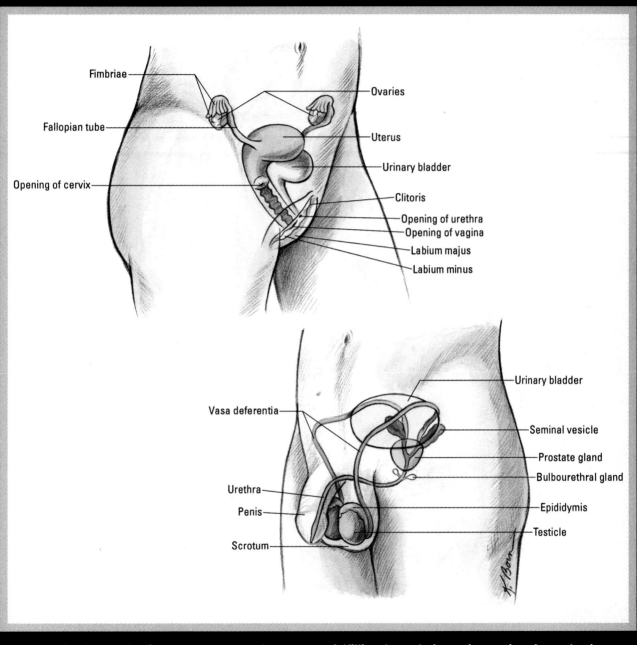

Female and male reproductive anatomy are complementary to fulfilling the evolutionary imperative of reproduction with variation. See Chapter 14.

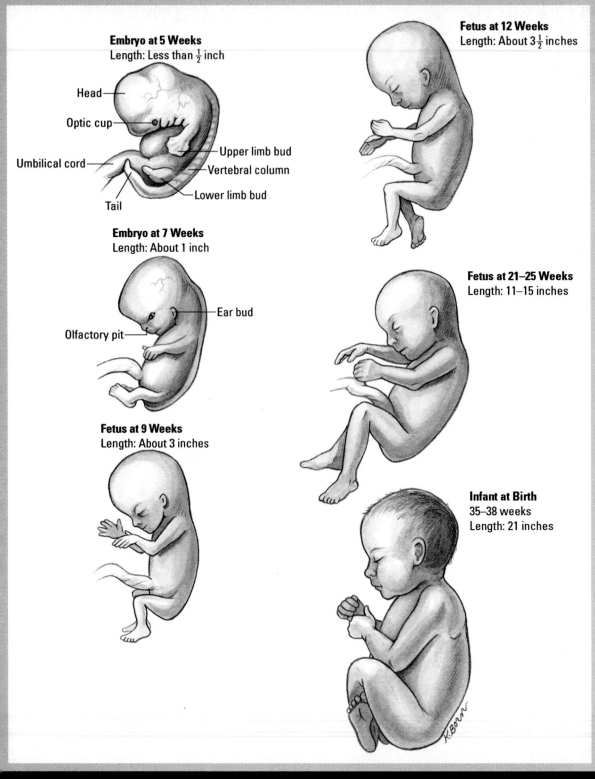

Embryo at 5 Weeks
Length: Less than ½ inch

Head
Optic cup
Umbilical cord
Tail
Upper limb bud
Vertebral column
Lower limb bud

Embryo at 7 Weeks
Length: About 1 inch

Olfactory pit
Ear bud

Fetus at 9 Weeks
Length: About 3 inches

Fetus at 12 Weeks
Length: About 3½ inches

Fetus at 21–25 Weeks
Length: 11–15 inches

Infant at Birth
35–38 weeks
Length: 21 inches

Rapid growth and differentiation in the fetus is fueled through the placenta. See Chapter 15.

Chapter 9

The Circulatory System: Getting Your Blood Pumping

In This Chapter

▶ Breaking down the heart's parts

▶ Discovering arteries, veins, and capillaries

▶ Following blood along its path through your body

▶ Looking at some circulatory system problems

*M*ore than any other system, the circulatory system has contributed strong imagery to people's daily language. "Heart" is a metaphor for love and for courage. People say they "nearly had a heart attack" to describe an experience of surprise or shock. Abstract but distinctive characteristic qualities are said to be "in one's blood." The blood itself runs cold, runs hot, and runs all over the place in informal speech and poetry. Scientifically speaking, emotions are more a matter of hormones than myocardium, and nobody's blood is any redder or any hotter than anyone else's. The heart is neither soft nor hard, but a muscular, fibrous pump, and blood is a complex biological fluid that must be kept moving through its specialized network of vessels. Ready to take a closer look?

Getting Substances from Here to There

The functions of the circulatory system are all related to transportation. Nearly every substance made or used in the body is transported in the blood: hormones, gases of respiration, products of digestion, metabolic wastes, and immune system cells. We discuss these transportation functions in the context of the other organ systems and only as they relate specifically to circulatory anatomy and physiology.

The blood also "transports" heat. Stimulated by the hormones of thermoregulation, blood flow can disperse heat to the environment at the body surface or conserve heat for essential functions in the body core.

Cardiac Anatomy

The circulatory system — or *cardiovascular system* — consists of the heart and the blood vessels. The heart's pumping action squeezes blood out of the heart, and the pressure it generates forces the blood through the blood vessels. The autonomic nervous system controls the rate of the heartbeat. One continuous layer of epithelium lines the entire circulatory system — the heart and all blood vessels.

Sizing up the heart's structure

The heart is shaped like a cone and is only about the size of your fist (see Figure 9-1). It lies between your lungs, just behind (posterior to) your sternum, and the tip (apex) of the cone points to the left. In most individuals, the heart is situated slightly to the left of center in the chest.

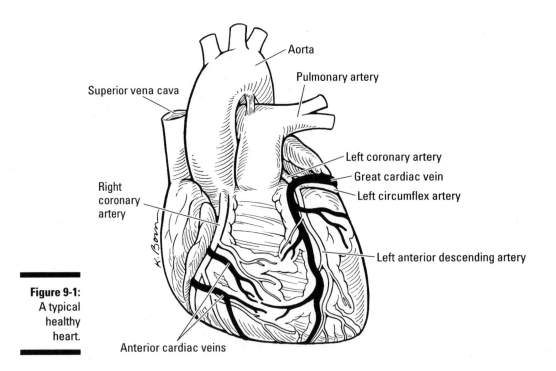

Superior vena cava

Aorta

Pulmonary artery

Left coronary artery

Great cardiac vein

Left circumflex artery

Right coronary artery

Left anterior descending artery

Figure 9-1:
A typical
healthy
heart.

Anterior cardiac veins

The four hollow spaces of the heart are called its *chambers*. You can check them out in the "Heart" color plate in the middle of the book. The heart is divided anatomically and functionally into left and right sides. Each side has one *atrium* and one *ventricle,* each with a separate function. A membrane called the *interatrial septum* separates the atria; a membrane called the *interventricular septum* separates the ventricles. The contraction of the heart muscle pumps blood into and out of all four chambers in a rhythmic pattern.

Between the chambers are several *valves* that allow measured quantities of blood to flow into the chambers and keep blood flowing in the right direction. The valves' names tell you their anatomical location or characteristics. The two *atrioventricular* (AV) *valves* lie between the atrium and the ventricle on each side; the *bicuspid valve* (BV) has two flaps; and the *tricuspid valve* (TV) has three flaps. The *semilunar valves* (SV) are shaped like half-moons. In Figure 9-2, the arrows show the direction of blood flow through the chambers of the heart.

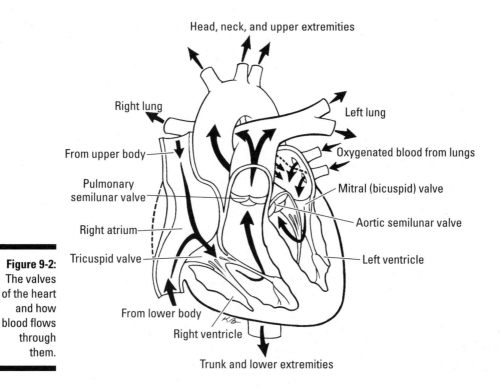

Figure 9-2:
The valves of the heart and how blood flows through them.

Head, neck, and upper extremities

Right lung

Left lung

From upper body

Oxygenated blood from lungs

Pulmonary semilunar valve

Mitral (bicuspid) valve

Right atrium

Aortic semilunar valve

Tricuspid valve

Left ventricle

From lower body

Right ventricle

Trunk and lower extremities

Examining the heart's tissues

The tissues of the heart perform the functions required to keep the double-pump working strongly and steadily. Like other hollow organs, the heart is made up of layers of endothelial and connective tissue.

✔ **Endocardium:** A layer of endothelial tissue that lines the inside of the chambers. This layer is continuous with the vascular endothelium, which we talk about in more detail in the "Starting with the arteries" section later in the chapter.

✔ **Myocardium:** The thick, muscular layer of your heart. The myocardium (literally, "heart muscle") is composed of cardiac muscle fibers that contract in a coordinated way to pump the blood out of the heart and into the aorta with enough force to carry it through the arterial system and out into the capillaries.

✔ **Epicardium:** The visceral layer of the *serous pericardium*. It's a conical sac of fibrous tissue that closely envelopes the myocardium and surrounds the roots of the major blood vessels. The epicardium secretes *pericardial fluid* into the *pericardial cavity,* which lubricates the tissues as the heart beats.

✔ **Pericardial cavity:** A fluid-filled space between the epicardium and the parietal layer of the serous pericardium. The fluid reduces friction between the pericardial membranes.

✔ **Parietal layer of the serous pericardium:** A serous membrane that's attached to the outermost layer of the heart, the *fibrous pericardium.* The fibrous pericardium is a thick, white sheet of fibrous connective tissue that anchors the heart and the major blood vessels, including the aorta, to the sternum and diaphragm. (Your heart is not just floating in your chest.) The parietal layer of the serous pericardium also secretes pericardial fluid into the pericardial cavity.

Supplying blood to the heart

Blood flows in and out of the heart every second of your life, and some of that blood needs to supply the cells of the heart itself with oxygen and nutrients. The coronary arteries supply oxygenated blood to the heart, while cardiac veins return deoxygenated blood to the pulmonary circulation loop.

✔ **Coronary arteries:** Two large coronary arteries and their many branches supply blood to the heart. Large arteries enter the heart on the left and right at the top. They're called the left and right coronary arteries because they sit atop and encircle the heart, looking like a crown (see Figure 9-3). The *right coronary artery* and its two major branches, the *marginal artery* and the *posterior interventricular artery,* primarily supply the right atrium and ventricle with oxygenated blood and nutrients. The *left coronary artery* and its two branches, the *anterior descending coronary artery* and the *left circumflex coronary artery,* primarily supply the left atrium and ventricle with oxygenated blood and nutrients.

✔ **Cardiac veins:** The cardiac venous system is similarly branched, often lying alongside the coronary arteries and their branches. Like everything about the heart, the cardiac venous system is composed of two parts, left and right. The *left cardiac venous system,* also called the *coronary sinus system,* receives deoxygenated blood from most of the superficial veins of the heart. The *right cardiac venous system* is composed of veins that originate on the anterior and lateral surfaces of the right ventricle and drain directly into the right atrium. The smallest vessels (the *venae cordis minimae*) drain the myocardium directly into the atria and ventricles.

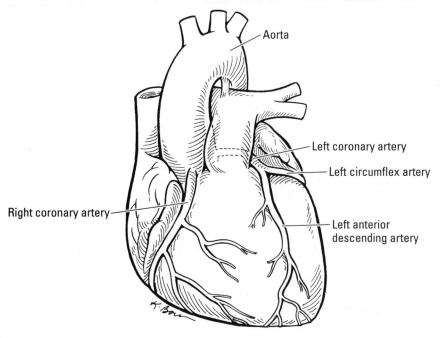

Aorta

Left coronary artery

Left circumflex artery

Right coronary artery

Left anterior descending artery

Figure 9-3:
The coronary arteries.

Looking at Your Blood Vessels

Your blood vessels comprise a network of channels through which your blood flows. But the vessels aren't passive tubes. Rather, they're very active organs that, when functioning properly, assist the heart in circulating the blood and influence the blood's constitution. The innermost layer of the heart and of all vessels is continuous — it's one very convoluted sheet of epithelium.

The vessels that take blood away from the heart are *arteries.* The vessels that bring blood toward the heart are *veins.* (The smallest ones are called *arterioles* and *venules,* respectively.) Generally, arteries have veins of the same size running right alongside or near them, and they often have similar names. (Turn to the "Arterial Components of the Circulatory System" color plate in the middle of the book for a diagram of the major arteries.) The arterial vessels (arteries and arterioles) decrease in diameter as they spread throughout the body. Eventually, they end in the *capillaries,* the tiny vessels that connect the arterial and venous systems. The venous vessels become increasingly larger as they converge on the heart. The smaller venules carry deoxygenated blood from a capillary to a vein, and the larger veins carry the deoxygenated blood from the venules back to the heart.

Starting with the arteries

Your arteries form a branching network of vessels, with the main trunk, called the *aorta,* arising from the left ventricle and splitting immediately into the *brachiocephalic trunk, left common carotid artery,* and *left subclavian artery,* which serve the head and upper limbs. The *descending aorta* — which serves the thoracic organs, abdominal organs, and lower limbs — gives off several branches. One is the *mesenteric arteries,* the main arteries of the digestive tract. It also gives off two *renal arteries,* which supply the kidneys; and the *common iliac arteries,* which supply the pelvis and lower limbs. The arteries branch into smaller and smaller vessels, becoming arterioles that end in capillary vessels.

Although an artery looks like a simple tube, the anatomy is complex. Arteries are made of three concentric layers of tissue around a space, called the *lumen,* where the blood flows (see Figure 9-4). The outside layer, the *tunica externa,* is a thick layer of connective tissue that supports the vessel and protects the inner layers from harm. The larger the artery, the thicker the connective tissue supporting it. (However, this type of tissue can become stiff with age.)

The next layer in, the *tunica media,* is a thick wall of smooth muscle and elastic tissue. This layer expands and contracts with every pulse wave (heartbeat), and sometimes it gives a little squeeze to increase the pressure and force the blood through.

The inner layer, the *tunica interna,* is a single-cell thick endothelial layer that lines the lumen and is continuous through all the organs of the circulatory system. The *vascular endothelium,* as this tissue is also called, is very active metabolically, releasing into the blood various substances that influence blood circulation and vascular health. It also specializes in transporting oxygen, nutrients, and other substances from the blood flowing in the lumen to the smooth muscle fibers of the tunica media.

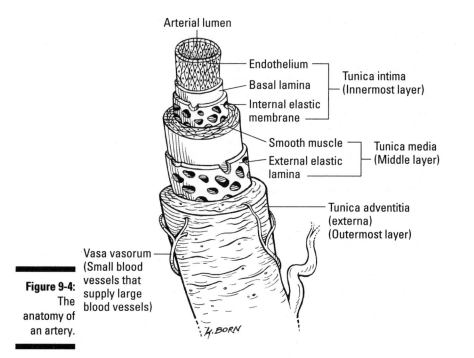

Arterial lumen

Endothelium — Tunica intima
(Innermost layer)

Basal lamina

Internal elastic
membrane

Smooth muscle — Tunica media
(Middle layer)

External elastic
lamina

Tunica adventitia
(externa)
(Outermost layer)

Vasa vasorum
(Small blood
vessels that
supply large
blood vessels)

Figure 9-4:
The
anatomy of
an artery.

H. BORN

Cruising through the capillaries

After passing from the arteries and the arterioles, blood enters the capillaries, which lie between larger blood vessels in *capillary beds.* A capillary bed forms a bridge between the arterioles and the venules.

Your capillaries are your smallest vessels:Oonly the single-cell thick epithelial layer surrounds the lumen. The *precapillary sphincters* of the *metarterioles* can tighten or relax to control blood flow into the capillary bed.

Lacking the structure and complexity of arteries and veins, capillaries rely on simple diffusion to get their job done: moving oxygen and nutrients from the blood to the cells and waste materials from the cells to the blood. *Capillary exchange* is the term used to describe these processes (see Figure 9-5).

The capillaries come into close contact with all the cells of your tissues. At the end near the arteriole, oxygen diffuses out of the red blood cells and nutrient molecules out of the plasma, across the capillary membrane, and directly into the tissue fluid. The oxygen and nutrients dissolved in the tissue fluid diffuse across the membrane of the adjacent cells. (See Chapter 3 for more on diffusion.)

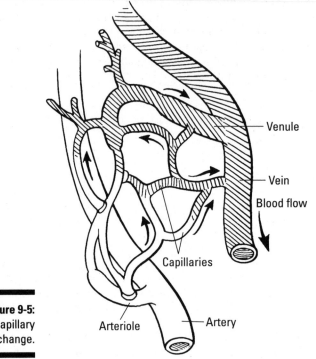

Figure 9-5:
Capillary
exchange.

Venule

Vein

Blood flow

Capillaries

Arteriole

Artery

At the venule end, the carbon dioxide and other waste materials diffuse out of the tissue fluid and across the capillary membrane into the blood. Then, the blood continues on through the venous system, and those waste materials are deposited in the proper locations on their way out of the body. The carbon dioxide diffuses out of the bloodstream in the lungs, so it can be exhaled and thus removed from the body, while other metabolic waste is filtered through the kidneys.

Capillary beds are everywhere in your body, which is why you bleed anywhere that you even slightly cut your skin.

Besides aiding in the exchange of gases and nutrients throughout your body, capillaries serve two other important functions:

✔ **Thermoregulation:** Precapillary sphincters tighten when you're in a cold environment to prevent heat loss from the blood at the skin surface. Blood is then shunted from an arteriole directly to a venule through a nearby *arteriovenous shunt.* When you're in a warm environment or you're producing heat through exertion, the precapillary sphincters relax, opening the capillary bed to blood flow and dispersing heat.

✔ **Blood pressure regulation:** When blood pressure (blood volume) is low, the hormones that regulate blood pressure stimulate the precapillary sphincters to tighten, temporarily reducing the total volume of the blood vessel system and thus raising the pressure. When blood pressure is high, they stimulate the sphincters to relax, increasing overall system volume and reducing pressure.

Visiting the veins

Small veins converge into larger veins, all merging in the *inferior vena cava* and *superior vena cava,* the largest vessels in the venous system. These major veins return blood from below and above the heart, respectively. The inferior vena cava lies to the right of, and more or less parallel to, the descending aorta. The superior vena cava lies to the right of, and more or less parallel to, the aorta.

In the lower body, the *internal iliac veins,* which return blood from the pelvic organs, and the *external iliac veins,* which return blood from the lower extremities, converge into the *common iliac veins.* The *renal veins* return blood from the kidneys. Both these major veins flow into the inferior vena cava.

Blood from the digestive tract travels in the *hepatic portal vein* to the liver (see Figure 9-6). Specialized cells in the liver move glucose molecules from the blood into storage. Phagocytic cells in the liver destroy bacterial cells that make it through the digestive process and remove toxins and other foreign material from the blood. Blood exits the liver through the *hepatic veins,* which flow into the inferior vena cava. The inferior vena cava empties into the right atrium.

Deoxygenated blood from the head and upper extremities drains into the *brachiocephalic veins.* The veins of the upper extremity — the *ulnar veins, radial veins,* and *subclavian veins* — also drain into the brachiocephalic veins. The *jugular veins* of the head and neck also drain into the brachiocephalic veins, which connect to the superior vena cava, which enters the right atrium.

After the blood from the right atrium has been pumped into the right ventricle, it's pumped into the lungs, where the blood is oxygenated, and then it flows back to the heart in the *pulmonary veins,* the only veins that carry oxygenated (red) blood.

Veins have a similar anatomy to arteries, although they tend to be wider and their walls thinner and less elastic. The tunica interna of a vein is also part of the continuous endothelial layer that lines the whole network. The tunica media has a layer of elastic tissue and smooth muscle, but this layer is much thinner in a vein than in an artery. The veins have virtually no blood pressure and thus they don't need a thick muscle layer to vary the vessel diameter or withstand fluid pressure. The outermost tunica externa is the thickest layer of a vein.

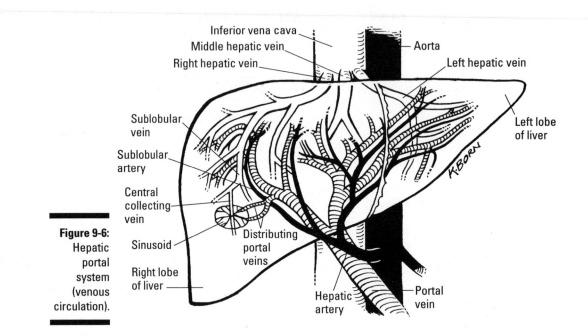

Figure 9-6:
Hepatic portal system (venous circulation).

Because veins don't have a thick muscle layer to push blood through them, they depend on contraction of skeletal muscles to move blood back to the heart. As you move your arms, legs, and torso, your muscles contract, and those movements "massage" the blood through your veins. The blood moves through the vein a little bit at a time. The larger veins have valves that keep blood from flowing backward. The valves open in the direction that the blood is moving and then shut after the blood passes through to keep the blood heading toward the heart.

Carrying Cargo: Your Blood and What's in It

Blood — that deep maroon, body-temperature-warm, and coppery-tasting liquid that courses through your body — is a vitally important, life-supporting, life-giving, life-saving substance that everybody needs. And every adult-size body contains about five quarts of the precious stuff.

Blood consists of many different types of cells in a matrix called *plasma*. That's what makes blood a connective tissue. The different types of cells —

red blood cells, white blood cells, and *platelets* — are referred to as *formed elements.*

Watering down your blood: Plasma

Plasma is about 92 percent water. The remaining 8 percent or so is made up of *plasma proteins,* salt ions, oxygen and carbon dioxide gases, nutrients (glucose, fats, amino acids) from the foods you take in, *urea* (a waste product), and other substances carried in the bloodstream, such as hormones and enzymes.

The plasma proteins, produced in the liver, have some important functions:

- ✔ **Albumin:** The smallest plasma protein and the most abundant, albumin maintains the *osmotic pressure* in the bloodstream within the homeostatic range.

- ✔ **Fibrinogen:** During the process of clot formation, fibrinogen is converted into threads of *fibrin,* which then form a meshlike structure that traps blood cells to form a clot.

- ✔ **Immunoglobulin:** This is another word for *antibody* — proteins that are created in response to an invading microbe (turn to Chapter 13 for more on the immune system).

Transporting oxygen and carbon dioxide: Red blood cells

Red blood cells (RBCs), or *erythrocytes* ("erythro" is the Greek word for "red"), are the most numerous of the blood cells and one of the most numerous of all cell types in your body. About one-quarter of the body's approximately 3 trillion cells are RBCs. RBCs are among the cell types that must be constantly regenerated and disposed of. In fact, you produce and destroy a few million RBCs every second!

The cytoplasm of RBCs is full to the brim with an iron-containing biomolecule called *hemoglobin.* The iron-containing heme group in hemoglobin binds oxygen at the respiratory membrane (Chapter 10 covers the respiratory system) and then releases it in the capillaries. This is the sole mechanism by which all your cells and tissues get the oxygen they need to sustain their metabolism. RBCs containing heme-bound oxygen are bright red, the familiar color of the arterial blood that flows from wounds. RBCs in the venous system have less heme-bound oxygen and are a dark red.

Squeezing blood from a bone

The process that forms blood cells is called *hematopoiesis.* In children, hematopoiesis occurs in the red bone marrow of the long bones such as the femur and tibia. In adults, it occurs mainly in the red bone marrow of the pelvis, cranium, vertebrae, and sternum. Special cells called *hematopoietic stem cells* divide and differentiate until they become specific blood cells. Turn to Chapter 3 for more about stem cells and cellular differentiation.

In brief, the process goes like this: A hematopoietic stem cell called a *hemocytoblast,* produces two lines of blood cell development. The *myeloid line* develops from the *myeloid stem cells* to form *erythrocytes, megakaryocytes,* and all *leukocytes* except lymphocytes. The *lymphoid line* develops from the *lymphoid stem cell* to form the *lymphocytes.* An *erythroblast* matures into an erythrocyte (RBC).

A *megakaryoblast* matures into a megakaryocyte, which fragments into platelets.

A myeloid stem cell can differentiate into any one of four types of WBCs: *basophil, eosinophil, neutrophil,* or *monocyte.*

After blood cells have matured in the red bone marrow, they enter the circulatory system, which transports them around the body.

How do the hematopoietic stem cells "know" which kind of cell to "become"? The answer appears to be that they don't. All the main types of cells are produced at random. Then, factors in the hematopoietic microenvironment (the red bone marrow) cause some cells to "choose to die," a cellular process known as *apoptosis.* This is how the bone marrow can regulate the populations of the different types of cells.

RBCs are so full of hemoglobin because they contain very little else. During differentiation, they lose their organelles, even their nucleus. Only their membranes retain their function.

An RBC has about a four-month life span, at the end of which it's destroyed by a *phagocyte* (a large cell with cleanup responsibilities) in the liver or spleen. The iron is removed from the heme group and is transferred to either the liver (for storage) or the bone marrow (for use in the production of new hemoglobin). The rest of the heme group is converted to *bilirubin* and released into the plasma (giving plasma its characteristic straw color). The liver uses the bilirubin to form bile to help with the digestion of fats.

At the same time as oxygen diffuses into a cell, carbon dioxide diffuses out, making its way in the interstitial fluid to the venous system. Some deoxygenated hemoglobin in the venous blood takes up carbon dioxide to form *carboxyhemoglobin.* At the respiratory membrane, carboxyhemoglobin releases the carbon dioxide and takes up oxygen again. Carbon dioxide is transported in several different ways in the blood to the respiratory membrane, where it enters the lung and is exhaled in the breath.

Plugging along with platelets

Platelets are tiny pieces of cells. Large cells in the red bone marrow called *megakaryocytes* break into fragments, which are the platelets. Their job is to begin the clotting process and plug up injured blood vessels. Platelets, also called *thrombocytes* ("thrombos" means "clot"), have a short life span — they live only about ten days.

Putting up a good fight: White blood cells

White blood cells (WBCs) are derived from the same type of hematopoietic stem cells as RBCs. However, they take different paths early in the process of differentiation. The WBCs, also called *leukocytes* (except for the T-lymphocytes), leave the red bone marrow and enter into circulation in their mature form. (Flip to Chapter 13 for more on WBCs and the immune system.)

Physiology of Circulation

You do it 100,000 times every day. Waking or sleeping, from a moment early in fetal development until the moment you die, the beat goes on. It may speed up slightly when you're working hard physically or you're excited or stressed, and some people can slow theirs down with meditation techniques. But for most people, it just plugs along, the same thing over and over.

What is it? The heartbeat, of course. The beating heart pushes blood around a double-circuit — out through your arteries, ultimately into your capillary beds, then across your capillary beds and into your veins, and then back through your veins. The blood passes through the heart to the lungs and then back to the heart and out through the arteries again. Each complete double-circuit takes less than one minute.

Putting your finger on your pulse

You can feel the rhythmic pulsation of blood flow at certain spots around your body, most commonly on the inside of the wrist or on the carotid artery of the neck. What you feel as you touch these spots is your artery expanding as the blood rushes through it and then immediately returning to its normal size when the bulge of blood has passed.

The entire cardiac cycle takes about 0.86 of a second, based on the average of 70 heartbeats per minute. If your cardiac cycles take less time, your heart is beating too fast *(tachycardia);* if there's too much time between your cardiac cycles, your heart is beating too slowly *(bradycardia).*

Generating electricity: The cardiac cycle

Certain structures of the heart, together called the *cardiac conduction system,* specialize in initiating and conducting the electrical impulses that induce your heartbeat, keeping it regular and strong in every part of the organ (see Figure 9-7). This is the heart's pacemaker.

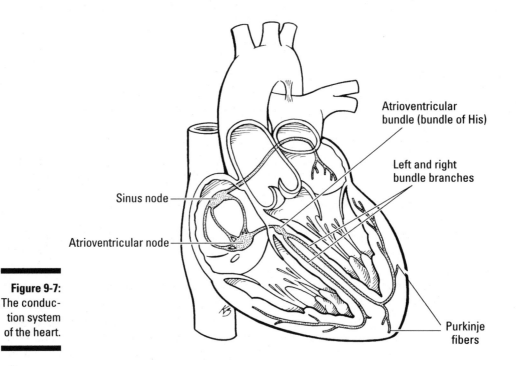

Figure 9-7:
The conduc-
tion system
of the heart.

Atrioventricular
bundle (bundle of His)

Left and right
bundle branches

Sinus node

Atrioventricular node

Purkinje
fibers

Anatomy of cardiac conduction

Five structures comprise the cardiac conduction system:

✔ **Sinoatrial node (SA node; pacemaker):** A small knot of cardiac muscle-like tissue (the pacemaker cells look like cardiac muscle cells, but they've lost the ability to contract) located on the back wall of the right atrium, near where the superior vena cava enters the heart. The SA node's cells initiate the electrical impulse that generates the heartbeat.

- **Arterioventricular node (AV node):** A similar small mass of tissue located in the right atrium but near the septum that divides the right and left atria from the ventricles. Its function is to relay the impulses it receives from the SA node to the next part of the conduction system.

- **Atrioventricular bundle (AV bundle):** A bundle of fibers that extends from the AV node into the *interventricular septum* (which divides the heart into right and left sides). The AV bundle transmits cardiac impulses.

- **Left and right bundle branches:** Where the septum widens, the AV bundle splits into the right and left bundle branches, each extending down under the endocardium and then up along the outside of the ventricles. These bundles carry cardiac impulses.

- **Purkinje fibers:** At the ends of the AV bundle branches are the Purkinje fibers, which deliver the impulse to the myocardial cells, causing the ventricles to contract.

The word *node* is used in many different contexts, and not only in anatomy. In general, it describes a local connecting point of component parts, literal and figurative. In cardiac anatomy, a *node* is a specialized type of tissue that looks like muscle and generates electrical impulses like nervous tissue.

The sequence of events

The cardiac conduction system is responsible for keeping the cardiac cycle going. If the cardiac cycle stops for too long, you experience serious consequences (see the "Cardiac disorders" section later in the chapter). Here's how the cycle goes:

1. **The electrical impulse is initiated in the SA node.** The right and left atria contract simultaneously, pumping blood into the right and left ventricles, respectively.

2. **The impulse passes to the AV node, which sends it to the AV bundle.** In the meantime, the right and left atria relax.

3. **The impulse passes into the right and left bundle branches and ultimately into the Purkinje fibers, causing the right ventricle to contract and to pump its deoxygenated blood into the pulmonary arteries and the lungs.** Simultaneously, the left ventricle contracts, pumping oxygenated blood into the aorta.

4. **The ventricles relax while the atria begin contracting, and the cycle starts all over again.**

The heart's electrical signals can be measured and recorded digitally to produce an image called an *electrocardiogram* (ECG or EKG). Three major waves of electric signals appear on the ECG, each showing a different part of the heartbeat. The first wave, called the *P wave,* records the electrical activity of the sinoatrial node and the contracting of the right and left atria. The second and largest wave, the *QRS wave*, records the electrical activity as the impulse spreads through the right and left ventricles and their resultant contraction. The third wave, the *T wave*, records the relaxation of the ventricles.

Problems with signal conduction due to disease or abnormalities of the conducting system can occur anyplace along the heart's conduction pathway. Abnormally conducted signals resulting in irregular heartbeat are called *arrhythmias* or *dysrhythmia*.

On the beating path: The circuits of blood through the heart and body

Remember, the heart is a double-pump, so it has two circuits: heart to lungs and back to heart, and heart to body and back to heart. These are called the *pulmonary circulation path* and the *systemic circulation path,* respectively (see Figure 9-8). Every drop of your blood travels around the double-circuit about once per minute.

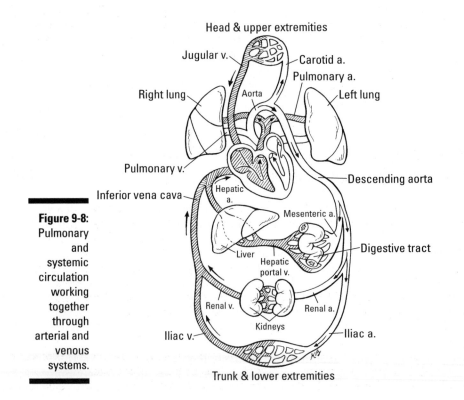

Figure 9-8: Pulmonary and systemic circulation working together through arterial and venous systems.

Pulmonary circulation

Deoxygenated blood enters the heart's right atrium from the largest veins in the body, the superior vena cava and the inferior vena cava. When the SA node initiates the cardiac conduction cycle, the right atrium contracts, pumping the blood into the right ventricle.

When the impulse is passed to the AV node and on to the AV bundle, the right bundle branch, and the Purkinje fibers, the right ventricle contracts, pumping blood into the pulmonary arteries, which take it to the lungs for gas exchange.

During the relaxation phase of the atria, the newly oxygenated blood flows into the left atrium.

The pulmonary arteries are the only arteries that carry deoxygenated blood.

Systemic circulation

When the SA node initiates the cardiac conduction cycle, the left atrium contracts, pumping the oxygenated blood into the right ventricle.

When the impulse is passed to the AV node and on to the AV bundle, the left bundle branch, and the Purkinje fibers, the left ventricle contracts, pumping blood into the aorta. From the aorta, it travels through the arteries and arterioles to the capillary beds, and then back to the heart through the veins.

During the relaxation phase of the atria, the deoxygenated blood flows into the right atrium.

Going up, going down, holding steady: Blood pressure

Blood pressure is a term used to describe the force of blood pushing against the wall of an artery. It's measured in millimeters of mercury at both the highest point (*systole,* when the heart is contracted) and the lowest point (*diastole,* when the heart is relaxed) in the cardiac cycle. The systole pressure is always higher than the diastole. The higher the systolic and diastolic values, the more pressure there is on the walls of the arteries. Two factors affect the blood pressure: the *cardiac output,* which is the amount of blood the heart pumps out per unit of time, and the *peripheral resistance,* a measure of the diameter and elasticity of the vessel walls.

The cardiac output is determined by the heartbeat rate and the blood volume put out from a ventricle during one beat. When either of these rises, the blood pressure rises. Heart rate is increased by physical exertion, the release of epinephrine (a hormone), and other factors. The blood volume is influenced by the action of ADH (antidiuretic hormone) and other mechanisms in the kidneys to control the amount of water that's removed from the urine and restored to the blood.

The arteries' diameter changes locally and continuously. The pressure of the pulse wave increases pressure on the vascular endothelium, inducing it to release molecules, mainly nitrous oxide (NO), that induce relaxation in the tunica media. The endothelium's ability to respond to the pulse-wave pressure is extremely important to vascular health. Resistance in the arteries to expansion as the blood rushes through raises the blood pressure.

As part of homeostasis, receptors in the arteries, called *pressoreceptors,* measure blood pressure. If the blood pressure is above the normal range, the brain sends out impulses to cause responses that decrease the heart rate and dilate the arterioles, both of which decrease blood pressure.

Not going with the flow

Not the least amazing thing about blood is its ability to stop flowing. The term for this is *hemostasis* (literally, "blood stopping," and not to be confused with *homeostasis*). Hemostasis is the reason why you didn't bleed to death the first time you cut yourself. When vessels are cut, blood flows, but only for a moment, just about long enough to clean the cut. As you watch, the blood stops flowing, and a plug, called a *clot,* forms. Within a day or so, the clot has dried and hardened into a scaly scab. Eventually, the scab falls off, revealing fresh new skin.

A blood clot consists of a plug of platelets enmeshed in a network of insoluble fibrin molecules. The *clotting cascade* is a physiological pathway that involves numerous components in the blood itself interacting to create a barrier to the flow. As soon as a blood vessel is injured, a signal is sent from the vascular endothelium to the platelets, also called *thrombocytes* (*thrombo* means "clot"), summoning them to the injury site. On exposure to the air, their membranes become sticky, and they start to adhere together. Proteins in the blood plasma, called *coagulation factors* or *clotting factors,* respond in a complex cascade to form fibrin strands, which strengthen the platelet plug. Within minutes, your blood is safe in your body, where it belongs.

A blood clot in the right place is something to be grateful for. But blood tends to start clotting whenever it's not flowing freely, and this tendency can cause problems in the peripheral vessels (the arteries and veins of the legs). Clots also form on the inner wall of blood vessels when the endothelium is injured by disturbances in blood flow (turbulence) or by free radicals in the blood. These tiny clots adhere to the wall, further disturbing flow and causing more turbulence and more injury. Atherosclerotic plaque may begin to form around the clot (see the next section for a discussion of atherosclerosis). Worst of all, perhaps, is when the clot, with the plaque attached, breaks off from the vessel wall and floats free (sort of) in the bloodstream. Sooner or later, this *embolus* lodges somewhere in a vessel, sometimes with sudden and fatal consequences.

Pathophysiology of the Circulatory System

The circulatory system endures constant mechanical stress (sheer stress, from the pressure of fluid flow), physical blockages large and small, and other physical and chemical assaults. The blood is a natural target for pathogens and parasites of all kinds.

Cardiac disorders

The heart muscle is subject to malfunctions of electrical and chemical signaling that disrupt its all-important rhythm.

Arrhythmia

An *arrhythmia* is any disorder of your heart rate or rhythm, such as the heart beating too quickly (faster than 100 beats per minute, called *tachycardia*), too slowly (fewer than 50 beats per minute, called *bradycardia*), or in an irregular pattern. Arrhythmia, also called *dysrhythmia,* can originate as a disorder of any of the structures of the cardiac conduction system. Any part of the heart can be affected. In general, arrhythmias that start in the ventricles are more serious than those that start in the atria.

Transient arrhythmia can be brought on by many factors, including caffeine, vigorous exercise, and emotional stress. Some arrhythmias are chronic conditions, treated with an implantable medical device called a *pacemaker.* Sometimes, ventricular arrhythmia can be life-threatening, especially when combined with tachycardia. This type of situation is a medical emergency that requires immediate treatment.

Myocardial infarction

Myocardial infarction (or "heart attack") is the irreversible damage of myocardial tissue caused by a blockage of blood flow to the myocardium — the occlusion of a coronary artery. The myocardial cells supplied by that artery become oxygen-depleted and die or become irreversibly damaged (infarction). Depending on the size of the infarct and its location, the severity of the follow-on effects on cardiac rhythm and the integrity of the myocardium itself can range from relatively mild to fatal.

Occlusion of a coronary artery may be from a blood clot (coronary thrombosis) following the rupture of an atherosclerotic plaque on the vascular endothelium.

Vascular disorders

Vascular disorders are malfunctions of the blood vessels. The following examples are disorders of the arteries. The veins, particularly the deep veins of the legs, are subject to damage and malfunction, too.

Hypertension and atherosclerosis

Hypertension, more commonly called "high blood pressure," means that blood is pushing too hard on the walls of the arteries. This can happen because cardiac output is high or because the arteries themselves have lost the ability to flex in response to the pulse-wave of blood.

As a fast-flowing stream damages the stream bank, the pressure of the blood flow exerts *shear stress* on the thin and vulnerable vascular endothelium. Platelets rush to the injury site and initiate the clotting cascade. A clot *(thrombus)* may form on the artery wall, intruding into the lumen and possibly blocking the artery.

The cumulative damage to arteries causes many problems: *ischemia* (reduced blood flow to tissues and organs); *angina* (sudden pain as blood squeezes through arteries); worsened shear stress because of turbulence in the blood flow; and *embolus,* a clot dislodged from the artery wall, which can be fatal.

Microinjuries in the artery walls lead to the formation of *arterial plaque.* Fats in the blood build up on the epithelial lining. Then, fibrous plaques begin to form on the fatty deposits. Next, calcium deposits become enmeshed in the fibrous plaques. *Atherosclerosis* is the narrowing and stiffening of the arteries due to arterial plaque. Atherosclerosis increases the risk of coronary artery disease (CAD) and myocardial infarction (MI).

Stroke

A *stroke* is damage to the brain caused by *ischemia* (reduced blood flow) or *hemorragia* (bleeding). Ischemia results from atherosclerosis in the arteries that supply the brain — the internal carotid arteries and the cerebral arteries. Bleeding may be caused by a ruptured vessel in the brain (an *aneurysm*). Whether ischemic or hemorragic, some brain tissue dies or is irreversibly damaged.

Disability following a stroke may be mild. The patient may even make an apparently complete recovery, as other parts of the brain may start "filling in for" the damaged areas by acquiring new capabilities. Or damage may be severely disabling, both physically and mentally. Frequently, people who experience stroke ultimately experience several more strokes of increased severity. Stroke is a common cause of death among the elderly.

When a stroke is disabling, the specific nature of the disability depends on which part of the brain is affected. Speech problems, from slight slurring to total *aphasia* (inability to speak), may result from damage to the speech control areas in the left frontal lobe. If the brain tissue that dies is in the center for vision, the stroke victim may become blind. Damage to motor-control areas cause problems ranging from *paresis* (impaired movement) in one eyelid to serious paralysis that affects one or both sides of the body.

Blood disorders

Any disruption or change in the composition and chemistry of blood is a disorder. Here you see how such changes can disrupt the delivery of oxygen, needed by all cells. You also see how parasites can divert the resources of the blood for their own processes.

Anemia

Anemia, the most common blood disorder, is a malfunction of RBCs resulting in low blood oxygen levels. The underlying cause may be related to any of the following:

- A low number of RBCs, which, in turn, can have a number of underlying causes related to cell differentiation and maturation in the bone marrow, or some process whereby RBCs are lost, such as chronic internal bleeding (called *occult bleeding*) or *hemolytic anemia* — when RBCs are destroyed at a higher rate than they're produced.

- A low quantity of hemoglobin in the RBCs, which may result from a deficiency of iron in the diet.

- Impaired oxygen-binding capacity in the hemoglobin molecules. Several diseases called *hemoglobinopathies* have this effect.

Sickle cell disease

Sickle cell disease (SCD) is a *hemoglobinopathy,* an inherited (genetic) disorder of the pathway for producing hemoglobin molecules. The abnormally shaped hemoglobin molecules that are produced have impaired oxygen-binding capacity. The RBCs containing abnormal hemoglobin have a life span even shorter than the life span for normal RBCs and often rupture in the blood. Free hemoglobin in the blood is toxic to the vascular endothelium's cells.

The abnormal hemoglobin distorts the shape of the RBCs, which causes blockages in the blood flow, setting off a *syndrome* (a clinical term for a group of related signs and symptoms) called a *sickle cell crisis*. A number of factors or events can trigger a sickle cell crisis, such as cold temperature, high altitude, physical or emotional stress, infection, and low blood oxygen level *(hypoxia)*. Blocked blood flow initiates a clotting cascade, causing pain and possibly necrosis (dead tissue) in the area, and damaging the artery wall. Damaged arteries leave the patient vulnerable to all the resulting disorders of atherosclerosis and hypertension; the prevalence of pulmonary hypertension, a very dangerous condition, is particularly high in the SCD patient population.

Like other hemoglobinopathies, SCD has a close association with geography. Those who carry the sickle-cell trait in their genome are descended from certain African populations. It's widely accepted that hemoglobinopathies, if not too severe, may confer some evolutionary advantage by the fact that they die sooner and are eliminated from the body, thus disrupting the parasite's life cycle.

Malaria

Malaria is a disease caused by infection with a pathogenic, single-celled organism named *Plasmodium*. The disease is transmitted from one person to another by the bite of a particular genus of mosquito. Parasite-infected RBCs from one person are injected into the bloodstream of the next person the mosquito bites. After a maturation period in the liver, the pathogen enters the blood and parasitizes the RBCs, disrupting their normal routine of picking up and dropping off oxygen. Within the RBCs, the parasites multiply, periodically undergoing amplification cycles and breaking out en masse to invade fresh RBCs. During such cycles, the patient suffers waves of fever, chills, nausea, and malaise.

Plasmodium has evolved several means of escaping immune surveillance, including wrapping itself in the cell membrane of the infected host liver cell. Another way the parasite evades immune detection is to escape from circulation before entering the spleen. It does this by causing a sticky protein to be secreted on the outside of the cell membrane of the RBC, causing it to stick to the vascular endothelium, with all the usual results of vascular damage and blood flow turbulence.

No one knows for sure how long malaria has been endemic in human populations, but it has been long enough for the human genome to evolve a genetic response in the form of altered hemoglobin molecules. Present in just the right amount, these abnormal hemoglobins can confer some protection from malaria.

Chapter 10

The Respiratory System: Breathing Life into Your Body

In This Chapter

▶ Understanding what the respiratory system does

▶ Checking out the parts of the respiratory system

▶ Drawing in some breathing knowledge

▶ Looking at some common respiratory system ailments

*V*ertebrates evolved in the sea from animals that obtained their oxygen by breathing water. Whatever it was that drove a fish to leave the watery environment, the challenge of breathing the gaseous atmosphere instead had to be met. All the major groups of land vertebrates (amphibians, reptiles, birds, and mammals) have succeeded by evolving the anatomy and physiology for breathing air. For mammals and birds, the physiological requirement for a high and steady body temperature (warm-bloodedness) put additional demands on the gas-exchange system. Early mammals evolved a system that has remained basically the same, with adaptations by each species, through all the major mammal groups. Rodents, canines, felines, cetaceans, bovines, and primates all have a tubelike airway, a moist gas-exchange membrane, and muscles under autonomic control. So take a deep breath and prepare to explore the respiratory system.

Functions of the Respiratory System

Air is the source of the oxygen your cells need for nearly all reactions. Your respiratory organ system manages the flow of air into and out of your body. Your respiratory system oversees a number of vital body functions, including:

✔ **Ventilating:** *Ventilation* — the act of breathing — isn't the same as *respiration* — the exchange of gases. You don't have to think about performing either task, but ventilation involves the physical movement of your diaphragm muscle, rib cage, and lungs to draw air in and push air out of the body. See the "Breathing: Everybody's Doing It" section later in this chapter.

✔ **Exchanging gasses:** The job of the respiratory system is to supply oxygen and remove carbon dioxide from the blood as it circulates through the body. Chapter 9 discusses how the blood gets into and out of the lungs in the process of *pulmonary circulation.* This is sometimes called *external respiration* to distinguish it from *cellular respiration,* which we discuss in Chapter 2. Gas exchange happens in the lungs, where the respiratory tissues and the circulatory tissues meet. See the "Respiratory membrane" section later in the chapter.

✔ **Regulating blood pH:** Maintaining blood pH within the homeostatic range requires coordination between the respiratory and urinary systems, with help from the endocrine system and, of course, the circulatory system itself.

✔ **Producing speech:** The ability of humans to consciously control breathing permits speech and singing.

Nosing around Your Respiratory Anatomy

The *respiratory tract* is the path of air from the nose to the lungs. It's divided into two sections: the *upper* respiratory tract (from the beginning of the airway at the nostrils to the pharynx) and the *lower* respiratory tract (from the top of the trachea to the diaphragm). Turn to the "Respiratory System" color plate in the center of the book to get an idea of what the internal structures of the respiratory tract look like.

The respiratory tract is one of the places in the body where cells are replaced constantly throughout life (see Chapter 2).

Nose

This is one time that it's okay to turn your nose up at your anatomy. Really. Point your nose up while looking in the mirror. (Yes, you have to put the book down.) See the two big openings? Those are your *nostrils,* and that's one of two places where air enters and exits your respiratory system. Now, see all those tiny hairs in your nostrils? Those little hairs serve a purpose. They trap dirt, dust particles, and bacteria. Okay. You can put your head down now. The rest of your respiratory parts are way inside of your body, so you can't see them in the mirror.

Just beyond your nostrils, the *nasal septum* separates your *nasal cavities.* Inside the nasal cavities, the three tiny bones of the *nasal conchae* provide more surface area inside the nose because they're rolled up (like conch shells). The cells of the *respiratory mucosa* that lines the inside of the nasal cavity have tiny *cilia* that move the dirt-laden mucus toward the outside of the nostrils.

The *lacrimal glands* secrete tears that flow across the eye's surface and drain through the openings in the corner of the eye *(lacrimal puncta),* into the *naso-lacrimal ducts* and the nasal cavities. That's why your nose runs when you cry.

Your *sinuses* are air spaces in your skull that lighten the weight of your head. The sinuses open into the nasal cavities so they can receive air as you breathe, and, like the nasal cavities, the sinuses are lined with mucous membranes.

Pharynx

Air passes through your *pharynx* on its way to your lungs. Along the way, it passes through and by some other important structures, like your larynx and tonsils.

Your pharynx is divided into three regions based on what structures open into it:

✔ **Nasopharynx:** The top part of your throat where your nasal cavities drain. If you press your tongue to the roof of your mouth, you can feel your *hard palate.* This bony plate separates your mouth *(oral cavity)* from your nose *(nasal cavities).* If you move your tongue backward along the roof of your mouth, you reach a soft spot. This spot is the *soft palate.* Beyond the soft palate is where your nasal cavities drain into your throat, your *nasopharynx.* Your soft palate moves backward when you swallow so that the nasopharynx is blocked.

Normally, the soft palate blocking the nasopharynx keeps food from going up into your nose. But when you're laughing and eating or drinking at the same time, your soft palate gets confused. When you go to swallow, it starts to move back, but when you laugh suddenly, it thrusts forward, allowing whatever's in your mouth to flow up into your nasal cavities and immediately fly out of your nostrils to the delight of everyone around you.

✔ **Oropharynx:** The middle part of your throat, frequently called "the back of the throat." It extends from the *uvula* to the level of the *hyoid bone.* It's the location of the *epiglottis,* a cartilage structure that guides materials passing though the mouth to the trachea or esophagus, as appropriate. It's why your food comes close to your windpipe but rarely goes in. We mention the special characteristics of the hyoid bone in Chapter 4. We describe the esophagus, part of the digestive tract, in Chapter 11.

✔ **Laryngopharynx:** The lower part of your throat adjacent to your larynx. The *larynx* (or *voice box*) is triangular. At the apex of the triangle is *thyroid cartilage,* commonly known as your "Adam's apple." If you could look down your throat onto the top of your larynx, you'd see your *glottis,* the opening through which air passes. When you swallow, a flap of tissue called your *epiglottis* covers your glottis and blocks food from getting into your larynx.

Inside your glottis are the *vocal cords* — gathered mucous membranes that cover ligaments. Your vocal cords vibrate when air passes over them, producing sound waves. Pushing more air over them increases the vibration's amplitude, making the sound louder. When you tighten your vocal cords, the glottis narrows, and your voice has a higher pitch.

Trachea

Your *trachea* (windpipe) is a tube that runs from your larynx to just above your lungs. Just behind your sternum, your trachea divides into two large branches called *primary bronchi* (singular, *bronchus*) that enter each lung.

The trachea and bronchi are made of smooth muscle and cartilage, allowing the airways to constrict and expand.

Lungs

Your lungs are large paired organs within your chest cavity on either side of your heart. Like the heart, they're protected by the rib cage. The lungs sit on top of the *diaphragm,* a powerful muscle that's fixed to the lower ribs, sternum, and lumbar vertebrae. The heart sits in a depression between the lungs, called the *cardiac notch.*

The right lung is somewhat larger than the left. Both lungs are separated into *lobes* (three on the right and two on the left). The lobes are further divided into segments and then into *lobules,* the smallest subdivision visible to the eye.

Pleural sac

Each lung is completely enclosed in the *pleural sac.* The pleural sac is similar to the pericardial sac in that it's made up of two membranes, the *parietal* pleura, attached to the thoracic wall, and the *visceral* pleura, attached to the lung's surface, with the pleural cavity between them. The pleural cavity contains a lubricating fluid called the *intrapleural fluid.*

Mucus: Gross, but necessary

Inquiring minds want to know: What is mucus? Where does it come from? What does it do?

Mucus (adjectival form, *mucous*) is a viscous fluid made up of water, salts (electrolytes), glycoproteins called *mucins,* enzymes, epithelial cells, and immune system components like *immunoglobulins* (antibodies) and *leukocytes* (white blood cells). In mammals, including humans, mucus is found in the structures of the respiratory, digestive, reproductive, and urinary systems. Many other animals secrete mucus, including many invertebrates, notably snails and slugs.

Mucus is produced in the *mucous membranes,* or *mucosae* (singular, *mucosa*). These membranes have cells and glands that secrete mucus. Mucosae line the airway (nose, trachea, bronchus, and bronchioles), the entire digestive tract, the ureters and urinary bladder, and reproductive tracts of both males and females. The average human body produces about a liter of mucus per day.

In the respiratory system, mucus oozes in a continuous stream from the mucosa that lines the entire airway. Cilia on the epithelial cells of the bronchi and bronchioles constantly sweep this slow stream of mucus back up the airway toward the throat (pharynx). In the nasal passages, mucus warms and moistens the inhaled air before it reaches the delicate *alveoli.* Mucus coats the hairs that line the nasal passages, trapping dust and irritating particles in its sticky mass and expelling them back outside the body. Its immune system components destroy airborne pathogens. Any particles that escape the nasal hairs are engulfed farther down the airway. The continuous movement of the respiratory mucus layer toward the oropharynx helps prevent foreign objects from entering the lungs during breathing. Hypersecretion of mucus is a common symptom in inflammatory respiratory ailments such as colds, the flu, respiratory allergies, asthma, and chronic bronchitis.

The digestive system is lined by mucosa for its entire length. Mucus serves to moisten and soften the digesting material and to lubricate its passage. A thick coat of mucus protects your stomach lining from the extremely acidic conditions inside. Mucus isn't digested in the intestinal tract, so mucus commonly appears in fecal matter.

In the female reproductive system, cervical mucus prevents infection. In the male reproductive system, the semen contains mucus. Both sexes secrete mucus-containing lubricating fluids during sexual intercourse. In other systems, mucus serves primarily to protect and lubricate surfaces.

Because of the adhesive force of the fluid interface between the parietal pleura and the visceral pleura, in essence, the lung is attached to the chest wall. Thus, as muscles associated with the thoracic wall contract and relax and the chest rises and falls, the lungs expand and contract.

The visceral pleura surrounds the *mediastinum,* the region that separates the left and right lungs and houses the heart, thymus, and part of the esophagus.

The intrapleural fluid completely surrounds the lungs. It keeps the pleural membranes moist and lubricated. Because this fluid has a pressure lower than atmospheric pressure (that is, little air is in the fluid), the lungs stay inflated.

The bronchial tree

After the primary bronchus enters the lung on each side, it splits into secondary and tertiary branches called *bronchi.* The tertiary bronchi divide into smaller branches called *brionchioles.* At the end of the smallest bronchioles are little structures that look like raspberries. These are the *alveolar sacs,* and each sac contains many *alveoli* (singular, *alveolus*). The alveoli's walls are composed of a simple squamous epithelium (designed to facilitate rapid diffusion) and elastic tissue that alternately stretches and constricts as you breathe.

Each of the approximately 300 million alveoli is wrapped with capillaries, whose wall, like the alveoli's walls, contains simple squamous epithelium, a tissue type adapted for the exchange of materials. The interface of the simple squamous epithelium of an alveolus and the simple squamous epithelium of a pulmonary capillary (along with its supporting connective tissue) is called the *respiratory membrane,* and that's where gas exchange actually occurs. Check out the "Structures of the Respiratory Membrane" color plate in the center of the book for a diagrammatic look at this physiologically crucial interface.

The lungs have a very large reserve volume relative to the oxygen exchange requirements when at rest. It's possible to live many years with only one lung.

Respiratory membrane

The term *respiratory membrane* is one way of referring to the area within the lung where the epithelial cells of the alveoli meet the capillaries and gas exchange takes place. Because of the folds and convolutions of the bronchioles and alveoli, the respiratory membrane has a large surface area. Rather, because gas exchange for a large, active, warmblooded animal requires a membrane with a large surface area, evolution has favored those with convoluted bronchioles and alveoli. Study Figure 10-1 carefully to get an understanding of the respiratory membrane's structure.

The respiratory membrane is where blood is reoxygenated in the process of pulmonary circulation, which we describe in detail in Chapter 9. The process of gas exchange at the respiratory membrane is almost exactly the same as the capillary exchange process, also described in Chapter 9, but at the respiratory membrane, oxygen flows through the alveolar wall and into the blood, and carbon dioxide flows out of the blood and through the alveolar wall into the air space. This is the opposite direction of the flow in the capillary beds.

The respiratory membrane is where carbon dioxide, the waste product of aerobic metabolism, is eliminated. It flows into the air in your lungs and you just breathe it out.

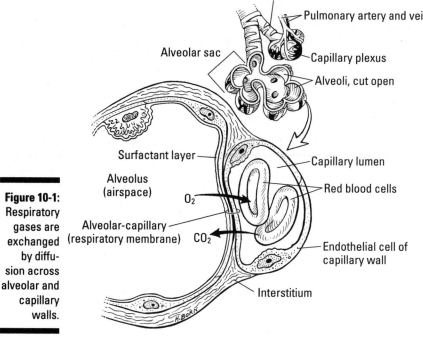

Respiratory bronchiole

Pulmonary artery and vein

Alveolar sac

Capillary plexus

Alveoli, cut open

Surfactant layer

Capillary lumen

Alveolus (airspace)

Red blood cells

O_2

Alveolar-capillary (respiratory membrane)

CO_2

Endothelial cell of capillary wall

Interstitium

Figure 10-1: Respiratory gases are exchanged by diffusion across alveolar and capillary walls.

Diaphragm

The *thoracic diaphragm* is a dome-shaped sheet of muscle separating the base of the lungs from the liver, and, on the left side, from the stomach and the spleen. The diaphragm pushes up beneath the lungs to control their contraction and expansion during ventilation. The motor fibers in the *phrenic nerves* signal to the diaphragm when to contract and relax. The diaphragm can also exert pressure on the abdominal cavity, helping the expulsion of vomit, feces, or urine.

Breathing: Everybody's Doing It

Breathing is essential to life, and thankfully, your body does it automatically. Air is alternately pulled into (inhaled) and pushed out of (exhaled) the lungs because of changes in the gas pressure in the alveoli. The change in pressure comes about because the alveoli are expanded when the chest cavity expands. In the following sections, we take a look at how your body breathes under different conditions.

Normal breathing

When you're sleeping, sitting still, and doing normal activity, your breathing rate is 12 to 20 inhalation/exhalation cycles per minute. That's 17,000 breaths a day or more.

Normal breathing *(eupnea)* is involuntary, which is why you never really forget to breathe. Breathing continues during sleep. In many cases, breathing continues even during a coma. Impulses to the diaphragm come through a pair of spinal nerves, called the *phrenic nerves.* They initiate the regular alternating contraction and release of the diaphragm. The rhythm of the impulse is controlled by the autonomic system in the brain stem. (You can control your breath voluntarily, but doing so involves the cerebral cortex.)

Inspiration (breathing in) is the result of the diaphragm's contraction. The diaphragm moves downward into the abdomen, expanding the lungs as it does so. This, of course, decreases the air pressure in the lungs. (Remember, the diaphragm is attached to the base of the lungs at the bottom and the thoracic wall at the sides.) The skeletal muscles of the ribs *(intercostal* muscles) help to expand the lungs by pulling the ribs up and out. Air is pulled in at the top of the airway (nose and mouth) and all the way down to take up the expanded space in the alveoli.

Taking measured breaths

To determine whether you're breathing properly and your lungs are functioning well, physicians can measure the amount of air you inhale and exhale, as well as how much air remains in your lungs after you exhale and how much air you're capable of holding.

The *tidal volume* (TV) is the volume of air inhaled and exhaled in a single breath during normal relaxed breathing. This amount is about half a liter.

Of course, if you breathe deeply, you can inhale and exhale more. The maximum volume of air that you can move through your lungs in one breath is your *vital capacity* (VC). By breathing really deeply, people usually can reach a VC of 4.8 quarts (4.5 liters) or slightly more.

The volume of air that you can force out beyond the TV is your *expiratory reserve volume* (ERV). The ERV averages about a liter.

The VC minus the TV and the ERV gives you the *inspiratory reserve volume* (IRV). This is how much more room you can make in your lungs above and beyond a normal inhalation. The IRV averages about 3.2 quarts (3 liters), but you can't actually breathe that way for any length of time.

Even after you force out all the air you can, you still have a little more left in your lungs. This is the *residual volume* (RV), which is 1.3 quarts (1.2 liters) on average.

Some of the air you inhale fills your pharynx, trachea, and nasal cavities and never actually gets down into your lungs. This volume is called *dead space.*

During normal breathing, *expiration* (breathing out) is a passive process that happens between impulses. Stretch receptors within the alveoli send nerve impulses to the respiratory center to release the contraction of the diaphragm. As the diaphragm relaxes, it moves back up out of the abdomen (because of abdominal muscle tone), the elastic tissue in the lungs snaps back (gets smaller), and the rib cage falls back (mostly because of gravity). This increases the air pressure in the lungs and pushes the air out (see Figure 10-2).

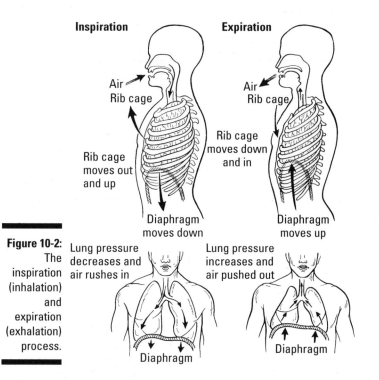

Figure 10-2:
The inspiration (inhalation) and expiration (exhalation) process.

Breathing under stress

Keep in mind as you read this section that "stress" just means that an extra physiological demand is being placed. Stress isn't necessarily negative — whether physical or emotional, stress can be painful or pleasurable, and is often both.

However much fun you are or are not having, stress increases metabolism. More oxygen is consumed, and more carbon dioxide is produced. *Chemoreceptor cells* in the carotid arteries and aorta detect an elevated level of carbon dioxide or hydrogen ions and alert your respiratory center. Both inspiration and expiration become active processes. You breathe more deeply and frequently. The intercostal muscles forcefully contract and push more air out of your lungs (this can't occur during rest). These processes restore homeostasis and support the elevated metabolism.

The forcible exhalation involved in coughing or sneezing is aided by the sudden contraction of the abdominal muscles, raising the abdominal pressure. The rapid increase in pressure pushes the relaxed diaphragm up against the pleural cavity, forcing air out of the lungs.

Controlled breathing

As far as has been determined, humans are the only animals who can bring their breathing under conscious control. Breathing also allows people to speak and sing, as well as moderate other physiological systems.

Holding your breath

You can stop breathing (hold your breath), at least for awhile, when unpleasant odors, noxious chemicals, or particulate matter is in the air around you, while you swim underwater, or just for the fun of it. The cerebral cortex sends signals to the rib muscles and the diaphragm that override the respiratory center signals — temporarily.

Holding your breath long enough to cause damage to your own brain from a lack of oxygen isn't possible. Metabolism and gas exchange continue as usual while you hold your breath. Carbon dioxide concentration increases in the blood. At a certain point, long before brain damage is even a possibility, the chemoreceptors that work with the respiratory center are stimulated to the point where they override the cerebral cortex. At the extreme, you lose consciousness. (The evolutionarily older brainstem puts the whippersnapper cerebral cortex into time out.) You exhale and the system rapidly returns to normal.

Speaking and singing

Speech, another uniquely human activity, requires breath control. The exhalation passes breath over the vocal cords, causing sound waves to be emitted, and the lips and tongue shape the sound waves into speech. The rate of exhalation is lower while you're speaking, controlled by the diaphragm, the intercostal muscles, and the abdominal muscles. Singing requires even more breath control than speaking.

Controlling other systems

The relationship between the autonomic nervous system and the respiratory system appears to be two-way. For example, anxiety prompts hyperventilation, and hyperventilation produces symptoms of anxiety. Consciously controlling the rate and depth of breathing, mainly by achieving awareness and control of the diaphragm, has been demonstrated to decrease anxiety and sympathetic nervous system activation.

Controlled breathing is a feature of many religious, spiritual, and physical disciplines in all traditions. Clinical benefits have been demonstrated for meditation and controlled breathing in a wide variety of conditions of the neural, cardiovascular, and pulmonary systems.

Air, air everywhere

Air. You need it every minute of your life. But what is it?

Air is the mixture of gasses that surrounds planet Earth. All *eukaryotic* (multicellular) organisms existing today, including, naturally, humans, have an intimate relationship with air. Air and multicellular organisms created each other in a process that began when life began and took a decisive turn when some bacteria "learned" to photosynthesize.

Nearly four-fifths (78 percent) of the mixture is a gas called *molecular nitrogen,* or N_2. This species is two atoms of nitrogen held together by a triple bond. The triple bond makes it unlikely that the atoms will separate and participate in any reactions. It is chemically inert. Nitrogen is an important element in many kinds of biochemicals, notably amino acids. (The root *amine* tells you that a compound contains nitrogen.) But animal physiology has not evolved a way to use the nitrogen in air to make proteins.

Still, from the point of view of human physiology, this four-fifths of air is just taking up space. Yes, exactly. These molecules separate the molecules of the other one-fifth of the volume of air, which is a gas called *molecular oxygen,* or O_2. Like molecular nitrogen, molecular oxygen is two atoms of the same element bonded together, but unlike N_2, the bonds between the oxygen atoms are weak. Those atoms will leave their partners (ionize) and go off with any passing hydrogen (often two at once!), carbon, or nitrogen ion that happens to be passing by. Both molecular oxygen and oxygen ions *(oxygen free radicals)* are so chemically active that they can be difficult to control. It's very dangerous stuff

to have around. The paradox is that humanity lives in a sea of it and has evolved a total and absolute dependence on it.

Molecular oxygen is a byproduct of photosynthesis. O_2 wasn't present in the atmosphere of the earth until certain bacteria evolved the anatomy and physiology (as these words are understood in the context of prokaryotes) for photosynthesis. Evolution favored organisms with the ability to harness light energy — until, that is, the amount of molecular oxygen byproduct in the air started posing a mortal danger to carbon-based objects (all organisms). Some evolutionary theorists have suggested that eukaryotic cells evolved in response to this selection pressure. The cell membrane *(phospholipid bilayer)* may have originated as a way to protect bacterial machinery from the disruptive effects of molecular oxygen while still benefitting from its exceptionally available energy.

Carbon dioxide, comprising about 0.4 percent of the total volume of the air, is mostly a byproduct of aerobic metabolism. The carbon-oxygen bonds are stronger than the oxygen-oxygen bonds, but not as strong as the nitrogen-nitrogen bonds; so carbon dioxide is moderately active chemically, but not enough so to support the processes of life.

Until recently, CO_2 was best known as the source of the carbon that is "fixed" (incorporated into biological molecules) by photosynthesis. Recently, it has become notorious as a pollutant in the air, released when oxygen disassembles the large carbon molecules in petroleum-based substances.

Pathophysiology of the Respiratory System

The structures of the airway and respiratory membrane are in constant contact with the air, with its constant threats of temperature extremes, desiccation, harmful chemicals and particles, and pathogens.

Hypoxemia

Fluctuations in oxygen concentration in tissues is part of normal physiology, such as at times of increased demand like during strenuous exercise or emotional stress. In a healthy body, homeostatic mechanisms kick in rapidly and effectively when oxygen levels fall below the normal range.

Hypoxemia (low oxygen concentration in the arterial blood), however, is a respiratory system disorder resulting from any condition that interferes with gas exchange at the respiratory membrane, such as an obstructed or blocked airway or alveolar scarring. Because all cells and tissues require an adequate oxygen supply to function optimally, hypoxemia harms all structures and inhibits all physiological processes to one extent or another.

A further complication of hypoxemia occurs when a low blood oxygen level stimulates the production of more red blood cells (RBCs). However, if iron is in short supply in the diet, which is a common condition worldwide, the body can't synthesize enough hemoglobin for all these RBCs, and many of them are dysfunctional, a condition called *polycythemia*. The blood is low in oxygen despite the excess of RBCs.

The other side of gas exchange is affected, too. Blood carbon dioxide levels rise, acidifying the blood and disrupting its normal function. Many enzymes, for example, are sensitive to pH.

Note: Don't confuse hypoxemia with *hypoxia,* a pathological condition of the circulatory system caused by atherosclerosis, arterial blockage, or anemia. The oxygen-poor air at high altitudes can cause hypoxia, too.

Airway disorders

Airway disorders are similar in their local and systemic effects. Many are chronic and all lead to hypoxemia, which may be mild, moderate, or severe.

Asthma

Asthma is classified as a chronic disorder involving a "reactive airway." That means that the airway (bronchi and bronchioles) becomes inflamed and swollen or constricted in reaction to certain triggers. The bronchial tubes may spasm. The mucosa lining the tubes may secrete excess or very thick mucus. Any or all of these conditions make breathing difficult.

Asthma is most often developed and diagnosed in childhood. Signs and symptoms of mild asthma include periods of coughing, shortness of breath, or wheezing. The symptoms are often brought on by exercise, but sometimes they occur without that trigger. The condition may progress to moderate asthma, with more frequent and severe bouts of shortness of breath, trouble breathing while resting, and an increased respiration rate, sometimes lasting several days. In severe asthma, respiratory distress brings hypoxemia. Asthma is a chronic condition and can be fatal.

Clinically, asthma is diagnosed as *intrinsic* (triggered by factors in the body) or *extrinsic* (triggered by factors from outside the body) and treated accordingly. Intrinsic asthma may set in after a severe respiratory tract infection. Other triggers include hormonal changes, emotional stress, and fatigue. Extrinsic asthma triggers are common irritants and allergens.

Bronchitis

Bronchitis is inflammation of the bronchi. *Acute bronchitis* may follow respiratory infection or exposure to irritating substances or cold temperatures. As a result, the body produces copious mucus, causing a persistent cough.

Chronic bronchitis is caused by long-term exposure to irritating substances, such as chemicals or cigarette smoke. The cilia of the respiratory mucosal cells are damaged and are less effective at cleaning the bronchial tissue of the respiratory mucus and the entrapped debris. Coughing develops to help the excretion of the mucus, but coughing irritates the bronchi further. The airway swells and constricts.

Lungs

In spite of all the respiratory system's defenses, inhaled pathogens, chemicals, and particles find their way to the lung. Once there, they can be difficult to push out again.

Pneumonia

The lung is an extremely hospitable environment for bacterial, viral, and fungal microbes. *Pneumonia* is a pathogenic infection in the lower respiratory system that results in mucus and pus building up in the airway. Pneumonia is classified by the location and extent of the infection as *bronchopneumonia, lobular pneumonia* (affecting part of a lobe), or *lobar pneumonia* (affecting the entire lobe).

Pathogens enter the respiratory system easily in inhaled air. A healthy respiratory system working with the immune system can eliminate almost all of them. As the body ages, both respiratory tissue and immune system functionality decline. Before the arrival of effective antibiotics in the pharmacological arsenal, bacterial pneumonia was a common cause of death in the elderly. (It even had a nickname, "the old man's friend," implying that a relatively quick death by pneumonia was a blessing.)

Tuberculosis (TB)

Pulmonary tuberculosis is an infectious disease of the lungs caused by the bacterium *Mycobacterium tuberculosis.* The bacterium may infect the lung without symptoms for a period of several months to many years. The infection may spread to the bones and other organs. The long asymptomatic period, during which the infected person can infect others just by breathing, is what makes TB such a stubborn public health problem.

In tuberculosis, areas of bronchial and lung tissue become inflamed and die, leaving a hole in the tissue from which air can leak. Air leaking out from the lung causes the lung to collapse.

M. tuberculosis evades efforts by the immune system to eliminate it. The immune system responds to its presence as it does for any infection: White blood cells and macrophages rush to the site of infection, and the macrophages engulf the bacterial cells and carry them to the lymph nodes (see Chapter 13). But the macrophages then behave oddly — they clump together, forming *tubercles.* Wherever the tubercles lodge, the surrounding tissue is killed, and scar tissue forms around the tubercle. Eventually, the lymph nodes become inflamed and may rupture, allowing the bacterium to spread to surrounding tissue. Because lymph nodes exist all over your body, TB can easily spread from your lungs to other areas.

Emphysema

Emphysema is a *chronic obstructive pulmonary disorder* (COPD). Commonly known as a smoker's disease, emphysema can also affect people who have had long-term exposure to pulmonary irritants, such as chemicals, asbestos, or coal. Eventually, the cumulative damage to the bronchioles causes them to collapse, trapping air inside them. The pressure of the trapped air can rupture the tiny alveoli, destroying the respiratory membrane in that area. The destroyed tissue may be replaced with inelastic scar tissue (fibrosis). The elasticity of the lungs is decreased, making it hard for a person with emphysema to breathe *(dyspnea).* As the disease progresses, the delivery of oxygen to the blood is compromised, resulting in the condition of hypoxemia.

Chapter 11

The Digestive System: Beginning the Breakdown

In This Chapter

▶ Explaining what the digestive system does

▶ Following the path of the digestive tract

▶ Focusing on the liver, pancreas, and digestive fluids

▶ Discovering some digestive system diseases

*J*ust how does that pile of steak, potatoes, and salad on your plate become the tissues of your body? The digestive organ system gets it halfway there. (The circulatory system does most of the rest.)

Living systems constantly exchange energy. Physiological processes — anabolic, catabolic, and homeostatic — require energy. Ultimately, that energy comes from the light energy that plants use to transform carbon in the atmosphere (as CO_2) into biological matter (as carbohydrate) in the process of *photosynthesis*. Humans get their energy by consuming this biological matter, either directly or by consuming other organisms that do. The digestive system takes apart this biological matter step by step and transforms it into a form that human cells can use. In this chapter, we explain the ins and outs of the process and the organs responsible for the chore.

Functions of the Digestive System

Digestion itself is a middle step. Other important functions of this organ system come before and after:

✔ **Ingesting:** Although all animals *ingest* — take something into the body through the mouth — only humans, and possibly some of the great apes, appear to enjoy food as they ingest it.

As we discuss in Chapter 7, the perception of subtle flavors is more closely connected to olfaction than digestion. The perception of the five basic flavors is also considered *neurosensory*. Perception in the mouth is more about texture and is closely related to food's protein and fat content. This concept is captured by the food industry term *mouth feel*. These sensory perceptions guide you in the selection of foods.

Sometimes, though, ingestion isn't a feast for the senses — nothing delicious is available. The body still needs calories, though, so whatever's available is chewed and swallowed just the same. (See the section "Starting with the mighty mouth" later in the chapter.)

✔ **Digesting:** Eating is fun, and ingestion is bearable, but neither provides biological molecules that your cells can use. That task is accomplished by the interaction of physical and chemical forces. The digestive tract is a muscular tube lined with chemical factories that operate under the direction of their own dedicated neural structures and under hormonal control (see the upcoming section "Structures of the Digestive System").

The digestive system processes everything down the same track, extracting fuel, biological molecules and monomers, and micronutrients from whatever you eat. (See the section "Moving through the intestines" later in the chapter.)

✔ **Exporting nutrients to the body:** The end products of digestion are biological molecules such as glucose that are absorbed across the digestive membrane into the blood and then distributed in the body (see Chapter 9 for an overview of the circulatory system).

✔ **Eliminating:** The elimination of digestive waste is part of digestion. Other organ systems have evolved to make use of the digestive system's structures to eliminate metabolic wastes of other kinds. As they say, one big pile is better than two little piles.

Structures of the Digestive Tract

The *digestive tract*, also called the *alimentary canal* ("alimentary" means food), or the *gastrointestinal (GI) tract*, is a tube through which ingested substances are pushed along for physical and chemical processing. The tube walls are made up of an outer fibrous layer, a muscular layer, a supportive connective tissue layer, and an inner layer (containing an epithelial lining), called the *digestive mucosa*. All the layers vary in thickness from one place to another along the digestive tract. The space inside the tube is the *lumen*, and its size varies, too.

The digestive system's gross anatomy (no pun here) is comparable to that of an industrial smelter. Some structures bring in raw materials; other structures extract, process, and ship out specific substances; and still other structures export the unused part of the raw materials back into the environment. The body uses both mechanical and chemical mechanisms to break down the raw materials and export products to the larger system (the economy in the case of the smelter; the organism in the case of the intestine). These efficient systems are organized linearly — things keep moving along in one direction at a steady pace.

Strictly speaking, the lumen isn't "inside" the body. Rather, the body itself is wrapped around a small piece of the environment — that is, the lumen. Neither the food that enters your mouth nor any of the partially digested substances that your digestive tract produces are inside the body, either. Fully digested biological molecules extracted from these substances and transformed into molecules that are usable by human cells cross out of the lumen and into the blood. At that point, they're inside the body.

As you read about the organs that break down food to nourish your body, refer to the "Digestive System" color plate in the middle of the book to see where the organs are located.

Examining the walls of the digestive tract

The upper third of the esophagus is made of skeletal muscle. Beginning in the middle third of the esophagus and extending to the *anal sphincter,* layers of smooth muscle line the digestive tract. As we discuss in Chapter 5, this smooth muscle contracts in pulsating waves, pushing the lumen's contents along in a single direction. This constant wave-like contraction is called *peristalsis.*

A mucous membrane lines the digestive system, running continuously from the mouth all the way through to the rectum. This membrane protects your digestive organs from the strong acids and powerful enzymes secreted in the digestive system. The membrane's innermost cells (next to the lumen) are among those cells that are continuously replaced (see Chapter 2 for more on cell renewal).

The digestive mucosa's secretions keep everything in the digestive tract moist, soft, and slippery, protecting the membrane and its underlying structures from abrasion and corrosion. The digestive mucosa contains tissues and cells that secrete other substances as well, including gastric acid, hormones, neurotransmitters, and enzymes. The digestive mucosa also contains an extensive network of lymphatic tissue, as we discuss in Chapter 13.

The digestive mucosa takes an active role in the final stage of digestion. It delivers the products of digestion from the small and large intestine to the blood for distribution through the body. Every molecule that enters the bloodstream passes through the digestive mucosa.

The digestive mucosa is called by different names in different parts of the digestive system. In the mouth, it's called the *buccal membrane;* in the stomach, the *mucous coat.* These different areas of the membrane secrete and absorb different substances and can therefore be considered separate structures. Mucus is secreted throughout its entire length.

Starting with the mighty mouth

Your mouth is the starting point of your digestive system, the gateway to your other digestive organs. Besides making eating a fun experience, your mouth *(oral cavity)* serves some important digestive functions.

Talking about your teeth and gums

Humans have 32 teeth — 16 on the top and 16 on the bottom. Your teeth tear and grind food into pieces small enough to swallow, and they come in four basic types: incisors for biting, canines for tearing (especially meat), and premolars and molars for grinding.

The *gingiva* (gums) hold teeth in position, and a binding material called *cementum* embeds your teeth's roots in your jawbone (see Figure 11-1). Blood vessels that run through the jawbone and up into the pulp of the tooth supply the teeth with blood. *Dentin,* a bonelike material, covers the pulp, and an extremely hard protective enamel covers the dentin.

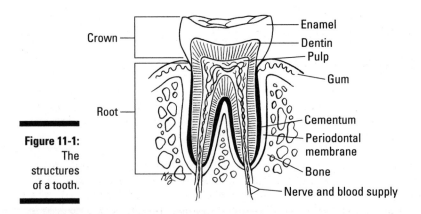

Figure 11-1:
The
structures
of a tooth.

Crown — Enamel — Dentin — Pulp — Gum — Root — Cementum — Periodontal membrane — Bone — Nerve and blood supply

Evolution appears to be favoring a smaller mouth in the human species. Many jawbones have difficulty accommodating the last molar in each row of teeth. Typically, these *wisdom teeth* erupt far later than the other permanent molars, usually in mid to late adolescence, and often cause jaw pain and dislocation of other teeth. Quite often, they don't erupt at all and remain "impacted" in the jawbone, which can also cause jaw pain.

Your tongue helps out

Your tongue is mainly skeletal muscle tissue. The muscle is covered on the upper surface by a mucous membrane, in which are embedded taste buds. (See Chapter 7 for more on taste buds.) The tongue muscles move the food around in your mouth to assist chewing. The mucus moistens and lubricates the *bolus,* the technical term for a mouthful of food in the process of being chewed.

Muscles attach your tongue to your skull bones, and a mucous membrane on the tongue's underside attaches your tongue to the oral cavity floor. That stringy piece of membrane that you see when you touch your tongue's tip to the roof of your mouth is the *lingual frenulum.*

The buccal membrane

The *buccal membrane* is that portion of the digestive mucosa that lines the inside of the mouth. Several *salivary glands* have ducts that course through the buccal membrane and secrete mucus and *salivary amylase,* a digestive enzyme, into the oral cavity. These glands often go into action before you take the first bite of your meal. A delicious aroma or even just the anticipation of eating something you enjoy can get those juices flowing.

The enzyme salivary amylase turns starch into sugar as you chew.

Pharynx and esophagus: Not Egyptian landmarks

The *pharynx,* better known as your throat, leads to the *esophagus,* the tube that extends from the mouth to the stomach. When you swallow, the bolus bounces off a piece of cartilage called the *epiglottis* and so is diverted from the *trachea* and into the esophagus (see Figure 11-2).

The esophagus has two sphincters, one at the top and one at the bottom, that control the movement of the bolus into and out of the esophagus. The *pharyngoesophageal sphincter,* composed of skeletal muscle, takes part in the muscular actions involved in swallowing. Peristalsis propels the bolus along the esophagus. The lower esophageal sphincter surrounds the esophagus just as it enters the stomach.

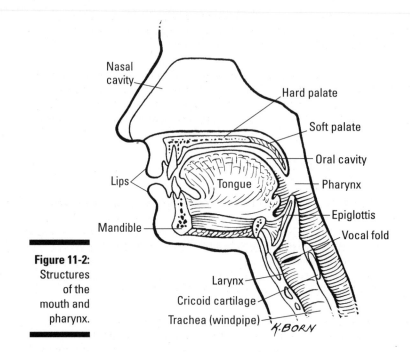

Figure 11-2:
Structures
of the
mouth and
pharynx.

Nasal cavity

Hard palate

Soft palate

Oral cavity

Lips

Tongue

Pharynx

Epiglottis

Vocal fold

Mandible

Larynx

Cricoid cartilage

Trachea (windpipe)

K. BORN

Stirring it up in your stomach

After passing through the bottom sphincter of the esophagus, the bolus drops into the stomach, the widest and most flexible part of the alimentary canal. Food remains in the stomach about two to six hours, during which it's churned in an acidic substance that the stomach secretes called *gastric juice,* ground up by thousands of strong muscle contractions, and offered in tiny pieces to protein-digesting enzymes. The "Stomach" color plate in the center of the book shows the various parts of the stomach.

The outside of the stomach is a tough connective tissue layer called the *serosa.* Beneath the serosa is the *muscular coat,* which has three layers of smooth muscle fibers — oblique, circular, and longitudinal — that contract in different directions. Stretch receptors in this layer send nerve impulses to the brain when the stomach is full. These two layers support the structure of the stomach as a hollow organ.

Beneath these connective tissue layers are two mucosal layers, the *submucosal coat* and the *mucous coat,* also called the *stomach lining.* Gastric glands in the mucosa secrete the components of gastric juice. The mucosal coat is corrugated, increasing the surface area inside the hollow. The folds are called *rugae.* As the stomach fills, the rugae smooth out, allowing the stomach to expand.

The stomach's muscular action is part of physical digestion, like chewing, swallowing, and peristalsis. But the stomach's contribution to chemical digestion is what really helps break down the food you eat.

As the stomach churns the bolus in the gastric acid, the material turns into an oatmeal-like paste called *chyme*. The chyme squirts into the small intestine through the *pyloric sphincter*, between the lower part of the stomach, called the *pylorus*, and the top of the small intestine, called the *duodenum*.

Moving through the intestines

The *intestine* is a long muscular tube (up to about 20 feet, or 6 meters) that extends from the pyloric sphincter to the anal sphincter. How does 20 feet of tubing fit into a relatively small space that's also crowded with other organs? It becomes narrow and convoluted. The intestines are classified as small and large based on their width, not their length (like hoses). The lumen of the small intestine is about 1 inch (2.5 centimeters) in diameter; the large intestine is about 2.5 inches (6.4 centimeters).

Overall, the intestine specializes in the import and export of biological substances of many kinds. As is usual for organs with import-export functions, the intestine has structures that maximize the surface area available for the exchange.

The intestine's muscular outer walls lie coiled closely together within the abdominal cavity, held in place by the fibrous sheets of the *peritoneum*. With two layers of smooth muscle tissue, longitudinal and circular, the intestine specializes in strong, sustained peristalsis.

The intestinal mucosa is continuous with the rest of the digestive mucosa. It's studded with specialized "work areas" that produce hormones, neurotransmitters, enzymes, and other substances integral to the digestive process.

The capillary beds that line the intestine define the interface of the digestive and circulatory systems. These capillaries are arrayed more or less continuously along the intestine's lumen.

The lumen is lined by the *villus* (plural, *villi*), a structure that's specialized for import and export processes and that's characteristic of tissues in body locations where substances are exchanged. *Villi* are fingerlike projections of the mucosa that multiply the surface area available for exchange, much like wharves and piers extending into a harbor increase the area for harbor activities.

Villi line the entire length of the small intestine, projecting out into the lumen. Each villus has its own assigned capillary for absorbing materials from the intestine into the blood (flip to Chapter 9 for more on the circulatory system). *Microvilli* are even smaller projections on the epithelial cells of the mucosa.

Some of these processes require *active transport* — the expenditure of some energy in the form of ATP. For more on active transport and ATP, see Chapter 3.

Investigating the small intestine

The small intestine does a lot of the physical work of the digestive system, beginning with peristalsis. The small intestine is also majorly involved in digestive chemistry.

The small intestine is an endocrine gland as well as a digestive organ, producing and secreting hormones that control digestion. Cells in the small intestine's walls secrete the hormones *secretin* and *cholecystokinin* (CCK), which stimulate the release of digestive fluids such as bile from the gallbladder and pancreatic juice from the pancreas.

The small intestine's walls are lined with secretory tissue that functions in chemical digestion. Cells in the duodenum's walls secrete digestive enzymes. *Brunner's glands* in the small intestine's lining secrete mucus and bicarbonate directly into the lumen to help neutralize the gastric juice in the chyme. (Most enzymes require a near neutral pH.)

The small intestine is divided into three structures along its 10-to-20-foot (3-to-6-meter) length: the *duodenum* (about 1 foot long, or .3 meters), the *jejunum* (about 3 to 6 feet, or 1 to 2 meters), and the *ileum* (about 6 to 12 feet, or 2 to 4 meters). It is approximately 1 to 2 inches (2.5 to 5 centimeters) in diameter.

The small intestine can measure around 50 percent longer at autopsy because of the loss of smooth muscle tone after death.

The pyloric sphincter controls the release of chyme into the duodenum. The rate of flow is limited by the ability of the duodenum to neutralize the strong acid. The processes of chemical digestion run furiously. The carbohydrates, proteins, and fats are broken down into molecules such as glucose, amino acids, fatty acids, and glycerol. Peristalsis moves the almost-completely-digested chyme along into the jejunum and ileum.

The body handles the two products of fat digestion, fatty acids and glycerol, a bit differently. Short-chain fatty acids are shuttled directly to the capillary through the villus. Long-chain fatty acids are transported through the villus to the lymphatic system. In the cells, long-chain fatty acids are assembled into compounds called *triglycerides.* Glycerol is absorbed by the liver and is either converted to glucose or used in glycolysis (breaking down glucose into energy).

By the time the chyme works its way through all three parts of your small intestine, the nutrients that your body needs have been absorbed into the blood. At the ileum, the indigestible matter passes into the large intestine.

The work of the large intestine

Chyme oozes from the small intestine to the large intestine (also called the *colon*), passing out of the ileum through the *ileocecal valve* into the *cecum,* the first portion of the large intestine. The material is now called *feces.*

The large intestine is about 6 feet long (almost 2 meters) and is positioned anatomically like a "frame" around the small intestine. Beyond the cecum, the large intestine moves upward as the *ascending colon,* across as the *transverse colon,* and downward as the *descending colon* and finally into the *sigmoid colon.*

In the large intestine, water is reabsorbed from the feces by diffusion across the intestinal wall into the capillaries. The removal of water compacts the indigestible material in the colon, forming the characteristic texture of the feces.

In addition to undigested food, the feces contain other bodily wastes to be excreted. The brown color of feces comes from the combination of greenish-yellow bile pigments, broken down hemoglobin, and bacteria.

Your intestines are home to unimaginably large numbers of bacteria, including hundreds of species. Trillions of tiny (prokaryotic) cells ingest some of the undigested material in your feces, producing molecules that have a well-known odor. (It's nothing to be embarrassed about, and nothing to be proud of, either.) Some of these bacteria produce beneficial substances like vitamin K, which is necessary for blood clotting. These substances are absorbed through the intestinal wall and transported into the blood via the capillaries. Check out Chapter 17 for more on microbes.

Passing through the colon and rectum

As the colon completes its work, peristalsis moves feces into the *rectum,* which is located at the bottom of the colon. Stretch receptors in the rectum signal to the brain the need to defecate (release feces) when the rectum contains about 5 to 8 ounces (142 to 227 grams). Pushed by peristalsis, the feces pass through the *anal canal* and exit the body through the anal sphincter.

Doing the Chemical Breakdown

The pancreas, liver, and gallbladder are often referred to as the *accessory organs of digestion.* They're not part of the digestive tract; they never come into contact with ingested material, and they take no part in the mechanical aspects of digestion. They produce and make available to the digestive tract's organs some of the chemical and biological substances that assist in digestion's chemical aspects.

The liver delivers

The *liver* is one of the most important organs, not just in digestion but in many other functions. The liver's digestive function is the production and transport of *bile,* one of the digestive chemicals.

Many of the terms related to the liver's structures and functions contain the prefix *hepato-,* meaning "liver."

Liver anatomy

Your liver is both the largest internal organ and the largest gland in the human body. A healthy adult human liver weighs about 3 to 3.5 pounds (1.4 to 1.6 kilograms). It's located under your diaphragm and above your stomach on the right side of your abdomen (see Figure 11-3). The liver is soft, pinkish-brown, and triangular, with four lobes of unequal size and shape: the *right lobe, left lobe, quadrate lobe,* and *caudate lobe.* The liver is covered by a connective tissue capsule that branches and extends throughout its insides, providing a scaffolding of support for the afferent blood vessels, lymphatic vessels, and bile ducts that traverse the liver.

The liver receives oxygenated blood through the *hepatic artery,* which comes from the *aorta.* It receives nutrient-rich blood through the *portal vein,* which carries blood from the capillaries of the small intestine and the descending colon (see Chapter 9 for more on the circulatory system). Three hepatic veins drain deoxygenated blood from the liver, exiting the liver at the top of the right lobe and draining into the *inferior vena cava.*

Each of the four lobes is made up of tiny lobules, about 100,000 of them in all. The *hepatic lobule* is the liver's functional unit (see Figure 11-4). Each lobule is made up of millions of hepatic cells and bile canals and is supported and separated by branches of the capsule. At the lobule's vertices are regularly distributed *portal triads* that contain a bile duct, a terminal branch of the hepatic artery, and a terminal branch of the portal vein. The *hepatocytes* are in a roughly hexagonal arrangement, with a vein in the center that carries the lobule's products out into the blood. On the surface of the lobules are ducts, veins, and arteries that carry fluids to and from them.

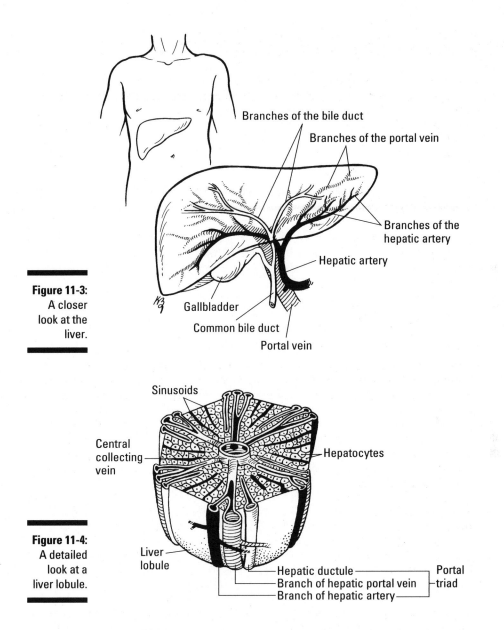

Figure 11-3:
A closer look at the liver.

Figure 11-4:
A detailed look at a liver lobule.

Liver regeneration

Among its other astounding powers, the liver possesses the ability to regenerate and regain its original size, structure, and function rapidly after partial resection surgery or massive injury. The human liver can fully regenerate from as little as 25 percent of its original tissue. This ability is unique among the major organs.

Live-donor transplantation, a procedure in which a healthy person donates a portion of his or her liver to a recipient whose liver is failing, has been performed successfully since 1989. Typically, the liver doubles in size in both the donor and recipient within only three to four weeks. Rapid replication of hepatocytes is the mechanism of growth.

Bile production and transport

The liver produces bile, a major factor in the digestion of fats and lipids of all kinds. The bile that some of the lobules produce is collected in bile *canaliculi,* which merge to form *bile ducts.* The *intrahepatic* (within the liver) bile ducts eventually drain directly into the duodenum through the *common hepatic duct.*

Bile can also be transported for storage into the gallbladder via the *extrahepatic* bile ducts. Your *gallbladder* is a pear-shaped sac tucked into the curve of your liver whose only function is to store bile and deliver it on demand to the duodenum. The bile flows through the common bile duct into the duodenum, near the entry point of the *pancreatic duct.*

Other functions of the liver

The liver functions in many other ways, affecting other organ systems. Here's a brief overview of its many functions:

- ✔ It processes and eliminates toxins. Toxic byproducts of some drugs, including alcohol, and other substances arrive from the digestive organs through the portal vein.

- ✔ It processes and eliminates metabolic waste. The liver removes dying red blood cells from the blood and converts the hemoglobin to *bilirubin* and other byproducts. These are delivered to the intestine and excreted with the feces. (The iron is recycled.)

- ✔ It stores glucose in the form of *glycogen* and reconverts it when blood glucose levels get low. This function is mediated by insulin and glucagon. (Turn to Chapter 8 for more on the endocrine system.)

- ✔ It stores vitamins and minerals.

- ✔ It produces many kinds of protein, including protein hormones, the plasma proteins, and the proteins of the clotting cascade (see Chapter 9) and the complement system (see Chapter 13), as well as the production of alpha and beta globulin (see Chapter 13).

Pancreas

The *pancreas* sits in the abdominal cavity next to the duodenum and behind the stomach. We discuss the endocrine function of the pancreas in Chapter 8.

The pancreas produces *pancreatic juice,* which is full of pancreatic enzymes that are important in digestion. Refer to Table 11-1 for information about pancreatic enzymes. Nearly every cell of the pancreas secretes pancreatic juice and passes it through the *pancreatic duct* into the duodenum.

Table 11-1	Pancreatic Enzymes	
Enzyme	*Targeted Nutrient*	*Result of Breakdown*
Trypsin	Proteins	Peptides (chains of amino acids)
Peptidase	Peptides	Individual amino acids
Lipase	Fats	Fatty acids and glycerol
Nuclease	Nucleic acids (DNA, RNA)	Nucleotides
Amylase	Carbohydrates	Glucose and fructose

Digestive fluids, enzymes, and hormones

Each part of the digestive system has its characteristic fluid, each a complex mixture of water, electrolytes, and biological substances with a specific role in digestion.

✔ **Mucus:** Every inch of the digestive tract has mucus glands whose secretions keep everything in the digestive tract moist, soft, and slippery, protecting the digestive membrane and its underlying structures from abrasion and corrosion. (For more on mucus, see the section "Examining the walls of the digestive tract" earlier in the chapter, and also flip to Chapter 10.)

✔ **Saliva:** Saliva (or spit) is a clear, watery solution that the salivary glands produce constantly in your mouth. You produce about 2 to 4 pints (1 to 2 liters) of spit every day. Saliva moistens food and makes it easier to swallow. It's also a component of the sense of taste — a food substance must be dissolved in the watery solution for its chemical signals to act on your taste buds (see Chapter 7).

Enzymes in saliva start starch digestion even before you swallow food. The combination of chewing food and coating it with saliva makes the tongue's job a bit easier — it can push wet, chewed food toward the throat more easily.

Saliva cleans the inside of your mouth and your teeth. The enzymes in saliva also help to fight off infections in the mouth.

✓ **Enzymes:** Thousands of enzymes are involved in digestion. Enzymes are specialized in their function — a given enzyme typically catalyzes one or only a few specific reactions. Digestive enzymes specialize in reactions that take specific molecules apart into component chemical entities. They can be broadly classified as *proteinases* and *peptidases, lipidases,* and various kinds of *carbohydrate-active* enzymes. Enzymes are part of the digestive fluids gastric juice and pancreatic juice.

The suffix *–ase* indicates an enzyme that breaks a molecule apart.

✓ **Gastric juice:** Gastric juice is secreted from millions of tiny gastric glands in the gastric mucosa and enters the hollow of the stomach through *gastric pits* on the mucosa's inner surface. Gastric juice contains *hydrochloric acid* (HCl), which is extremely acidic and kills bacteria that may have entered the body with food. It also contains the powerful proteinase *pepsin,* which can work only in this highly acidic environment.

✓ **Pancreatic juice:** Pancreatic juice contains many types of digestive enzymes. Refer to Table 11-1 for some details.

✓ **Bile:** Bile, also called *gall,* is a very alkaline, bitter-tasting, dark-green to yellowish-brown fluid produced by the liver. Bile may remain in the liver or be transported to the gallbladder for storage before being expelled into the duodenum.

The physiological function of bile is to emulsify fats — that is, to create an environment in which lipid-based substances can be mixed in a watery matrix for transportation and to make them available for chemical reactions to break them down. Bile's high alkalinity helps neutralize the strongly acidic chyme that comes into the duodenum from the stomach. Another purpose of bile is to help absorb the fat-soluble vitamins K, D, and A into the blood.

The color of bile comes from the *bilirubins* of red blood cells (RBCs) dismantled in the liver and pancreas. These pigments are deposited in the bile for elimination through the digestive tract. They play no part in chemical digestion.

✓ **Hormones:** The hormone *gastrin* regulates the secretion of HCl, mucus, and *pepsinogen.* As long as gastrin is flowing, your stomach continues to secrete gastric juice.

Cells in the small intestine's walls secrete the hormones secretin and cholecystokinin (CCK), which stimulate the release of bile and pancreatic juice, respectively.

✓ **Buffers:** To lower the extreme acidity (raise the pH) of gastric acid and create a more hospitable environment for most digestive enzymes, the small intestine and the pancreas secrete *sodium bicarbonate,* the same compound found in baking soda.

Pathophysiology of the Digestive System

This section gives you some info on several common diseases and disorders of the digestive system.

Diseases of the oral cavity

Bits and pieces of food that remain in the mouth promote the growth of the normal bacteria present in the mouth. The bacteria in the mouth make themselves right at home, secreting a gelatinous matrix called a *plaque* into the spaces between the *gingiva* (gums) and the teeth. Within about a day, the secreted material hardens into *calculus* or *tartar* (both terms mean, basically, "hard stuff"). The accumulation of plaque and the overgrowth of the bacteria cause *gingivitis* (inflammation of the gingiva). Gingivitis is a major factor in tooth loss. Many otherwise healthy people live with this chronic, low-level inflammation, a risk factor for cardiovascular and other diseases, as we discuss in Chapter 13.

Over time, the acid byproducts of the bacteria's metabolism erode the teeth's enamel, creating a "cavity." If the erosion continues, pathogenic bacteria can make their way into a tooth's pulp, a condition called a *tooth abscess*.

Disorders of the stomach and intestines

The stomach and intestines take food in one end and push it out toward the other, with potential for trouble at every point along the way.

Constipation

Constipation results when the large intestine absorbs too much water out of the feces, which makes the feces dry, hard, and a bit painful on exit. Almost everyone experiences constipation at one time or another, and some people suffer chronic constipation. Underlying causes can be dietary (generally, too little fiber or too little water), a lack of exercise, certain foods and beverages, certain drugs, or the slowing of intestinal activity that can come with aging. The most common cause of constipation is ignoring your body's signal to defecate. The feces remain in the colon too long, and too much water is absorbed into the intestinal lining, drying out the feces.

Diarrhea

An excessive amount of water remaining in the feces causes *diarrhea*. Something is preventing a normal amount of water from being absorbed through the large intestine and back into the blood. Although a bout of diarrhea is merely an inconvenience to adults in the developed world, it's a major health concern in developing countries, particularly for children. Each year, diarrhea kills 1.5 million children around the world, and it's one of the most common causes of death of children under 5 years old, according to the World Health Organization.

One possible cause of diarrhea is pathogenic bacteria that infect the large intestine (not the beneficial bacteria that normally inhabit the large intestine) through contaminated food. The rate of peristalsis increases in an attempt to eliminate the pathogen quickly, without the water being reabsorbed.

Another cause of diarrhea is stress, working through the hormonally-driven mechanisms of the *flight-or-fight response* (see Chapter 8). This mechanism evolved to help organisms survive acute threats, such as the apparition of a predator. Among the effects of adrenaline, the most important hormone in this mechanism, is the stimulation of peristalsis. In life-threatening situations, or those perceived as life-threatening, the lower intestines may expel their contents suddenly, presumably to jettison unproductive weight and permit faster running or harder fighting.

However, chronic stress can stimulate the chronic release of adrenaline, moving feces out of the large intestine too fast, impairing the reabsorption of water, and causing chronic diarrhea.

Appendicitis

Your *appendix* is a little sac attached to the cecum at the beginning of the large intestine. During the transfer of chyme from the small intestine to the large intestine, some material may flow into the appendix. Normally, this material makes its way out. But if it doesn't come out, depending on the material and how long it remains, the appendix can become inflamed or infected by intestinal bacteria, causing *appendicitis*. Fortunately, appendicitis is severely painful; the nature and location of the pain is diagnostic for the condition. (Figure out where your appendix is, just in case. This life-threatening condition is quite common and is always a medical emergency.)

In the worst case, or if left too long untreated, the appendix swells and bursts. The boundary between the large intestine and the peritoneum is breached, causing life-threatening peritonitis and shock.

Gastric and duodenal ulcers

Gastric or *duodenal ulcers* are lesions in the mucosa of these tissues. Breaks in the thick protective mucus layer permit the highly acidic gastric juice to contact the lining's cells, causing pain and further tissue damage.

For many years, emotional stress was considered to be the underlying cause of ulcers. Physicians and physiologists reasoned that the parasympathetic nervous system of a person under stress sent signals to secrete more gastric juice than was needed, leading to excess acid. A large industry was built around gastroenterologists prescribing antacid drugs to ulcer patients. Then, in the late 1980s, a group of Australian physicians and researchers demonstrated that these ulcers were the result of a bacterial infection. The species *Helicobacter pyloris* (translation: screw-shaped rod in the stomach) is moderately infectious and is present in many people; only sometimes are the bacteria able to penetrate the mucous layer and embed in the epithelium. (They move through the mucus and embed in the epithelium by twisting, like a screw. In most stomachs, they don't make it past the mucus layer.) A standardized antibiotic regimen is usually successful in eliminating the infection; the ulcers resolve as the mucosal lining is continuously replaced.

Bowel syndromes

Bowel syndromes are of two types: *noninflammatory* and *inflammatory*. We discuss both types in the following sections.

Irritable bowel syndrome (IBS)

In *irritable bowel syndrome* (IBS), irritation of the tissues results in a change in peristalsis. The rhythm of peristaltic contraction may either speed up or slow down. The patient may suffer diarrhea, constipation, or both. Stress reduction and a high-fiber diet usually are in the treatment plan. IBS is characterized as noninflammatory because autoimmunity and the inflammation response aren't induced.

A syndrome is a collection of symptoms rather than a disease.

Crohn's disease

Crohn's disease is an inflammatory bowel disease, meaning that the intestinal lining becomes inflamed. (The inflammation response is part of the syndrome.) The mucosal layer, the muscular layer, the *serosa* (the covering tissue), and even the lymph nodes and membranes that provide blood supply to the intestine can be affected. As the intestinal lining swells, ulcers, *fissures* (cracks), and *abscesses* (pus-filled pockets) can form. Crohn's usually affects the large intestine, and frequently the ileum as well.

In the disease's early stages, sufferers have diarrhea and pain in the lower right side of the abdomen. Later, areas of lymph follicles, called *Peyer's patches,* form on the small intestine's lining. Those follicles make the intestinal wall thicken and stiffen as fibers are deposited, which narrows the hollow space inside the intestine.

Diarrhea becomes severe because the water from feces can't be absorbed through the diseased lining. As segments of bowel become more diseased, they begin to stick to other diseased segments, which shortens the bowel. Complications such as *fistulas* (abnormal openings between the skin and an organ or between two organs) or obstructions can occur. Patients with Crohn's disease often have an *anal fistula* (an opening between the diseased bowel and the skin near the anus or between the bowel and the vagina or bladder).

Crohn's sufferers develop nutritional deficiencies because of poor nutrient absorption. Usually, people with Crohn's lose some beneficial intestinal bacteria, including those that synthesize vitamin B12. This chronic deficiency can lead to a condition called *pernicious anemia*.

Treatment for Crohn's includes dietary changes, rest, stress reduction, vitamin supplements, and medications to reduce inflammation and pain. Surgery sometimes is necessary, including removal of the large intestine and ileum of the small intestine.

Ulcerative colitis

Ulcerative colitis (*colitis* is inflammation of the colon) is a fairly common inflammatory bowel disease in which the intestinal lining becomes inflamed. (The inflammation response is part of the syndrome.) Ulcers form in the mucosal and submucosal layers of the large intestine, starting in the rectum and moving upward. The colon's lining swells and produces a lot of mucus and pus. Abscesses can form in the lining, and the tissue surrounding them can become irritated, or damaged, or die. Ulcerative colitis can become a life-threatening condition.

Because of all this damage, the feces are filled with blood and mucus. If the blood loss is severe enough, *anemia* (relative lack of red blood cells) can develop. As the disease progresses, the colon's lining thickens and develops scar tissue, so absorption of nutrients and electrolytes is reduced.

The cause of ulcerative colitis is under study. A problem with T lymphocytes appears to negatively affect the epithelial lining of the colon. Sometimes an infection may start the process. Although stress doesn't cause ulcerative colitis, it can bring on an attack.

Diseases of the accessory organs

The organs of chemical digestion are subject to the malfunction of the powerful chemicals they themselves produce, as well as to infectious diseases, nutritional deficiencies, and imbalances originating in other organ systems. Malfunction in any of these organs can have far-reaching effects in physiology.

Systemic symptoms of liver disease

Because the liver has so many important functions in physiology, malfunction of this large organ can show up as signs and symptoms in other organ systems.

When the liver has failed, the bilirubin from aged RBCs isn't eliminated properly from the body. Some may be deposited on the skin, where it causes intense itching, the most commonly reported symptom of liver failure. Yellow coloration of the *sclera* (whites) of the eyes is a well-recognized sign of liver disease, from which the common term *jaundice* (yellowing) is derived. The bilirubin may enter the urine at the kidney, giving the urine a dark color diagnostic of liver problems. The feces are pale because the metabolite of bilirubin that gives feces their brown color isn't being produced and delivered to the large intestine.

Among the proteins made in the liver are most of the components of the *clotting cascade* (see Chapter 9) and those of the *inflammation cascade* (see Chapter 13). Impaired production of these proteins can result in bruising and excessive bleeding. Decreased concentrations of plasma proteins, especially *albumin,* soon brings on *edema* (fluid retention) in the abdomen, legs, and feet. A general loss of nutrients may result in chronic fatigue.

The liver is capable of considerable self-regeneration. These problems may self-correct by homeostatic mechanisms when the cause of the liver malfunction is eliminated.

Viral hepatitis

An inflammation of the liver is called *hepatitis.* Among the most common causes of hepatitis are the viral infections commonly named "hepatitis viruses A through E," in order of their scientific discovery. The study of hepatitis viruses is an area of active biomedical research, and more hepatitis viruses may be identified and named. The viruses themselves aren't related, but they all produce a similar set of symptoms. In the *prodromal stage* (immediate onset of infection, lasting about two weeks), the afflicted suffer from a sick feeling known as *generalized malaise,* as well as nervous system problems such as an altered sense of taste or smell, or light sensitivity. In the clinical stages, the liver is inflamed and enlarged, leading to abdominal pain, tenderness, indigestion, and accumulation of bilirubin.

The immune system discovered eons ago what medicine became aware of only recently: These viruses are all different. The immune system produces specific *immunoglobulins* (antibodies) against each virus. Some advanced diagnostic tools used in clinical diagnosis and public health identify the virus by identifying exactly which viral antibodies are present in the patient's blood.

Some of the viruses that infect the human liver have effectively resisted the efforts of the immune system to eliminate them. Chronic infection can bring long-term morbidity, including extensive scarring of the liver, and early death.

You've got gall (stones)

Gallstones start as crystals of cholesterol forming in the gallbladder. Like pebbles in a jar, they reduce the gallbladder's storage capacity. Worse, they can block the common bile duct, causing bile pigments to back up into the blood, a condition called *obstructive jaundice*. A laparoscopic laser technique called *lithotripsy* can obliterate the gallstones.

Painful pancreatitis

Inflammation of the pancreas *(pancreatitis)* may be mild or severe, acute or chronic. Pain associated with mild pancreatitis is centered around the navel and doesn't lessen with vomiting. In severe pancreatitis, the pain is an unrelenting, piercing pain in the middle of the abdomen.

Acute pancreatitis can easily go on to become a condition called *edematous pancreatitis* (fluid accumulation) or a condition called *necrotizing pancreatitis* (death of cells and tissues in the pancreas). The cause of both is the same: Digestive enzymes produced in the pancreas are blocked from exiting into the intestine. Inflammatory changes in the ducts usually underlie the blockage.

Without the enzymes, digestion may slow or be incomplete, which may harm the body's ability to maintain homeostasis. But worse, if the condition persists, the digestive enzymes are eventually turned loose on the tissues of the pancreas itself, destroying the body's ability to produce and deliver the digestive enzymes ever again. Unlike the liver, the pancreas does not regenerate.

Sometimes, a bout of pancreatitis impairs the body's ability to produce insulin, and diabetes results. (Turn to Chapter 8 for more on diabetes.)

Chapter 12

The Urinary System: Cleaning Up the Act

. . . .

In This Chapter
▶ Understanding what the urinary system does
▶ Breaking down the urinary system's parts
▶ Ruminating on the role of urine
▶ Seeing how your body maintains homeostasis
▶ Describing some urinary system disorders

. . . .

Generally speaking, everything that enters your body through the digestive system must be excreted back into the environment in some form eventually. Most of the waste products of cellular metabolism and many other substances exit your body via your urinary system. Urine is the body's primary waste product, and urination (the release of urine to the environment) is the final step of metabolism.

In this chapter, we take you on a tour of the urinary system, reveal the makeup of urine, explain how the kidneys help your body maintain homeostasis, and look at some of the problems that affect the urinary system.

Functions of the Urinary System

Each class of land animals has faced the challenges presented by an environment where water may often be scarce and the demands of thermoregulation include eliminating the byproducts of a high-energy metabolism. The functions of the human urinary system are similar to those of typical reptiles, birds, and mammals, with some differences in the details.

✔ **Doing the dirty work:** Cellular metabolic wastes are toxic, and would cause harm to the cell if they were allowed to accumulate. Cells continuously export their metabolic wastes into the extracellular fluid. As extracellular fluid is absorbed into the venous blood during capillary exchange, the waste substances enter the circulatory system. Carbon dioxide is excreted in the lungs in the pulmonary circulation cycle. Other byproducts of cellular metabolism travel in the arterial (oxygenated) blood to the renal artery and into the kidney. The function of the urinary system is to remove these toxic byproducts from the blood and then eliminate them from the body. (See the section "Putting out the trash: Kidneys" later in this chapter.)

The term _excretion_ can mean the movement of material from the inside to the outside of a cell, as well as to the release of material from the inside to the outside of the body.

✔ **Making and expelling urine:** _Urine_ is a watery solution produced by your kidneys. It's the matrix through which the water-soluble waste products of metabolism, mainly nitrogenous metabolic byproducts, are removed from the blood and eliminated from the body. It also helps maintain the chemical equilibrium of the blood because substances harmful to the chemical equilibrium of the blood can be dumped there.

✔ **Balancing water content:** Around half of your body weight is water, contained in intracellular and extra cellular fluid, the plasma of your blood, your lymph fluid, and the fluids in your digestive tract and mucous membranes. (Turn to Chapters 3, 10, 11, and 13 for details about these fluids.) Your kidneys precisely regulate the release and retention of water to maintain blood volume, composition, and chemical variables within homeostatic ranges (see the "Maintaining Homeostasis" section later in the chapter).

✔ **Performing endocrine functions:** The kidney is one of the most important endocrine organs. As we discuss in Chapter 8, the adrenal endocrine glands are located on top of the kidneys but outside the kidney capsule, so they're not part of the kidneys. Many of the hormones produced and secreted by these glands act on the kidneys.

   • **Regulating RBC production:** The hormone _erythropoietin,_ which regulates RBC (red blood cell) production _(erythropoiesis)_ in the bone marrow, is produced in the _peritubular capillary endothelial cells_ in the kidney and liver (see the "Structures of the Urinary System" section later in the chapter). Erythropoietin also has other physiological functions related to wound healing and recovery (flip to Chapter 9 for information on blood cell production).

   • **Regulating bone growth:** _Calcitriol,_ the physiologically active form of vitamin D, is synthesized in the kidneys from a precursor molecule synthesized in the liver. Calcitriol regulates, among other things, the concentration of calcium and phosphate in the blood, which promotes the healthy mineralization, growth, and remodeling of bone (Chapter 4 has more on bone growth).

Other mechanisms of excretion

Although most metabolic waste is excreted through the urinary system, some other organ systems have a role as well:

✔ Your lungs excrete carbon dioxide, a waste product of cellular respiration, in your exhaled breath.

✔ Your skin excretes some water, salts, and fats in the form of sweat.

✔ Your colon (large intestine) excretes the physical remains of dead cells and chemical components of cells not being reused or recycled, in addition to its role in digestion.

✔ Your liver doesn't excrete wastes directly into the environment. However, it has a large role in breaking up dead cells into chemical constituents and routing different substances for recycling or excretion through the kidneys and colon. Cells of the lymphatic system package up the cell bodies of invader organisms, which are also eliminated through the liver and colon.

Structures of the Urinary System

The urinary system is quite compact. Unlike some other organ systems, you can identify the point where it begins and another point, not too far away, where it ends. See the "Urinary System" color plate in the center of the book.

Like the alimentary canal, the *urinary system* is essentially a system of tubes through which a substance passes, undergoing a series of physiological processes as it does so. The familiar tissue layers — an outer fibrous covering, a muscular layer, and a mucous layer lining the interior surface — are seen throughout the urinary system, beginning at the ureter. The mucus protects the tissues from the caustic urine.

The male and female urethrae are adapted to interact with the respective reproductive systems and therefore differ in some aspects of anatomy and physiology.

Putting out the trash: Kidneys

The urinary system begins with the *kidneys,* fist-sized paired organs, reddish-brown in color, located just below the ribs in your lower back. The kidney is shaped like an elongated oval; actually, it's shaped just like a kidney bean. (Coincidence?) The inner curve of the kidney has a large concavity (depression) called the *hilum* where a number of vessels enter or exit, including the ureter, renal artery, renal vein, lymphatic vessels, and nerves. A connective tissue membrane called the *peritoneum,* as well as adipose (fat) tissue, attach your kidneys to your posterior abdominal wall. (The term *renal* comes from the Latin "ren," meaning "kidney.")

Beneath the peritoneum, a lining of collagen called the *capsule* encloses the kidney. Fibers of this layer extend outward to attach the organ to surrounding structures.

The kidneys are attached to the dorsal side of the abdominal cavity, on the outside of the cavity. The kidneys are said to "float." Your back muscles on either side of your vertebral column help to protect your kidneys, as do your lowest ribs. Still, a hard blow to the back can dislodge a kidney fairly easily.

Renal blood supply

The *renal artery,* a large branch of the abdominal aorta, brings blood into the kidney for filtering. The renal vein drains filtered blood out of the kidney. The kidneys receive about 20 percent of all blood pumped by the heart each minute.

Kidney tissues

Under the capsule, the kidney's various tissues are arranged in more or less concentric layers. The outermost layer, just beneath the capsule, is the *cortex.* Beneath the cortex is the *medulla,* a series of fan-shaped structures comprising a membrane that is convoluted (folded) into conical structures, called *renal pyramids,* that secrete urine at their tips into sac-like structures. The innermost layer is the *renal pelvis,* which channels the urine from these sacs into the *ureter.* See the "Kidney and Nephron" color plate in the middle of the book to see the locations of these structures.

Nephron

Microscopic in size (about a million of them exist in each kidney), the *nephron* is the kidney's filtering unit. Each nephron has two parts: the *renal corpuscle* and the *renal tubule.* The renal corpuscle also has two parts: the *glomulerus* (plural, *glomeruli*), a special kind of capillary bed derived from arterioles branching from the renal artery, and the *glomerular capsule* or *Bowman's capsule,* a double-walled epithelial cup that partially encloses the glomerulus.

Leading away from the capsule, the nephron loops up, forming the *proximal convoluted tubule,* or PCT. The tube then straightens, pushing down into the medulla and looping up again (the *loop of Henle*). At the top, it loops again into the *distal convoluted tubule* (DCT). This tubule connects with a collecting duct that carries the urine through the renal pyramids into a series of sac-like structures in the renal pelvis.

The nephrons are surrounded by the *peritubular capillaries,* which perform an important role in direct secretion, selective reabsorption, and the regulation of water. See the "Selectively reabsorbing" section later in the chapter, and take a look at the "Kidney and Nephron" color plate in the middle of the book to see the nephron's individual parts.

Make sure you keep the *kidney capsule* and the *glomerular capsules* straight. The kidney has one kidney capsule on the outside and a million microscopic glomerular capsules inside.

Holding and releasing

All complex animals have evolved their ways of removing metabolic waste and regulating water balance. Only mammals have evolved an organ, the bladder, for sequestering urine so that it may be excreted voluntarily. Whatever the evolutionary imperative that caused the development of this part of the mammalian urinary system, some groups have developed chemical signaling mechanisms that utilize it. For humans, the ability to control the time and place of urination may be crucial for their ability to live in dense groups.

The *urinary tract* is the name of the part of the urinary system from the top of the ureter to the external urethral opening, the route of urine out of the body.

Ureters

The *ureters* are tubes that transport the urine from a kidney to the bladder. The ureter emerges from the renal pelvis.

The walls of the ureter are similar in structure to those of the intestines: The muscular layer contracts in waves of peristalsis to move urine from the kidney to the bladder.

Bladder

The *bladder* is a hollow, funnel-shaped sac into which urine flows from the kidneys through the ureters. The bladder has a capacity of about 20 ounces (about three-fifths of a liter). It lies in the pelvic cavity, just behind the pubic bones and centered in front of the rectum. In females, it's in front of the uterus and vagina.

Like other organs in the urinary and digestive systems, the bladder is made up of an outer protective membrane, several layers of muscles arranged in opposing directions, and an inner mucosal layer. The muscle layers contract to expel urine into the urethra. The mucosa is made up of a special kind of epithelial tissue called *transitional epithelium* in which the cells can change shape from columnar to squamous to accommodate larger volumes of urine (see Figure 12-1). Pressure (stretch) receptors in the muscle layer send impulses to the brain when the bladder is becoming full. Flip to Chapter 3 for more details about epithelial tissue.

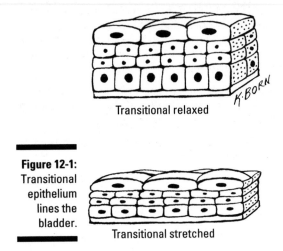

Transitional relaxed

Figure 12-1:
Transitional
epithelium
lines the
bladder.

Transitional stretched

Urethra

The *urethra* is a tube that carries urine from the bladder to an opening (orifice) of the body for elimination. The *mucosa* (epithelium) of the urethra is composed of transitional cells as it exits the bladder. Farther along are stratified columnar cells, followed by stratified squamous cells near the external urethral orifice. Mucus from small glands in the mucosa help protect the epithelium from the corrosive urine.

The male and female urethrae are adapted to interact with the respective reproductive systems and therefore differ from each other in some aspects of anatomy and physiology.

In both males and females, a sphincter at the proximal end of the urethra (between the bladder and the urethra) retains urine in the bladder. This sphincter, called the *internal urethral sphincter,* is made of smooth muscle and is under the control of the autonomic nervous system. It opens to release urine into the urethra.

Where the urethra traverses the *urogenital diaphragm* (the floor of the pelvis), there's a sphincter made of striated fibers (skeletal muscle) called the *external urethral sphincter,* which is under voluntary control.

✔ **Female urethra:** In females, the urethra is about 1.5 inches (3.8 centimeters) long, from its point of emergence from the bladder to the urethral orifice. It runs along the anterior wall of the vagina and opens between the clitoris and the vaginal orifice. The external sphincter is located just inside the exit point.

✔ **Male urethra:** In males, the urethra is about 8 inches (20 centimeters) long, from its point of emergence from the bladder to its opening at the tip of the penis, called the *urethral meatus.* As it passes through the prostate gland and the penis, the male urethra is divided into three named sections, based on anatomical structures:

- The *prostatic urethra* contains the internal sphincter and passes through the prostate. Openings in this region allow for the passage of sperm and prostatic fluid into the urethra during orgasm.

- The *membranous urethra* contains the external sphincter. It is only about 1 inch (2.5 centimeters) in length.

- The *cavernous urethra,* or *spongy urethra,* runs the length of the penis on its ventral surface, ending at the urethral meatus.

The Yellow River

Urine is a bodily fluid with specific functions, just like blood and lymph. But unlike those, you see urine in the normal course of everyday events. Here's everything you always wanted to know about how it gets to be its own sweet self.

Composition of urine

Urine is about 95 percent water, in which is dissolved urea, other nitrogenous compounds, and varying concentrations of electrolytes, plus assorted inorganic and organic compounds. Its odor comes from ammonia and other substances derived from ammonia, like urea. See Table 12-1 for more about the nitrogenous constituents of urine.

Table 12-1 Nitrogenous Components of Urine and Their Sources	
Nitrogenous Component	*Source*
Urea	A byproduct of the breakdown of amino acids
Creatinine	A byproduct of the metabolism of the amino acid *creatine,* present in large quantities in muscle cells
Ammonia	A byproduct of the breakdown of proteins by bacteria
Uric acid	A byproduct of the breakdown of *nucleotides*

Urine is yellow because it contains *urobilinogen,* a compound formed from the breakdown of red blood cells (see Chapter 9). The normal color range of urine is some shade of yellow, from nearly clear to dark amber, depending mostly on the level of hydration. When you're abundantly hydrated, lots of water goes into the urine, making it more diluted and therefore paler. When you're less than optimally hydrated, the kidneys put less water into the urine. The urine becomes more concentrated and appears darker.

Ingested food, beverages, and pharmaceutical products influence the composition of urine. Urine contains *hippuric acid,* produced by the digestion of fruits and vegetables, and *ketone bodies,* produced by the digestion of fats. Some foods, beverages, and drugs impart color and odor to urine. Some physiological conditions and disorders affect the composition of urine, and urine has long been analyzed for clinical signs.

To maintain the blood pH within the homeostatic range (7.3 to 7.4), the kidney can produce urine with a pH as low as 4.5 or as high as 8.5.

Filtering the blood

The glomerulus and the glomerular capsule make up the area of interface between the circulatory system and the kidney. That's where your body filters blood.

The blood pressure in these capillaries is higher than in other capillaries, and it pushes the blood up against the glomerular wall with some force. Because it's a process that relies on mechanical force (fluid pressure), glomerular filtration separates particles based on one criterion: molecule size.

The thin, permeable glomerular wall acts as a filtration membrane. Water passes through into the capsule, bringing along small-molecule solutes, including wastes and toxins like urea and creatinine and useful small-molecule substances like glucose, amino acids, and electrolyte ions. Large solutes, like proteins and nutrients, remain in the plasma inside the glomerular vessel and continue on in the circulating blood. The watery solution (filtrate) passes into the top of the renal tubule at the capsule.

Selectively reabsorbing

An important aspect of urine production is the restoration of glucose, amino acids, and electrolyte ions, not to mention water, from the filtrate to the blood. As the filtrate passes along the nephron, it's acted upon by different types of specialized tissue. Processes in the following specialized tissues are responsible for the composition of urine.

The color yellow

Urinalysis offers a fairly complete assessment of the state of the body. However, some valuable clues can still come from the ancient physician's method — just looking at urine's color.

Dark yellow urine often indicates dehydration.

Very pale urine indicates "over-hydration." The body takes care of this excess hydration through homeostatic mechanisms, such as forming diluted urine. Water poisoning, in which a large quantity of water is consumed over a short period, is possible but extremely rare. The copious urine that's produced to eliminate the excess water causes electrolytes to be depleted, leading to problems in multiple organ systems.

Consumption of asparagus can turn urine greenish. Consumption of beets can impart a reddish or pinkish color.

Excess B vitamin excretion may cause urine to be a yellow to light orange color. This could indicate a metabolic disorder or may simply be the effect of vitamin B supplements. Very high-dose B vitamin supplements may turn urine fluorescent yellow or greenish.

The metabolites of some medications, too many to list, can impart interesting colors to urine.

Bloody urine is termed *hematuria* and is potentially a sign of a bladder infection, dysfunctional glomeruli, or carcinoma.

Dark orange to brown color can be a sign of any of numerous diseases and disorders.

Black or dark-colored urine, referred to clinically as *melanuria,* may be caused by a melanoma.

Reddish, pinkish, or brown urine may be a sign of *porphyria,* a complex disorder. Most often, it's a sign of the consumption of beets.

Microvilli

Microvilli line the nephrons, increasing the surface area within the tubule where substances can enter and leave the filtrate.

PCT

Na+ (sodium ion) undergoes active transport from the filtrate to the blood at the PCT. Where Na+ goes, Cl– just naturally follows. This situation creates an osmotic pressure gradient that pulls water out of the filtrate and into the blood. Approximately 65 percent of the water and salts are returned to the blood at the PCT.

Some substances, including K+ and some hormones, are secreted from the adjacent *peritubular capillaries* into the PCT. This process is termed *direct secretion.*

Loop of Henle

At the loop of Henle, electrolytes and urea move out of the nephron into the medulla. The filtrate remaining in the nephron, now *urine,* drains through the DCT into the collecting duct and then into the renal pelvis.

Medulla

The solutes that enter the medulla from the loop of Henle draw water into the medulla. Capillaries in the medulla restore some of the water to the blood.

DCT

Active transport passes amino acids, glucose, and electrolytes (those useful substances that happen to be small enough to pass through the filtration membrane) back into the blood at the DCT. Water resorption at the DCT occurs under the control of ADH.

Collecting duct

Resorption of water occurs in the collecting duct. As salts are reabsorbed, water follows.

The amount of water reabsorbed from the filtrate back into the blood depends on the hydration situation in the body. However, even in cases of extreme dehydration, the kidneys produce around 16 ounces (about half a liter) of urine per day just to excrete toxic waste substances. Numerous other substances are returned to the blood at specific points, conserving them and maintaining the blood's chemical environment. Over 99 percent of the filtrate produced each day can be reabsorbed. The processes are called *selective reabsorption* because the details are different for (chemically) different types of substances.

Expelling urine

Urine drips down the nephron's collecting ducts to the renal pelvis. From there, it moves along the ureter and into the bladder. As the urine accumulates, the pressure receptors in the lining send signals to the brain. The first message is sent when the bladder is about half full (around 6 to 8 ounces, or 177 to 237 milliliters). At a volume of 12 ounces (354 milliliters), the messages become stronger, and it becomes difficult to control the external urethral sphincter.

When the time is right to empty the bladder, the brain sends an impulse through the autonomic nervous system to open the internal urethral sphincter and contract the bladder muscles. Urine flows out of the bladder, through the urethra, and out of the body.

Maintaining Homeostasis

The chemistry of life, stunningly complex and precise, requires a tightly controlled environment to proceed optimally. The kidneys, under hormonal control, are key organs for maintaining the chemical homeostasis of blood and other body fluids. The endocrine system has a huge range of subtle and interacting mechanisms for controlling kidney function. (See Chapter 8 for more on the endocrine system.)

Fluid balance and blood pressure

Aspects of fluid balance that involve the blood include blood volume and the amount and nature of the electrolytes in the plasma. These factors are linked, as we discuss in the following sections.

Blood volume (water content)

The concentration of *electrolytes* (ions), such as Na+, Cl−, and K+, helps determine the overall blood volume because the kidneys push water into the blood or withdraw water from the blood to bring the concentrations of these electrolytes within homeostatic ranges.

The blood volume is an important factor in circulation. For values above the optimum range, the greater the blood volume, the harder the heart must work to pump the blood and the greater the fluid pressure in the arteries. For values below the optimum range, the less the blood volume, the weaker the pressure filtration in the glomeruli (as well as the effects of inadequate circulation in all tissues and organs). However, the urinary system, when functioning properly, maintains blood chemistry (electrolyte concentration in the plasma), even if that requires putting some strain on other systems.

The adult body contains about 23 pints (11 liters) of extracellular fluid, constituting about 16 percent of body weight, and about 6 pints (almost 3 liters) of plasma, constituting about 4 percent of body weight. Plasma and extracellular fluid are very similar chemically and, in conjunction with intracellular fluid, help control the movement of water and electrolytes throughout the body. Some of the important ionized components of interstitial fluid are protein, Mg+, K+, Cl−, Ca+, and certain sulfates.

Hormonal mechanisms to control blood volume

The *renin-angiotensin system* (RAS) (also called the *renin-angiotensin-aldosterone system* [RAAS]) is called into action when the presence of toxins in the blood signals low blood pressure in the glomeruli. Specialized kidney cells secrete the enzyme *renin,* also called *angtiotensinogenase,* which sets off a series of reactions in different organ systems, eventually resulting in the production of the hormone *angiotensin II,* a powerful vasoconstrictor. Angiotensin II also stimulates the secretion of the hormone aldosterone from the adrenal cortex, which causes Na+ to be reabsorbed from the nephrons into the blood. Where salt goes, water follows. The movement of water into the blood increases blood volume, which, along with the vasoconstriction, raises blood pressure and enhances the effectiveness of the glomerular filtration.

Some medications for *hypertension* (high blood pressure) act by disrupting the production of an enzyme that catalyzes the crucial reaction in this pathway that produces angiotensin II. These are the so-called *angiotensin-conversion-enzyme inhibitors* (or *ACE inhibitors*). Without angiotensin II, no vasoconstriction takes place, and the kidneys don't receive the signal to increase the water content of the blood.

The pituitary gland secretes ADH when the hypothalamus senses dehydration (from a lack of water consumption or a loss of water through sweating, diarrhea, or vomiting). Sodium is reabsorbed by the nephrons but not enough water follows, so ADH causes more water to be reabsorbed from the urine, which decreases the amount of urine and helps maintain normal blood volume and pressure.

Atrial natriuretic hormone (ANH) is a polypeptide hormone secreted by heart muscle cells in response to signals from sensory cells in the atria that blood volume is too high. ANH prevents the kidneys from secreting renin. In fact, the overall effect of ANH is to counter the effects of the RAS. ANH acts to reduce the water, sodium, and adipose (fat molecules) loads on the circulatory system, thereby reducing blood pressure.

Regulating blood pH

The homeostatic range for blood pH is narrow. The optimum value is around 7.4 (slightly alkaline). *Alkalosis* (increase in alkalinity) is life-threatening at 7.8. *Acidosis* (increased acidity) is life-threatening at 7.0.

Neutral pH is 7.0, the value for water.

Acids and, to a lesser extent, *alkalis* (bases) are products or byproducts of metabolic processes. The digestion of fats produces fatty acids. Capillary exchange acidifies the blood because the metabolic waste carbon dioxide is combined with water in the red blood cell, forming carbonic acid. Muscle activity produces lactic acid. Some acids are ingested in food and drink. The kidneys respond to changes in blood pH by excreting acidic or basic ions into the urine.

Your body has buffering processes in which the kidneys are involved. A *buffer* is a type of chemical that binds with either acid or base as needed to increase or decrease the solution pH. Buffers are made in cells and available in the blood.

Three mechanisms work together to maintain tight control over blood pH:

- ✔ Minor fluctuations in pH are evened out by the mild buffering effects of substances always present in the blood, such as the plasma proteins.

- ✔ When sensors in the kidneys detect that the blood is too acidic, they induce the breakdown of the amino acid glutamine, releasing ammonia, a basic substance, into the blood. When the ammonia arrives at the kidney, it's exchanged for Na+ in the urine and eliminated.

- ✔ By far the most important buffer for maintaining acid-base balance in the blood is the carbonic-acid-bicarbonate buffer. The body maintains the buffer by eliminating either the acid (carbonic acid) or the base (hydrogen carbonate ions). Carbonic acid concentration can be reduced within seconds through increasing respiration — the excretion of carbon dioxide through the lungs increases pH. Hydrogen carbonate ions must be eliminated through the kidneys, a process that takes hours.

The *pH scale* measures the concentration of hydrogen ions (H+) in a solution. The scale ranges from 0 (extreme acidity, high concentration of hydrogen ions) to 14 (extreme alkalinity, low concentration of H+, and consequently, high concentration of OH–). Water has a neutral pH of 7.0. Solutions with balanced amounts of these two components have a pH in the neutral range.

Pathophysiology of the Urinary System

When things go wrong in the urinary system, the healthy functioning of the entire body is affected.

Kidney pathologies

The kidney is a complex and delicate organ, crucial to physiology but vulnerable to injury. Because of the kidneys' importance in homeostasis, kidney problems can be underlying problems in other disorders, especially those of the circulatory system.

Calculating kidney stones

The proper name for a kidney stone is a *renal calculus,* meaning "pebble" or "hard." A kidney stone is literally a little pebble made from chemicals in the urine. Kidney stones may result from urine that's chronically concentrated because of chronic inadequate hydration or excess sodium. When urine is more concentrated, solutes, such as uric acid, tend to precipitate (fall out of a solution and clump together). Crystals of uric acid may develop into kidney stones, or they may travel to your joints, causing *gout* (turn to Chapter 4 for more about gout). Small stones, up to about 0.1 inch (.3 centimeters) diameter, can flow more or less quietly out through the urinary tract. Larger stones can cause considerable pain, sometimes for several days, when passing, and sometimes damage the ureters or urethra. The largest stones, 0.3 inch (.8 centimeters) or greater, frequently become *impacted* — lodged in the urinary tract and unable to be passed through the urine stream. An impacted kidney stone may cause a backup of urine in the kidney, ureter, bladder, or urethra (see the "Urinary tract obstruction" section later in the chapter).

Alcohol hangover

Among its many physiological effects, alcohol inhibits the release from the pituitary of ADH. As we discuss earlier in the chapter, ADH regulates the kidney's action in restoring water to the blood. Without ADH, too much water is excreted in the urine, and blood volume falls below the optimum range. The resulting dehydration causes dizziness, headache, and other symptoms associated with a hangover. Homeostatic mechanisms (negative feedback and endocrine response) restore fluid balance over a period of hours after the consumption of alcohol ceases, especially if adequate water is consumed.

Urinary tract pathologies

Obstructions, blockages, and colonization by life forms with their own agendas plague all tubal systems, biological and inanimate. At least your body can fight back against these threats to the smooth flow of urine in the urinary tract.

Urinary tract obstruction

The urinary system's various tubes are vulnerable to an obstruction that blocks the flow of urine. The smaller the tube, the greater the risk of obstruction. The obstructing object is sometimes a stone (see the "Calculating kidney stones" section earlier in the chapter). Impacted kidney stones require immediate medical attention to avoid the risk of infection or more severe kidney damage.

Urinary tract infection

The urinary tract is subject to infection, known as UTI. Numerous pathogenic bacteria are adept at entering the body through the external opening of the urinary tract (urethra) and establishing infection in any structure of the urinary system. A bladder infection is known as *cystitis,* a kidney infection is *pyelonephritis,* and a urethra infection is *urethritis.*

UTIs are very noticeable and uncomfortable, so patients often seek treatment immediately, before the infection spreads higher up into the urinary system. The kidney is particularly vulnerable to damage from invading bacteria. If the pathogen is able to invade the bloodstream through the close connection with the kidney, the infection, then called *septicemia,* can be extremely serious. Septicemia can lead to *septic shock,* which can be life-threatening.

Urethral blockage

As men age, the prostate gland enlarges, usually starting around age 50. If the prostate gland enlarges too much, it can press on the urethra and block the flow of urine. The blockage can lead to UTIs in men, and it often causes painful urination (technically called *dysuria*). As the bladder becomes stressed from the blocked urethra, urine may leak out, or, more commonly, the urge to urinate becomes more frequent.

Incontinence across the continents

Incontinence is the inability to control the release of urine. There are four types of incontinence:

- ✔ **Stress incontinence:** Urine leaks out when pressure is put on the bladder, such as when a person runs, coughs, lifts something heavy, or laughs. This type is relatively common among women who have given birth several times because the muscle supporting the bladder weakens and the sphincter muscle of the urethra stretches.

- ✔ **Urge incontinence:** Also called *overactive bladder,* this type of incontinence is a sudden involuntary contraction of the bladder's muscular wall that brings on an urgent need to urinate, possibly accompanied by a sudden loss of urine. Often, these contractions occur regardless of the amount of urine in the bladder. This condition affects about 1 in 11 adults, particularly older adults. Underlying causes may be nervous system diseases, bladder tumors, bladder infection, and bladder irritation.

- ✔ **Overflow incontinence:** In this condition, patients never feel the urge to urinate, the bladder never empties, and small amounts of urine leak continuously. This type is prevalent in older men with an enlarged prostate and is rare in women. The upper portion of the urethra passes through the prostate, so when the gland becomes enlarged, it may obstruct the passage of urine through the urethra.

- ✔ **Total incontinence:** Some structural problems, spinal injuries, or diseases can cause a complete lack of bladder control. The sphincters of the bladder and urethra don't operate, and urine flows directly out of the bladder.

Chapter 13

The Immune System: Living in a Microbe Jungle

In This Chapter

▶ Figuring out what the immune system does

▶ Lymphing along

▶ Exploring immune system cells, molecules, and mechanisms

▶ Looking at some immune system disorders

*Y*our immune system is all that stands between you and a planetful of invasive microorganisms that regard you, metaphorically speaking, as a large serving of biological molecules that could feed their own processes. Your body calls on the immune system to protect itself from invading microbes, such as bacteria and viruses, other foreign cells, and your own cells that have gone bad (such as cells that have become cancerous).

Functions of the Immune System

The *immune system* consists of a variety of components: the *lymphatic system,* a body-wide network of vessels and organs through which flows an important body fluid called *lymph;* a variety of very peculiar cell types; and several types of biological molecules, some of them just as peculiar. These all have specialized functions for attacking and eliminating microbial invaders:

✔ **Confronting marauders:** Whether you're well or ill, your immune system is always alert and active. That's why none of the bacteria, fungi, parasites, and viruses that are present by the uncountable millions in the air you breathe, on surfaces you touch, and on your food, are eating you. Sure, you have lots of ways of keeping the invaders from entering your body in the first place. But disease organisms have evolved their own devious means to get past your defenses. (That's why they're "diseases" and not just "bugs.") Plenty get through. When your immune system is functioning properly, most don't stay for long. The immune system hunts them, destroys them, and wraps their cellular remains for elimination from the body.

✔ **Stopping renegades:** The second major function of the immune system is to recognize and destroy cells of your own body that have "gone rogue," potential seeds for cancerous growth. Cells go rogue every day. Most cancers can take hold only when the immune system malfunctions and fails to eliminate them, as happens with aging.

Patrolling the border

Your body has several ways of keeping out invaders, both living and nonliving:

✔ **Skin:** This barrier keeps out innumerable invaders. Glands in the skin secrete oils that make the barrier even more effective. The *stratum germinativum (basale)* and *stratum spinosum* contains special cells that connect the *integumentary* system and the immune system, called *epidermal dendritic (Langerhans) cells.* For more on the skin's barrier functions, flip to Chapter 6.

✔ **Eyes and mouth:** The flushing action of tears and saliva helps prevent infection of the eyes and mouth.

✔ **Mucous membranes:** The mucosal lining of the respiratory and digestive systems traps microbes in icky sticky goo. Some of the mucosa's cells have cilia on the surface facing the lumen of the tract. These tiny, hairlike organelles move foreign matter and microbes trapped into the mucus to where they, and the mucus, are eliminated.

✔ **Gastric juice:** This acidic substance secreted into the stomach destroys most bacteria that are inadvertently ingested. See Chapter 11 for more on gastric juice.

✔ **Gut:** The beneficial bacterial species that live in the intestine can prevent the colonization of pathogenic bacteria by secreting toxic substances or by competing with pathogenic bacteria for nutrients or attachment to tissue surfaces.

Loving Your Lymphatic System

The *lymphatic system* plays a crucial role in circulation by draining the fluids that pour out into the extracellular space during capillary exchange and returning them to the blood. The lymphatic system is more than a drainage network, though. It removes toxins, helps transport fats, and stabilizes blood volume despite environmental stresses. Possibly its most interesting functions are those related to its role in the immune system, fighting biological invaders. To understand the extent of the lymphatic system, turn to the "Lymphatic System" color plate in the center of the book.

Lymphing along

In Chapter 9, you read about the circulatory system as a transportation network that brings substances to cells and takes their waste products away. In Chapters 8, 10, and 11, you find out how organ systems "make use of" the blood to distribute their anabolites to the target tissue. During capillary exchange, much of this cargo is pushed out through the membranes of the capillary cells: oxygen, ions, glucose and other nutrients, proteins, hormones, and so on, all in a watery solution. This section is about the watery solution.

The watery solution, called *interstitial* (extracellular) fluid, is the fluid in between cells, about 2 to 4 pints (1 to 2 liters) total volume at any given moment. (*Interstitial* means "in-between spaces.") It is essentially the same fluid you met in Chapter 9, introduced as plasma, the fluid matrix of blood. (We discuss the role of interstitial fluid in cell metabolism in Chapter 3.)

Like plasma, interstitial fluid is continuously flowing: The pressure of the heartbeat pushes this watery solution across the capillary cell membrane and out into the interstitial space, a total of about 50 pints (24 liters) per day. Most of it is reabsorbed into the blood at the venous end of the capillaries. The rest goes on a detour through the lymphatic system. The fluid coursing through the lymphatic system is called *lymph*. After passing through the lymphatic system, the fluid rejoins the circulatory system through two large veins, and it is plasma again.

Plasma, interstitial fluid, and lymph are all the same watery solution of plasma proteins, electrolytes (ions), and various dissolved and undissolved substances. The fluid circulates around the body. The name it's called by depends on where it is.

Structures of the lymphatic system

The structures of the lymphatic system resemble those of other organs and systems that function to move fluids around. The lymphatic system has its own tubes, pipes, connectors, reservoirs, and filters. It lacks its own pumping organ but, like venous circulation, makes use of skeletal muscle action for this purpose.

Lymphatic vessels

The *lymphatic vessels* are the tubes that carry lymph. They form a network very similar to the venous system. You could even think of the lymphatic system as an alternative venous system, because the lymph that the vessels transport comes out of the arterial blood and is delivered back into the venous blood. Like the venous system, the lymphatic system's vessels start small (the lymph capillaries) and get larger (the lymphatic vessels), and even larger (the lymphatic ducts). Like veins, lymphatic vessels rely on skeletal muscle action and valves to keep the fluid moving in the right direction. The structure of the lymphatic vessel wall is similar to that of the veins, but thinner. Lymph vessels are distributed through the body, more or less alongside the blood vessels.

A few parts of the body don't have lymph vessels: the central nervous system (CNS), n the epidermis, bones and most cartilage, and the teeth.

Lymphatic ducts

The largest of the lymphatic vessels, the *lymphatic ducts,* drain into two large veins. The right lymphatic duct, located on the right side of your neck near your right clavicle, drains lymph from the right arm and the right half of the body above the diaphragm into the *right subclavian vein.* The thoracic duct, also called the *left lymphatic duct,* which runs through the middle of your thorax, drains lymph from everywhere else into the *left subclavian vein.*

Lymph nodes

Lymph nodes are bean-shaped structures located along the lymph vessels (see Figure 13-1). Dense clusters of lymph nodes are found in the mouth, pharynx, armpit, groin, all through the digestive system, and other locations. Each lymph node is encapsulated (covered) by a fibrous connective tissue *capsule. Afferent* lymphatic vessels cross the capsule on the convex side, bringing lymph into thenode. The node's *efferent vessel,* which carries the filtered lymph out of the node, emerges from the indentation on the concave side of the capsule, called the *hilus* (as in the kidney).

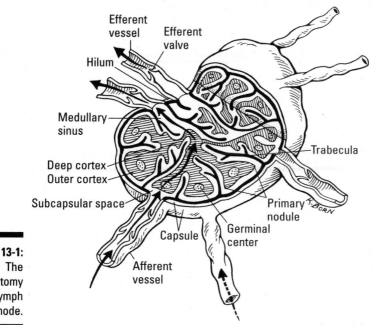

Efferent
vessel
Efferent
valve
Hilum
Medullary
sinus
Trabecula
Deep cortex
Outer cortex
Subcapsular space
Primary
nodule
Capsule
Germinal
center
Afferent
vessel

Figure 13-1:
The
anatomy
of a lymph
node.

On the inner side, the capsule sends numerous extensions that divide the node internally into structures called *nodules*. A nodule is filled with a meshlike network of fibers to which *lymphocytes* and *macrophages* (another immune system cell type) adhere. As the lymph flows through the node, pathogens, cancerous cells, and other matter in the lymph are engulfed and destroyed by macrophages or marked for destruction by B lymphocytes. The cleaned-up lymph travels toward the venous system in the efferent vessels.

The lymph nodes also provide a safe and nurturing environment for developing lymphocytes. (See the "Lymphocytes" section later in the chapter.)

Lymph nodes are sometimes mistakenly called *lymph glands.* They don't secrete anything, so they don't meet the definition of glands.

Swelling and tenderness in the lymph nodes, especially of the pharynx, is a symptom of an infectious disease. Swollen lymph nodes ("swollen glands") is not itself a disease or pathological condition. It's the immune system doing its job.

The splendid spleen

The *spleen* is a solid organ, located to the left of and slightly posterior to the stomach. It's roughly oval in shape, normally measuring about 1 by 3 by 5 inches (3 by 8 by 13 centimeters) and weighing about 8 ounces (23 grams). Essentially, its structure is that of a really large lymph node, and it filters blood in much the same way the lymph nodes filter lymph, removing pathogen cells along with exhausted RBCs and many kinds of foreign matter.

The spleen's structure is similar to that of other organs, like the kidney, the liver, and the lymph nodes. The spleen is enveloped by a fibrous capsule. The spleen has a *hilus,* a spot where several different vessels cross the capsule. The spleen's hilus contains the *splenic artery, the splenic vein,* and an *efferent lymph vessel,* a similar configuration to the lymph node. Note that the spleen has no role in filtering lymph and no afferent lymph vessels.

Inside, the spleen is divided into functional subunits by outgrowths of the capsule's fibrous tissue. Within each subunit, an arteriole is surrounded by material called *white pulp* — lymphoid tissue that contains lymphocyte production centers. Farther toward the outer edges of each compartment, similar masses called *red pulp* surround the arteriole. The red pulp is a network of channels filled with blood, where most of the filtration occurs. (It's also the major site of destruction of deteriorating RBCs and the recycling of their hemoglobin.) Both white pulp and red pulp contain *leukocytes* that remove foreign material and initiate an antibody-producing process.

Give us a "T"

A T cell, that is. The *thymus gland* overlies the heart and straddles the trachea, sitting just posterior to the sternum. It produces *thymosin,* a hormone that stimulates the differentiation and maturation of T cells. (See the "Lymphocytes" section later in the chapter for more on T cells, and flip to Chapter 8 for more on hormones.) The thymus is relatively large in childhood; it decreases in size with age.

Identifying Immune System Cells

Immune system cells are special in many ways. In shape and size, they're far from the compact epithelial or muscle cell types. Immune system cells have about a dozen distinctive shapes and many different sizes, and some have the ability to transform themselves into other, even weirder forms and to multiply extremely rapidly. See Table 13-1 for an overview of the different types of immune system cells.

Table 13-1		Cells of the Immune System	
Cell Type	*Origin*	*Function*	*Comment*
Neutrophil	Myeloblast	Phagocytizes bacteria.	40–70% of total WBCs.
Basophil	Myeloblast	Initiates the inflammation response by releasing histamine and heparin at the site of injury.	1% of total WBCs.
Eosinophil	Myeloblast	Destroys antigen-antibody complexes.	1–4% of total WBCs.
Monocyte	Myeloblast	Matures into a macrophage, which phagocytizes bacteria and viruses.	4–8% of total WBCs; largest of WBCs.
Macrophage	Monocyte	Phagocytizes antigens and dead cells; stimulates the production of other WBCs.	Monocytes produce macrophages in large quantities in the early stages of the inflammation response.
B lymphocyte (B cell)	Lymphoid stem cell	Produces antibodies.	Combined, B and T lymphocytes account for 20–45% of total WBCs.
T lymphocyte (T cell)	Lymphoid stem cell; matures in thymus gland	Several different subsets of T cells have been discovered, each with a distinct function.	First to respond to infections.
NK cell	Lymphoid stem cell	Destroys cancerous cells and cells infected by viruses.	Granular.
Mast cell	Precursors in the red bone marrow	Inflammation.	Responds to allergens.

Over the past few decades, immunologists and cell biologists have discussed immune system cells in terms of several classification systems that have not held up well on further investigation. Forming and discarding theories and classification systems based on new knowledge is expected and welcome in immunology, as in any science. The structure and physiology of immune system cells will be under study for the foreseeable future. The following discussion of specific types of immune system cells gives you some idea about some established concepts. Keep this limited goal in mind, especially when checking out Table 13-1.

Looking at leukocytes

Immune system cells are called *leukocytes* ("white cells"), because they appear white in color under a microscope. (That's "white" as in *pus*.) Although all blood cells, red and white, develop from *hematopoietic stem cells* in the red marrow, the leukocytes contain no hemoglobin and no iron. Unlike RBCs, all leukocytes retain their nuclei, organelles, and cytoplasm through their life cycle. Many fewer leukocytes are produced than RBCs, by a factor of around 700.

Also called *white blood cells* (WBCs), leukocytes are present everywhere and functioning at all times. You notice their presence in the acute phase of certain diseases — the immune response, not the invader directly, produces the well-known symptoms of flu. They function not only in the blood (really, in the plasma), but also in the interstitial fluid and the lymph. They're never far from a site of injury or infection because they're everywhere. When a splinter pierces your finger, a contingent of local WBCs arrives at the site instantaneously.

Sometimes it's difficult to remember that these bizarre warrior cells with their amazing superpowers are your cells — are *you* — just like your skin cells and your blood cells. Discussing them is difficult without resorting to language that makes them seem like a quasi-military force from outside your body. These cells are acutely aware (metaphorically speaking) that "they" are "self" (you). In fact, that's the primary distinction that matters to them: *self* or *nonself.* The overarching mission of a leukocyte is to protect self from other *biotic* (living) nonself, destroying the invaders where possible and necessary and establishing more or less mutualistic relationships with other life forms where possible and necessary. It's hard not to picture a disciplined army; but remember it's a metaphor.

Lymphocytes

The *lymphocytes* are one group of leukocytes (WBCs). (We discuss the function and some of the life cycle of lymphocytes in the "Structures of the lymphatic system" section earlier in the chapter). The group includes the B cells and T cells (small, agranular lymphocytes), as well as NK cells (large granular lymphocytes). To simplify: Activated B cells, produce antibodies; some T cells destroy antigens; and NK cells attack cancerous cells and cells infected by viruses. The surfaces of lymphocytes are covered with receptors, which are molecules that fit with a specific *antigen*. (Check out the "Examining Immune System Molecules" section later in the chapter for an explanation of antibodies and antigens.)

All lymphocytes originate in the red marrow from the same type of hemato-poietic stem cell. B lymphocytes and NK cells leave the marrow fully differen-tiated and enter the circulatory and lymphatic systems. T cells travel to the thymus gland to complete their differentiation in an environment rich in the hormone *thymosin*. Then they move to the medulla of a lymph node, where they further differentiate into one of several different cell types, each with its own function in the immune response: helper T cells, *cytotoxic* (cell killing) T cells, or suppressor T cells.

In the granular-versus-agranular dichotomous classification, NK cells are gran-ular. The term *granular* means the cytoplasm contains granules of proteins. In the case of NK cells, the granules contain proteins that signal to the cancerous or infected cell that it's time to surrender or self-destruct, a process called *apoptosis* (pronounced *ay*-po-*toh*-sis).

Phagocytizing leukocytes

Several different types of WBCs perform their function in part by processes involving *phagocytosis* (the digestion of foreign material). (We explain this process in further detail in the "Phagocytosis" section later in this chapter.)

Neutrophils

Neutrophils are the most numerous of the WBCs (40 percent to 70 percent of the total number) and are continuously present and active in the circulatory and lymphatic systems. Neutrophils squeeze through the capillary walls and into infected tissue, where they phagocytize the invading bacteria. They also function to limit the populations of the beneficial bacterial species in the respiratory and digestive passages. (See the "Degranulation" section later in the chapter for a brief description of another mechanism of neutrophils.)

Monocyte and macrophage

Monocytes aren't really stem cells, but they have some functions in common with stem cells — they exist to produce other specialized cells on demand. Monocytes divide to produce two other kinds of immune cells, *macrophages* and *dendritic cells*. In a homeostatic state, the monocytes replenish these cells as necessary. In response to inflammation-response-related stimuli, monocytes travel to the site and begin to turn out vast numbers of its daughter cells.

Macrophages (literally, "big eaters") are large phagocytic cells that target antigens and dead self cells. In the early stages of the immune response, macrophages initiate the mass production of other types of WBCs.

Dendritic cells take macrophagy to a whole new level beyond the scope of this book.

Examining Immune System Molecules

As noted earlier in the chapter, leukocytes are difficult to classify either by structure or by function because the structure and physiology of leukocytes are stupefyingly complex and astoundingly flexible. Most leukocytes, even those that phagocytize, also produce chemicals of many kinds that modulate the functional activities of many other cell types. Immune response involves a lot of cell-to-cell communication, and leukocytes make and use proteins, enzymes, hormones, and neurotransmitters. Some of these molecules are familiar from the physiology of other organ systems. Others are special creations of the immune system and have no other functions.

Histamine

Histamine is a nitrogen compound with several physiological functions but is best known for its role in local immune responses. Histamine is produced by basophils and by mast cells found in nearby connective tissues as part of the inflammatory response. It plays a major role in many allergic reactions. As part of the inflammation response, histamine dilates small blood vessels, activates the vascular endothelium, and increases blood vessel permeability to WBCs and inflammatory proteins. It also irritates nerve endings, leading to itching or pain. Histamine is found in virtually all animal body cells.

The itchy bump on your skin after a mosquito bite is caused not by the bite but by the histamine that's released to initiate the process by which the antigens that the mosquito introduced are destroyed.

Complement system proteins

The *complement system* supports the activity of antibodies to clear pathogens. The system is composed of about 26 proteins. The process is similar in many ways to the clotting cascade in response to blood vessel injury (see Chapter 9). In fact, some of the very same proteins are involved.

The complement system must be activated by one of two very specific mechanisms, involving either antigen-antibody complexes, called *specific immune response,* or antigens without the presence of antibodies, called *nonspecific immune response.*

The functions of complement proteins include

- Making bacteria more susceptible to phagocytosis
- Directly lysing some bacteria and foreign cells
- Producing chemotactic substances (signaling molecules)
- Increasing vascular permeability
- Causing smooth muscle contraction
- Promoting mast cell degranulation

Antibodies

An *antibody* is a type of protein molecule with immune function.

Antibodies, also called *immunoglobulins* (abbreviated *Ig*), are produced in activated B cells in response to the presence of an *antigen* — a cell or molecule that your immune system recognizes as nonself. An antigen can be part of a nonself cell, like a protein on the surface of a bacterial cell or virus; a cancerous (self) cell; or an allergen or toxin. The healthy human body has thousands and thousands of antibodies, each specific to one antigen, and is capable of producing large quantities of some of them on demand.

The membranes of lymphocytes are covered with receptors for thousands of different antigens, including antigens they haven't encountered. When a B cell encounters a new antigen (a membrane receptor binds it), the B cell multiplies prodigiously and (almost) all the new cells are devoted to the production and release of antibodies specific to that antigen. The antibodies circulate, efficiently binding and disabling their target. The antigen-antibody complex calls in the phagocytes and may activate the complement system. To protect you from further infection, antibodies saturate your tissues.

Antibody specificity

What makes one antibody specific to one antigen? In a word, shape. Remember, antibodies are proteins, and like those other useful proteins, the enzymes, they do their jobs by binding very tightly to a counterpart molecule of very specific configuration. Antibody molecules are Y-shaped, with a *binding site* on each of the short arms. A binding site, in turn, has a specific and intricate shape. An antibody can bind only an antigen that has the complementary shape. A common metaphor for this is a lock-and-key mechanism: Just as a key has a specific shape and can fit only one lock, the antibody can bind only antigens that match its shape. When an antibody binds its antigen, the antigen may be inactivated. Other immune system processes then remove the *antibody-antigen complex* from the body.

IgG antibodies

Immunoglobin molecules come in five classes, named, with great imagination, IgA, IgD, IgE, IgG, and IgM. IgG antibodies are the most important class. They're the type most involved in the secondary immune response. These proteins circulate in the blood and the other body fluids. They directly attach to microbes and bacterial toxins, and they step up phagocytosis. IgG molecules are relatively small and move easily across cell membranes.

Lab tests that detect the presence of specific antibodies in the plasma or serum can be used to diagnose some diseases, particularly bacterial and viral infections. The presence of the antibody indicates that the immune system has met up with the antigen at some time. The class of the antibody indicates when that may have been. Antibodies of the IgM class indicate recent infection; IgGs indicate that the original infection was in the past.

Immune System Mechanisms

The recent and continuing development of technological tools for microbiology and molecular biology has permitted observations of the previously unimagined details of the immune system's mechanisms. Our conceptual understanding of them follows a little behind. The following sections introduce you to the subtlety and complexity of immune system mechanisms.

Phagocytosis

Phagocytosis is the simplest of immune response mechanisms and probably older than multicellular life: The invader or foreign matter is just surrounded and digested. Phagocytosis is probably the body's most frequently used mechanism, too, because the most numerous of the leukocytes, the neutrophils (up to 70 percent of the total number), work this way. (Refer to the "Phagocytizing leukocytes" section earlier in the chapter.)

What's all the pus about?

Pus, like mucus, is yucky. But pus is proof that your body is fighting an invader and that your immune system is doing its job.

You could say that pus is a "product" of the immune system. This thick, white-to-yellow goo, sometimes shot with blood, appears at a site of injury or infection. (What would adolescence be without at least a few pus-filled pimples on the skin?) Its color and texture come from the dead phagocytes that make up most of it — WBCs that have done their job and then died with the cellular remains of thousands and thousands of invaders wrapped up inside. Pus also contains some dead tissue, some blood, and some lymph. Your body eliminates pus through the colon or through the skin's pores.

Degranulation

Various types of WBCs have *granules* in their cytoplasm. Granules aren't chemicals but little packets of chemicals, such as histamine, cellular toxins, enzymes, and other proteins. Cell biologists have identified several different types of granules with very specific chemical contents.

A process called *degranulation*, triggered by immune system processes, moves the granules out of the cell and breaks them apart, releasing the chemicals into the interstitial space. After they've been absorbed into the interstitial fluid, these chemicals carry out a range of specialized immune functions. Some destroy invaders directly. Some regulate immune system processes.

Granules in *eosinophils* play a crucial part in the immune response to enteric parasites through their release of toxic proteins (our own natural pesticides!). Eosinophil numbers increase during allergic reactions and parasitic infections. (Eosinophils also perform a limited phagocytic function in the destruction and elimination of antigen-antibody complexes.)

Neutrophils, discussed above as phagocytizing cells, contain granules in their cytoplasm that release many powerful substances. Because neutrophils are the most numerous of the WBCs, their granulocytic properties are very important in the immune response, especially to bacterial infection.

The degranulation of *basophils* releases histamine and *heparin* (an anticoagulant). This is a source of the histamine found at the site of inflammation and allergic reactions. Like eosinophils, basophils play a role in both parasitic infections and allergies.

Mast cells are present in most tissues, characteristically surrounding blood vessels and nerves, and are especially numerous near the boundaries between you and the outside world, such as in the skin and in the mucosa of the respiratory and digestive systems. Mast cells are granular cells that play a key role in the inflammatory process. When activated, a mast cell rapidly releases the contents of its granules and various hormonal mediators into the extracellular space. They're involved in allergic reactions, anaphylaxis, and autoimmunity.

Inflammation is swell

The last time you got a thorn or a splinter in your hand or foot, you may have noticed that the site of entry became red, hot, swollen, and tender. These are signs and symptoms of an *inflammation response.* The inflammation response is a basic way the body reacts to infection, irritation, or injury, a mechanism for removing the injurious object and initiating the healing process. Inflammation has recently been recognized as a type of *nonspecific immune response.*

These symptoms have been recited since antiquity: *rubor, calor, tumor,* and *dolor.* A fifth, *functio laesa* (loss of function), was omitted from the original list a millennium or so ago when people noticed that this symptom was present in any number of conditions.

When the splinter punctures your skin, the injured cells release mediator chemicals, particularly *histamine* and *bradykinin,* that initiate the inflammation response. Histamine also activates the complement system. The induction of the first complement protein stimulates the production of others, which stimulate still others, continuing in a rapid but controlled chain reaction until the full-blown inflammation response is raging. Immune system cells, mainly monocytes and other phagocytic cells, rush to the site to fight any microbes that were carried in on the thorn or entered through the skin opening.

When histamine is released, the chemical bradykinin is released along with it. Bradykinin causes the nerves to send pain messages to the brain. Thanks, bradykinin, we guess.

Immunity

It's the upside of infection and disease, at least some infections and diseases. After your immune system has dealt with certain antigens, it produces the condition of immunity from the associated pathogen. *Immunity* is the condition of being able to resist infection by a particular pathogen.

Physicians and physiologists recognize two types of immunity: *antibody-mediated immunity* and *cell-mediated immunity*.

Antibody-mediated immunity

Antibody-mediated immunity, also called *humoral immunity,* is closely related to the B cell response that we discuss in the "Antibodies" section earlier in the chapter. For some antigens, descendants of the B cells that produce the antibodies during the first encounter (the *primary response*) remain active at an imperceptibly low level for years, sometimes decades, so that when the antigen is encountered subsequently, the antibody factory can go into immediate mass production. These cells are called *immune memory cells* and their response is called the *secondary response.*

During primary response, signs and symptoms of disease may rage while the lymphocytes are gearing up to produce antibodies. Because of the rapidity of the secondary response, you may not have symptoms and you may not even be aware of being exposed to the antigen. Antibody-mediated immunity is the reason you typically don't get the same flu twice in one year.

The so-called *passive immunity* conferred on the fetus is antibody-mediated immunity received secondhand from the mother. Her IgGs pass through the placenta.

The term *humoral immunity* is possibly the last vestige of the ancient physiological concept of the four "humors" of the body. The "humor" (fluid) in this case is blood.

Cell-mediated immunity

Cell-mediated immunity is the job of T lymphocytes. It's most effective in removing virus-infected cells but also participates in defending against fungi, protozoans, cancers, and pathogenic bacteria. It plays a major role in the rejection of transplanted organ.

Immunization

Immunization is the process of inducing immunity to specific antigens by inoculation. *Inoculation* is the introduction of an antigen into the body to stimulate the production of antibodies. The antigen is introduced, often by injection but sometimes orally or intra-nasally, in a preparation called a *vaccine* that contains a sample of the pathogen either killed or in an *attenuated* (alive but weakened) form. Essentially, vaccination induces a very mild primary response (so mild you may not be aware of it) so the secondary response can be activated rapidly when you encounter the antigen again in the environment. Researchers have developed vaccines for many infectious diseases and are developing more all the time.

Pathophysiology of the Immune System

Malfunction or failure in any part of the immune system is a threat to homeostasis and continued existence in any organism. But other animals just don't have the same problems as humans do. The peculiarities of immune-related diseases in humans are due, in part, to human culture, which permits so many to survive to experience the age-related decline in immunity and requires continuous close contact between individuals that can sustain endemic or promote epidemic microbial diseases.

The immune system and cancer

Speaking of cancer as one disease is no longer possible, but all cancers are similar in that they represent a failure of the immune system to detect and destroy malignant cells. Often, the underlying cause of immune malfunction is simply the aging of the immune system. *Malignant cells,* by definition, divide and reproduce rapidly and in an uncontrolled way. The immune system destroys those cells as fast as it can. A young and vigorous immune system may eliminate all the malignant cells. Eventually, however, the immune system slows down and makes mistakes. A malignancy that arises may steadily gain ground and finally defeat it.

The immune-compromised patient is vulnerable to cancers of many kinds. The high incidence of an otherwise rare cancer called *Kaposi sarcoma* was among the first clues to emerge of the deadly new epidemic in the early 1980s and a strong clue that the target of the new virus was the cells of the immune system (see the "HIV and AIDS" section later in the chapter).

An organ transplant recipient must be on a regimen of immunosuppressant drugs his or her whole life. Besides other debilitating side effects, these drugs make the patient vulnerable to cancers of many kinds. Some patients suffering from autoimmune diseases are treated with immunosuppressants, too, and have the same vulnerability.

Immune-mediated diseases

Immune-mediated diseases are conditions that result from abnormal activity of the immune system. *Autoimmune diseases* are disorders caused by the immune system mounting an attack on its own cells. *Allergy* is, essentially, an overreaction of the immune system to a harmless substance in the environment.

Autoimmune diseases

Like a real-life horror movie, the awesome coordinated power that stands ready to engage any and all biological invaders in deadly combat attacks and destroys the body's own tissues. Autoimmune diseases are an active area of basic and clinical research, but on a fundamental level, what causes the immune system to turn on its "self" remains obscure. In most cases, a combination of factors is probably at work. For example, a viral infection may activate (or deactivate the suppression of) a genetic error. Certain autoimmune disorders affect many more women than men, so hormone activity is probably a factor.

Autoimmune disorders are many and various in terms of the pathophysiology. An autoimmune disease can be relatively benign, like *vitiligo,* in which the immune system destroys *melanocytes* (pigment-making cells), resulting in white patches of skin on different parts of the body. Other autoimmune diseases are far more serious. They can affect any part of the body, including the heart, brain, nerves, muscles, skin, eyes, joints, lungs, kidneys, glands, digestive tract, and blood vessels. Clinical experts disagree among themselves about whether certain conditions should be classified and treated as autoimmune conditions. The list of "accepted" autoimmune disorders numbers several dozen, some of which we mention in other chapters of this book.

Allergy

An allergic reaction is acquired (brought about by exposure to a triggering substance called an *allergen*) and rapid. Mild allergies are common in all human populations. Severe allergic reactions can be life-threatening.

Exposure to the allergen causes IgE antibodies to set off an excessive activation of certain types of WBCs (mast cells and basophils), which release excessive histamine. Histamine causes swelling of mucous membranes, such as in the nose and throat. The swelling causes nasal congestion and that annoying itch in the throat. Congestion and swelling can trap bacteria in the nasal cavities and lead to sinus infections or ear infections. The inflammatory response can lead to such symptoms as eczema, hives, and hay fever. Allergies are a significant factor in asthma.

Anaphylaxis is a rapid, severe, whole-body allergic reaction. Anaphylaxis results from an acquired hypersensitivity to an allergen. An initial exposure, called the *sensitizing dose,* to a substance like bee sting toxin or a protein in a food produces no symptoms but sensitizes the person's immune system to the allergen. A subsequent exposure, called a *shocking dose,* sets off anaphylaxis. *Anaphylactic shock* is anaphylaxis associated with vasodilation (dilation of the blood vessels) over the whole body, which results in low blood pressure and severe broncho-constriction to the point where the person has difficulty breathing. Respiratory failure, shock, and heart arrhythmias can lead rapidly to death. The treatment is immediate administration of epinephrine. (Turn to Chapter 8 for information about epinephrine.)

Chronic inflammation

The inflammation response is an important mechanism in the body's natural defense system against infection and disease. *Chronic inflammation,* on the other hand, is a disease. In chronic inflammation, the mechanisms that destroy invaders are turned on the body's own tissues.

Even low-level inflammation, such as occurs in a moderate case of gingivitis, can cause problems, and not just at the inflammation site. The proteins of the inflammatory response and the complement system can travel in the blood and harm cells and tissues anywhere in the body. Chronic inflammation is now widely recognized as an underlying disorder that contributes to many and diverse conditions, including cardiovascular disease; neurological diseases such as clinical depression and Alzheimer's; diabetes; many kinds of cancer; and even premature labor and preterm birth. The list of diseases and disorders that are now recognized to have an inflammatory component is getting longer all the time.

Infectious diseases

Some microbes not only commandeer your body for their own purposes; they use your body as a platform from which to launch invasions into the bodies of all your closest associates. This is a brief look at two different types of chronic infectious viral diseases.

HIV and AIDS

HIV is a species of virus that attacks cells of the human immune system, specifically the helper T cells. The immune system mounts a response, as it does to any other infection, and it may fight the virus to a standstill for years. But, as far as is known, the immune system is unable to eliminate the virus entirely. Like herpes viruses (see the next section), HIV hides out within cells. But as reactivated herpes viruses damage the nerve cells they hide out in, HIV damages the immune system cells, inhibiting the functioning of the immune system itself.

The diagnosis of *acquired immune deficiency syndrome* (AIDS) is made partly on the patient's status regarding certain infections that a healthy immune system has no difficulty beating back (opportunistic infections). Eventually, the response to pathogens and malignant cells is inadequate, and the patient succumbs to an opportunistic infection, cancer, or other disease.

Herpes viruses

Herpes viruses are a leading cause of human viral disease, second only to influenza and cold viruses. People living to middle age usually have antibodies to most of the eight known human herpes viruses, whether or not they're aware of having been infected.

The immune system is capable of suppressing the herpes viruses but not of eliminating them. After a patient becomes infected by a herpes virus, the infection remains in the body for life. Following the primary infection, the virus may migrate to the *ganglia* (nerve bundles) and establish a latent infection, which may reactivate at any stage. Reactivation is frequently, but not always, associated with further disease. Immunocompromised patients are at risk for serious disease and death from reactivated herpes viruses, which are prominent among the opportunistic infections that have been the actual cause of death in many AIDS patients.

The *Varicella-Zoster virus,* the herpes virus that causes chickenpox, is usually acquired in childhood (unless the child has been vaccinated), and more than 90 percent of the population of the United States carries antibodies. The virus is spread by respiratory aerosols (inhaled particles) or by direct contact with the skin lesions of an actively infected patient.

For several days following initial infection, the virus sits in the respiratory mucosa, where it infects macrophages and lung cells. At this stage, no symptoms appear. The virus spreads to lymphocytes and monocytes, and then to epithelial sites throughout the body. The virus reaches the surface of the skin, and lesions, typically hundreds, form on the skin and mucosa, usually most pronounced on the face, scalp, and trunk. The disease is more severe in older children and adults and can be very severe in immune-compromised patients.

Reactivation of the virus may occur in late life. The recurrence of viral replication is accompanied by severe pain in areas innervated by the latently infected ganglia. Symptoms include chronic burning, itching pain and increased sensitivity to touch *(hyperesthesia)* called *post-herpetic neuralgia* or *shingles.* The pain may last months or years. Reactivation can affect the eye and the brain via certain cranial nerves.

Part IV
Life's Rich Pageant: Reproduction and Development

"That's a telecast of parade balloons used in the Macy's Thanksgiving Day Parade. Your ultrasound images are over here."

In this part . . .

*H*ere's where you find out where babies come from and what happens after they're born. The reproductive function gets its own section, not because the reproductive organ system is separate from other physiological processes (it's actually deeply enmeshed with the endocrine system, and its organs need nourishment, oxygen, waste removal, and immune defense, just like all the others), but because we can't discuss development in the same terms as an organ system. Discussing reproduction and development together is logical.

One aspect of the anatomy and physiology of reproduction is different from the body's other systems. The other systems are all dedicated to the survival of the body they're part of. Reproduction, in contrast, does nothing to enhance physiological well-being and actually poses severe threats to the organism's survival. Yet people are driven to risk all for a chance to throw their parents' genes forward one more generation.

Chapter 14

The Reproductive System

In This Chapter

▶ Unscrambling the human egg

▶ Getting the gametes together

▶ Responding to pregnancy's changes

▶ Reviewing reproductive system problems

*L*ike all animals, humans have an instinctive knowledge of mating. However, only humans have a need to understand the processes of mating and reproduction. This chapter gives you information about the anatomy and physiology of reproduction. Look elsewhere for information about dating and mating rituals.

Functions of the Reproductive System

The anatomy and physiology of the reproductive system is dedicated to supporting your role in continuing the human species (whether or not you choose to fulfill that role). Or, to look at it another way, the reproductive system is dedicated to making sure your admirable characteristics are present in the next generation. Or, to look at it from the "selfish-gene perspective," your reproductive system is the means by which some genes replicate themselves and fight on in the never-ending battle for continuing existence.

The following list gives you an overview of what the reproductive system is responsible for:

▶ **Making gametes:** The *gametes* are made within the organs of the female and male reproductive systems. Also called the *sex cells,* gametes are of two kinds: The *ova* (singular, *ovum*) are the female gametes and the *sperm* (singular, *sperm*) are the male gametes. Specialized cells generate the gametes in a cell-division process called *meiosis* (see the "Meiosis" section later in the chapter). At the cell level, the processes are essentially identical in female and male bodies. At the tissue, organ, system, and organism levels, the processes are very different. (We discuss the various processes throughout this chapter.)

The terms *spermatozoon* (singular; literally "seed of an animal") and *spermatozoa* (plural) are the excruciatingly correct technical terms, used now almost exclusively in formal scientific writing.

✔ **Moving gametes into place:** However differently ova and sperm are made, if the reproductive system is to succeed in its purpose, one ovum and one sperm must make their way to the same place at the same time under the right conditions for them to fuse. Many of the reproductive system's tissues and organs chaperone the gametes from the place and time of their production to another place, where they're most likely to encounter their destiny.

✔ **Gestating and giving birth:** Only the female reproductive system has organs for gestating a fetus and giving birth. (See the "Pausing for Pregnancy" section later in the chapter for a detailed discussion of pregnancy.)

✔ **Nurturing the newborn:** The female reproductive system has tissues and organs specialized for nourishing the newborn for the first few months of life, until the older baby is capable of digesting other food.

Producing Gametes

The process of meiosis includes the sequence of cell-level events that result in the formation of sex cells (gametes) from somatic cells. (Flip to Chapter 3 for the details of cell division and differentiation.) Meiosis is the only cellular process in the human life cycle that produces *haploid* cells.

Somatic cells are *diploid*, meaning that each cell nucleus contains two complete copies of the DNA that came into being in the zygote. Sex cells (gametes) are *haploid*, meaning that each cell nucleus contains one copy of the DNA of the mother (somatic) cell. When two gametes fuse to form the zygote, each contributes its DNA to the new zygote, which is, therefore, diploid.

Meiosis

A cell divides by *mitosis* for purposes of asexual reproduction and for growth, development, replacement, and repair, as we discuss in Chapter 2. (Chapter 3 also has details about the cycle of cell growth and division.) A cell divides by *meiosis*, on the other hand, to produce gametes — that is, for purposes of sexual reproduction. The process of meiosis is similar in its mechanics to the process of mitosis, but there are several key distinctions between the two.

The most obvious difference is that meiosis has two parts, called *meiosis I* and *meiosis II*. Each part proceeds in a sequence of events similar to that of mitosis (prophase, metaphase, anaphase, and telophase). In mitosis, the mother cell is diploid, and both daughter cells are also diploid, each having one complete and identical copy of the mother cell's genome. In contrast, meiosis, when it functions optimally, results in four haploid daughter cells. What's more, the four haploid genomes are all different.

The early stages of meiosis (prophase I in Figure 14-1) includes a mechanism called *crossing-over* or *recombination* for exchanging genes between chromosomes. The result is that the cell that becomes the gamete (one of the four haploid products of meiosis) is carrying chromosomes that are completely unique and not identical to the mother cell's chromosomes. Like a lot of topics in cell biology, the complexity of these processes is beyond the scope of this book.

Note that replication of DNA does occur in meiosis, during the interphase that precedes the onset of meiosis I. After two sequential dichotomous divisions, the two complete copies are distributed among the four daughter cells.

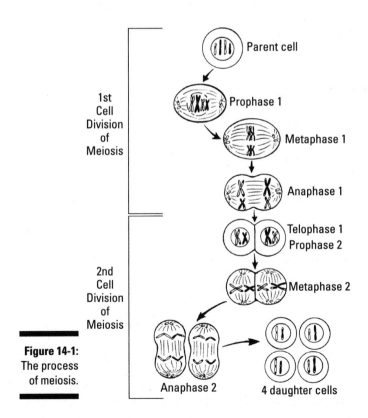

Figure 14-1:
The process of meiosis.

1st Cell Division of Meiosis

Parent cell

Prophase 1

Metaphase 1

Anaphase 1

Telophase 1
Prophase 2

2nd Cell Division of Meiosis

Metaphase 2

Anaphase 2

4 daughter cells

Meiosis includes a number of mechanisms intended to ensure that each gamete has exactly one complete and correct copy of each gene. Any omission, duplication, or error is very likely to be fatal to the gamete or the embryo.

Female gametes: Ova

A mature ovum (see Figure 14-2) is one of the largest cells in the human body, about 120 micrometers in diameter (about 25 times larger than sperm) and visible without magnification. The ovum contains a haploid nucleus, ample cytoplasm, and all the types of organelles usually found in the somatic cell, all within a plasma membrane. The plasma membrane is enclosed within a glycoprotein membrane called the *zona pellucida,* which protects the zygote and pre-embryo until implantation.

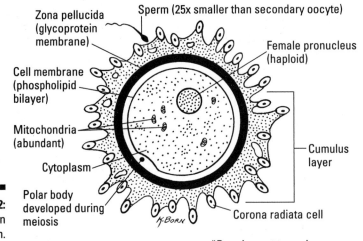

Zona pellucida (glycoprotein membrane)

Sperm (25x smaller than secondary oocyte)

Female pronucleus (haploid)

Cell membrane (phospholipid bilayer)

Mitochondria (abundant)

Cytoplasm

Cumulus layer

Figure 14-2: The human ovum.

Polar body developed during meiosis

Corona radiata cell

*Drawing not to scale

Oogenesis (the development of ova) in humans begins in embryonic and fetal development with specialized somatic cells called *oogonia.* A few of these cells head down the path of meiosis, producing cells called *primary oocytes.* However, this meiosis is suspended at the prophase I point until the female reaches puberty. At birth, the human female has about 700,000 oogonia and primary oocytes in suspended meiosis.

The primary oocyte undergoes meiosis I, producing two cells, called a *secondary oocyte* and the *first polar body.* Most of the primary oocyte's cytoplasm moves to the secondary oocyte. The first polar body undergoes meiosis II, and its daughter cells degenerate.

The cells released from the ovary at ovulation are secondary oocytes. If the secondary oocyte is not fertilized, it degenerates without undergoing meiosis II.

When (or if) a sperm initiates fertilization, the secondary oocyte immediately undergoes meiosis II, producing the ovum (plus a second polar body, which degenerates). Following fertilization, the ovum contains the sperm nucleus, and after approximately 12 hours, the two haploid nuclei fuse, producing the zygote.

Male gametes: Sperm

A mature sperm has three parts: a head that measures about 5 x 3 micrometers, containing a haploid nucleus; a short middle section; and a long flagellum. The sperm is adapted for traveling light — it has very little cytoplasm (see Figure 14-3). The head is covered by a structure that contains enzymes that break down the ovum's membrane to allow entry. The middle contains mitochondria and little else. Mitochondria produce the energy that fuels the sperm's highly active flagellum, which propels the sperm through the female reproductive tract.

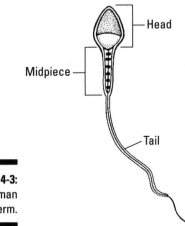

Figure 14-3:
The human sperm.

The process of sperm development *(spermatogenesis)* from meiosis to maturation takes place inside the *testes.* Specialized cells called *spermatogonia* divide by mitosis to produce another generation of spermatogonia. Mature spermatogonia, called *primary spermatocytes,* divide by meiosis, producing four haploid gametes.

Similar to the case with females, males are born with spermatogonia in their *seminiferous tubules,* which remain dormant until puberty. During puberty, hormonal mechanisms pull the spermatogonia out of dormancy.

In contrast to oogenesis, which is cyclic, spermatogenesis is continuous beginning at puberty and continuing lifelong in some men. In contrast to the one-per-month gametogenesis in females, males produce astronomical numbers of sperm. Each ejaculation produces about 1 teaspoon of semen, which contains about 400 million sperm in a matrix of seminal fluid. Mature sperm can live in the epididymis and vas deferens for up to six weeks.

Determining sex

An important difference between males and females is that in females, all chromosome pairs are made up of two identical-looking strands, whereby in males, the strands are different from each other. This difference is easily visible under a high-power microscope: One of the pair is "normal" length (about the same length as all the other chromosomes) and the other is markedly shorter than all other chromosomes. The first is called the X chromosome, and females have one set of them in all their somatic cells. The second is called the Y chromosome, and males have a mismatched pair (one X and one Y chromosome) in all their somatic cells. After meiosis in the female, all ova have one X chromosome. After meiosis in the male, each sperm has either an X or a Y. The fusion of an ovum with an X sperm produces a female (XX) zygote. The fusion of an ovum with a Y sperm produces a male zygote.

The Female Reproductive System

Among mammals, which include humans, the female body is specialized for reproduction to a much greater extent than the male body. The "Reproductive System (Female and Male)" color plate in the center of the book shows the female reproductive system in detail. Following is a brief discussion of the organs of the female reproductive system.

Organs of the female reproductive system

The organs of the female reproductive system are concentrated in the *pelvic cavity.* Many of the female reproductive organs are attached to the *broad ligament,* a sheet of tissue that supports the organs and connects the sides of the uterus to the walls and floor of the pelvis.

Ovaries

The *ovaries* are two almond-shaped structures approximately 2 inches (5 centimeters) wide, one on each side of the pelvic cavity. They house groups of cells called *follicles*.

The ovaries are the primary sex organs because they're the site of *oogenesis,* the process of oocyte maturation. The ovaries also have a major role in endocrine signaling, especially the production and control of hormones related to sex and reproduction.

Beginning at the female's puberty, the process of ovulation begins. The oogonia that have been dormant in her ovaries since early in her fetal development are hormonally activated, and secondary oocytes are released at a rate of approximately one per month from *menarche* (the first menstrual period of her life) to *menopause* (the last) — that is, from her early teen years to her late 40s or early 50s. The human female ovulates about 400 times during her lifetime.

Uterus

The *uterus* or *womb* nourishes and shelters the developing fetus during gestation. It's a muscular organ about the size and shape of an upside-down pear. The walls of the uterus are thick and capable of stretching as a fetus grows.

The lining of the uterus, called the *endometrium,* is built and broken down in the *menstrual cycle,* which we discuss in the section "Cycling approximately monthly" later in the chapter. A portion of the endometrium *(deciduas basalis)* becomes part of the placenta during pregnancy.

The *uterine cervix* (neck) is a cylindrical muscular structure about 1 inch (2.5 centimeters) long that rests at the bottom of the uterus like a thimble. It controls the movement of biological fluids and other material (not to mention, occasionally, a baby) into and out of the uterus. Normally, the cervix is open ever so slightly to allow sperm to pass into the uterus. During childbirth, the cervix opens wide to allow the fetus to move out of the uterus.

Fallopian tubes

The *fallopian tubes* run from the ovary to the uterus. They are not connected to the ovaries; they just kind of hang over them. These tubes transport the released ovum to the uterus during the monthly cycle, or the *pre-embryo* (early-stage fetus) to the uterus in the event of conception.

Vagina

The *vagina* is the part of the female body that receives the male penis during sexual intercourse and serves as a passageway for sperm to enter into the uterus and fallopian tubes. The vagina is about 3 to 4 inches (8 to 10 centimeters) in length. The uterine cervix marks the top of the vagina.

During childbirth, the vagina must accommodate the passage of a fetus weighing on average about 7 pounds (3 kilograms), so the vagina's walls are made of stretchy tissues — some fibrous, some muscular, and some erectile. In their normal state, the vagina's walls have many folds, much like the stomach's lining. When the vagina needs to stretch, the folds flatten out, providing more volume.

Vulva

In females, the external genitalia comprise the *labia majora, labia minora,* and the *clitoris.* Together, these organs are called the *vulva.* The term *labia* (singular, *labium*) means "lips." The labia of the vulva are loose flaps of flesh, just like the lips of the mouth (called *labia mandibulare* and *labia maxillare,* by the way). The labia protect the vagina's opening and cover the pelvis's bony structures.

Here are some details about the three parts of the vulva:

- **Labia majora:** These large folds of skin — one fold on each side — cover the smaller labia minora. The labia majora extend from the *mons pubis* (pubic mound) back toward the anus. The mons pubis contains fat deposits that cover the pubic bone. Following puberty, pubic hair covers the mons pubis and the labia majora.

- **Labia minora:** These hairless folds of skin lie underneath the labia majora and cover the opening of the vagina. The labia minora are attached near the vaginal orifice (opening) and extend upward, forming the foreskin that covers the clitoris.

- **Clitoris:** This part of the vulva located above the vagina's opening and above the urethra has a shaft and glans tip, just as a penis does, and it's extremely sensitive to sexual stimulation. The clitoris contains erectile tissues that fill with blood during sexual stimulation. Because the tissue of the labia minora cap the clitoris, the swelling and reddening is also obvious in the labia minora. Stimulation of the clitoris can lead to orgasm in the female. Although females don't ejaculate, females do experience a building and release of muscular tension. Female orgasm causes the muscle tissue that lines the vagina and uterus to contract, which helps to pull the sperm up through the reproductive tract.

Breasts

Like other female mammals, female humans have mammary glands (breasts) that produce a substance we call *milk* for the nutrition of relatively helpless infants with high calorie requirements. Besides nutrition, breast milk boosts the infant's immune system.

The breast contains about two dozen lobules that are filled with *alveoli* and contribute to ducts (see Figure 14-4). The ducts merge at the *nipple*. Inside the alveoli are milk-producing cells. During puberty, the lobules and ducts form, and adipose tissue is deposited under the skin to protect the lobules and ducts and give shape to the breast. During pregnancy, hormones increase the number of milk-producing cells and increase the size of the lobules and ducts.

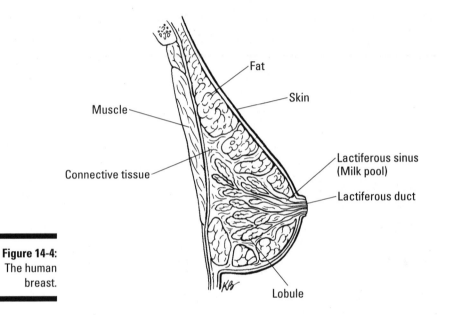

Muscle

Connective tissue

Fat

Skin

Lactiferous sinus
(Milk pool)

Lactiferous duct

Lobule

Figure 14-4:
The human
breast.

After the infant is born, the mother's pituitary gland secretes the hormone *prolactin,* which causes the milk-producing cells to create milk, and *lactation* begins. The infant suckles the milk out of the ducts through the nipple. Lactation continues as long as a child nurses regularly.

The hormone *oxytocin* is strongly involved in milk release (let-down reflex). Stimulation of the nipple prompts the secretion of oxytocin from the mother's pituitary gland. Oxytocin expels milk from the lobules by causing them to contract, just as it stimulates uterine contractions to expel the fetus. This hormone has also been strongly correlated with neuro-emotional phenomena, such as family bonding.

The accepted theory of the evolution of mammary glands is that they originated in early mammals as apocrine sweat glands. The prominence of human breasts has prompted a century and a half of speculation about whether they have functions related to the maintenance of the pair bond between the infant's parents, a significant factor in the infant's survival to reproductive age.

Cycling approximately monthly

The *menstrual cycle (monthly cycle)* consists of both the *ovarian cycle* and the *uterine cycle,* both of which are approximately 28 days in duration (see Figure 14-5). These cycles run concurrently to prepare the ovum and the uterus, respectively, for pregnancy.

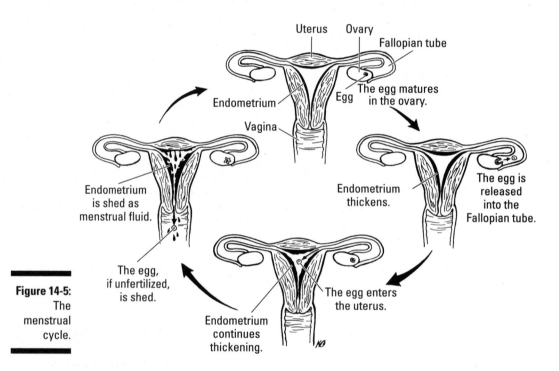

Figure 14-5: The menstrual cycle.

By convention, the first day of menstrual bleeding is counted as Day 1 of the menstrual cycle. Menstrual bleeding begins at a point in the cycle when the levels of estrogen and progesterone are at their lowest. However, the entire menstrual cycle is directed by several hormones, not just estrogen and progesterone.

Looking at the ovarian cycle

The 28-day ovarian cycle is the most important part of the menstrual cycle because it's responsible for producing the hormones that then control the uterine cycle. (See the next section, "Clocking the uterine cycle.") From Day 1 to Day 13, triggered by low estrogen level, *follicle-stimulating hormone* (FSH) stimulates the development of a follicle and *luteinizing hormone* (LH) stimulates the maturation of an oocyte in one of the ovaries. When the follicle is developed enough, it begins to secrete estrogen. When the level of estrogen reaches the appropriate level, a negative feedback mechanism involving the *hypothalamus* (the "master gland") slows the secretion of FSH and LH. When the follicle is fully mature and the oocyte is ready to be released, FSH and LH secretion peaks. On Day 14, the oocyte is released *(ovulation)*. An oocytes lives for only 12 to 24 hours after ovulation.

At the time of ovulation, the anterior pituitary gland, which has been secreting FSH and LH simultaneously, secretes a surge of LH that causes the follicle from which the oocyte was released to become a *corpus luteum* (yellow body). The corpus luteum secretes the hormone *progesterone*, which triggers the hypothalamus. When the corpus luteum has secreted a sufficient amount of progesterone, the hypothalamus stops the anterior pituitary gland from secreting any more LH. At that point, the corpus luteum begins to shrink (about Day 17). When the corpus luteum is gone (about Day 26), the levels of estrogen and progesterone are at the lowest levels of the cycle (sometimes causing symptoms of premenstrual syndrome), and menstruation starts (about Day 29, or Day 1 of the new cycle).

Like any cycle, the whole process starts over. When the level of estrogen is low during menstruation, the hypothalamus detects the low level and secretes *gonadotropin-releasing hormone* (GnRH), which prompts the anterior pituitary gland to release its gonadotropic hormone — FSH — so that another follicle is stimulated to develop a new oocyte that secretes estrogen. Now, you're back to the first paragraph of this section.

Clocking the uterine cycle

The 28-day uterine cycle, which aims to prepare the uterus for a possible pregnancy, overlaps with the ovarian cycle.

- **Days 1 to 5:** The first 5 days of the uterine cycle is when the level of estrogen and progesterone are lowest — the period of menstruation. The low level of sex hormones fails to prevent the tissues lining the uterus (the *endometrium*) from disintegrating and shedding. As the hormone levels drop, blood vessels spasm, cells undergo *autolysis* (self-destruction), tissues tear apart from the uterus wall, and blood vessels rupture, causing the bleeding that occurs during a period. The blood and tissue (menstrual flow) passes out of the uterus through the cervix and then out of the body through the vagina.

- **Days 6 to 14:** During this *proliferative phase,* estrogen production is highest. The developed follicle secretes estrogen, which makes the endometrium regenerate fresh tissue. The tissues lining the uterus and the glands in the uterine wall grow and develop an increased supply of blood. All these changes are preparation for nourishing an embryo and supporting a pregnancy, should the oocyte, which is released on day 14, become fertilized and implant in the wall of the uterus. (See the section "Pausing for Pregnancy" later in this chapter.)

✔ **Days 15 to 28:** During this *secretory phase,* the corpus luteum secretes an increasing level of progesterone, which further thickens the endometrium, and the glands of the uterus secrete a thick mucus. If the egg becomes fertilized, the thickened endometrium and mucus help to "trap" the fertilized egg so it implants properly in the uterus. If the egg doesn't become fertilized within a day or two, the corpus luteum begins to shrink because it won't be needed for a pregnancy. As the corpus luteum shrinks, the progesterone and estrogen levels decline, which causes the endometrium to "shred and shed" just before menstruation.

Early on in a pregnancy, the corpus luteum serves as a source of estrogen and progesterone until the placenta develops and can secrete estrogen and progesterone on its own.

Winding down the cycle

Physiologically, menopause essentially reverses the hormonal pathway of adolescence. When a woman enters menopause, her ability to reproduce ends — ovulation stops, and she no longer can become pregnant. She may also experience hot flashes and sweat baths if faulty signals from the parasympathetic nervous system disrupt the body's ability to accurately monitor its temperature. Other body processes slow down, including cellular metabolism and the replacement of the structural proteins in the skin, which leads to wrinkles. A woman's bones may also weaken when the breakdown of bone tissue occurs faster than the buildup of bone tissue during bone remodeling.

Menopause is one of the unique aspects of human physiology. Not that reproductive cycling declines and stops as the female ages. That happens in many mammal, bird, and reptile species (although relatively few individuals in any species live long enough to experience the decline of their reproductive capabilities).

The unique part is that human females often live a substantial proportion of their life span beyond their reproductive capacity. (A woman in her 80s has lived around 40 percent of her life after menopause.) Much research into this phenomenon is concentrated at the interface of biology and culture. One theory holds that an adult female with no offspring of her own to feed tends instead to feed her grandchildren or other children in her community. Children with grannies eat better, the theory goes, improving their chances of surviving to reproductive age and pushing her genes into one more generation.

The Male Reproductive System

The male reproductive system produces sperm and moves them into the female reproductive system. On a rare occasion (relative to the astronomical number of sperm that the average human male produces), a sperm fertilizes an egg. All the billions and billions of other sperm a man produces in his lifetime have a limited life span — about six weeks.

The female gamete is released from the ovary as a secondary oocyte. Only when fertilization is initiated does the ovum (egg) come into being.

If *reproduction* is defined as ending with the creation of a new organism, that's the end of the male's reproductive function. If *reproduction* is defined as including the nurturance of the new organism until it is itself ready for reproduction, the anatomy and physiology of the male may well be substantially devoted to the task for decades, along with those of the female. For the purposes of this chapter, we use the more limited definition.

Some people are confused about the meaning of "evolutionary success" for the individual (you, for example). Truly, it's not how many zygotes arise from your gametes, or even how many offspring are born from these zygotes, but how many of these offspring survive to reproduce. Simply, evolutionary success is rated not by the number of your children (and certainly not by your level of sexual activity or number of opposite-sex partners) but by the number of your grandchildren.

The organs of the male reproductive system

The organs of the male reproductive system produce gametes, called sperm, and transfer them to the female reproductive system. (Refer to the "Reproductive System (Female and Male)" color plate in the center of the book for a look at the male reproductive system.) In contrast to some other organ systems, and especially to the reproductive system of females, the male reproductive organs are located in an exposed location on the periphery of his body. The heart, lungs, kidneys, and the female's ovaries and uterus are all located beneath protective layers of skin, muscle, and other connective tissue and suspended in a fluid matrix, all of which support homeostasis in these organs. The exposed location of the male reproductive organs is common in mammals but unusual in other vertebrates.

Ask about the location of the testes outside the male body and you'll likely get a simple answer about sperm requiring a slightly lower temperature than body core temperature for development. Sorry, but that's not an explanation. That's a description of an evolutionary mechanism. The question is: Why did evolution favor males, whose gametes must be produced outside their bodies? Like a lot of intriguing questions, that one is beyond the scope of this book. Just remember to be careful about accepting simplistic answers to complex questions in anatomy and physiology.

Testes and scrotum

The *testes* (singular, *testis*) are paired organs that produce sperm and hormones. Like the ovaries of the female reproductive system, the testes are the site of gamete production and therefore the *primary sex organs.*

The testes contain fibrous tissue that forms compartments in the testis. Inside the compartments are long, coiled *seminiferous tubules,* the site of *spermatogenesis,* the process of sperm development from meiosis to maturation. The walls of the seminiferous tubules are lined with thousands of *spermatogonia* (immature sperm). The seminiferous tubules also contain *Sertoli cells* that nourish the developing sperm and regulate how many of the spermatogonia are developing at any one time.

The *epididymis* (plural, *epididymides*) is a long, cordlike structure that lies atop each testis. The epididymis is continuous with (that is, "becomes") the *vas deferens,* which is the final place of maturation for sperm. The vas deferens is a tube that connects the epididymis of each testis to the penis.

The testes are held within the *scrotum* beneath and outside of the abdomen. The scrotum contains smooth muscle that contracts when the scrotal skin senses cold temperatures and pulls the scrotum (and thereby the testicles) closer to the body to keep the sperm at the right temperature. The scrotum's inside muscle layers are an outpouching of the pelvic cavity. The scrotum's outside skin is continuous with the skin of the perineum and groin.

Prostate gland

Several other structures secrete the substances that make up the ejaculatory fluid that provides a matrix for the propulsion of sperm into the female reproductive tract. Among these are the *prostate* and the *seminal vesicles.* The prostate also contains some smooth muscles that help expel semen during ejaculation.

Penis

The penis consists of a shaft and the *glans penis* (tip). The tube-shaped *urethra* runs through the shaft of the penis, and the *glans penis* contains the *urethral orifice.* The semen is ejaculated through the urethra and urethral orifice. *Foreskin* (also called *prepuce*) covers the glans penis. The foreskin of newborn males is often removed in a surgical procedure called *circumcision.*

During sexual arousal, erectile tissue enables the penis to be inserted into the female vagina, delivering the sperm to the vicinity of the secondary oocyte (if one is available).

The urethra and urethral orifice also function in the urinary system as the tube through which urine leaves the body. However, semen doesn't contain urine. At the point of ejaculation, a sphincter closes off the bladder to keep urine, which is acidic, from mixing with the sperm, which live in a basic environment.

Seminal fluid and ejaculation

The *seminal vesicles,* glands located at the juncture of the bladder and vas deferens, have ducts that allow the fluid they produce to sweep the sperm from the vas deferens into the urethra.

Next, the *prostate gland* adds its fluid, which contains mainly citric acid and a variety of enzymes that keep semen liquefied. The prostate gland surrounds the urethra just below where it exits from the urinary bladder.

The *bulbourethral glands,* also called *Cowper's glands,* sit within the floor of the pelvis near the bulb of the penis on either side of the urethra. These two small glands have ducts leading directly to the urethra.

These three types of glands — the seminal vesicles, the prostate gland, and the bulbourethral glands — secrete fluids that have several functions:

✔ They're slightly basic with a pH of 7.5, just the way sperm like their environment to be.

✔ They nourish the sperm by providing the sugar fructose so that the sperm's mitochondria can make enough energy to move its tail and travel all the way to the egg.

✔ They contain *prostaglandins,* which are chemicals that make the uterus reverse its downward contractions. When the uterus contracts, the sperm are pulled farther upward into the female's reproductive tract.

As the glands add the secretions, forming the semen, pressure builds up on the structures of the male reproductive tract. When the pressure has reached its peak, the semen is expelled out of the urethra through the penis. Peristaltic waves (like those that occur in the digestive tract; see Chapter 11) and rhythmic contractions move the sperm through the vas deferens and urethra. The term for this discharge is *ejaculation* — part of orgasm in males, as is the contraction and relaxation of skeletal muscles at the base of the penis. As the muscles contract rhythmically, the semen comes out in spurts.

Pausing for Pregnancy

Pregnancy is established in two stages: fertilization of the secondary oocyte and implantation of the blastocyst in the uterus. Development of the embryo after implantation is the subject of Chapter 15. The female body makes many and various adaptations to pregnancy and delivery, which we examine in the following sections.

Steps to fertilization

Ovulation sends a secondary oocyte from the follicle of the ovary into the uterine tube. Then, within an appropriate time, heterosexual intercourse results in the ejaculation of semen into the vagina. Some few million sperm make their way through the cervix, up through the uterus, and into the uterine tube to the vicinity of the waiting secondary oocyte.

To achieve fertilization, one sperm must penetrate the secondary oocyte's membrane, which initiates meiosis II, and its nucleus must fuse with that of the ovum. At this point, the secondary oocyte is fertilized, the ovum is developed, and, with the fusion of the nuclei, the zygote comes into being.

The probability of any act of intercourse resulting in fertilization is actually quite low because many complicating factors exist. The timing of intercourse relative to ovulation is crucial. The released secondary oocyte is viable for only a matter of hours; sperm live a little longer in the female reproductive tract (12 to 72 hours). The environment within the female reproductive tract may be more or less hospitable to the sperm, depending on the female's hormone levels and other physiological processes. Even when a single sperm has made contact with the secondary oocyte, fertilization is not assured.

Implantation

Following fertilization, the zygote divides immediately. Several more cell division cycles take place as the pre-embryo moves down the uterine tube. Experts believe that many pre-embryos die at this stage, sometimes because of genetic or developmental abnormalities. Only if the pre-embryo arrives at the uterus and properly embeds itself into the endometrium is pregnancy established.

A successfully implanted pre-embryo, now call a *blastocyst,* begins immediately to take over its mother's body. It begins to produce a hormone called *human chorionic gonadotropin* (hCG), which maintains the *corpus luteum,* elevates levels of progesterone and estrogen, and inhibits menstruation.

The presence of hCG can be detected chemically in the urine of a pregnant woman within 10 to 14 days after fertilization. It may be detected before that symptomatically by the mother as a sensation of being nauseated in every cell of her body.

Adapting to pregnancy

The maternal body responds to pregnancy with many anatomical and physiological changes to accommodate the growth and development of the fetus.

Most structures and processes revert to the nonpregnant form (more or less) after the end of the pregnancy. See Chapter 15 for details about how the fetus grows in the uterus.

Uterus

During pregnancy, the uterus grows to about five times its nonpregnant size and weight to accommodate not only the fetus but also the placenta, the umbilical cord, about a quart of amniotic fluid, and the fetal membranes. The size of the uterus usually reaches its peak at about 38 weeks gestation. During the last few weeks of pregnancy, the uterus has expanded to fill the abdominal cavity all the way up to the ribs. The size of the expanded uterus and the pressure of the full-grown fetus may make things difficult for the mother.

The placenta acts as a temporary endocrine gland during pregnancy, producing large amounts of estrogen and progesterone by 10 to 12 weeks. It serves to maintain the growth of the uterus, helps to control uterine activity, and is responsible for many of the changes in the maternal body.

Near the end of pregnancy, the uterine cervix softens. Enlarged and active mucus glands in the cervix produce the *operculum*, a mucus "plug" that protects the fetus and fetal membranes from infection. The mucus plug is expelled at the end of the pregnancy. Additional changes and softening of the cervix occur at the onset of labor.

Ovaries

Hormonal mechanisms prevent follicle development and ovulation in the ovaries.

Breasts

The breasts usually increase in size as pregnancy progresses and may feel inflamed or tender. The areolas of the nipples enlarge and darken. The areola's *sebaceous glands* enlarge and tend to protrude. By the 16th week (second trimester), the breasts begin to produce *colostrum*, the precursor of breast milk.

Other organ systems

Pregnancy affects all organ systems as they support the growth and development of the fetus and maintain homeostasis in the female. Here are a few important physiological consequences of pregnancy:

- The other abdominal organs are displaced to the sides as the uterus grows.

- Decreased tone and mobility of smooth muscles slows peristalsis and enhances the absorption of nutrients. An increase in water uptake from the large intestines increases the risk for constipation. Relaxation of the cardiac sphincter may increase regurgitation and heartburn. Nausea and other gastric discomforts are common.

✔ Increases occur in blood volume, cardiac output, body core temperature, respiration rate, urine volume, and output from sweat glands.

✔ Immunity is partially suppressed.

✔ Spinal curvature is realigned to counterbalance the growing uterus. Slight relaxation and increased mobility of the pelvic joints prepares the pelvis for the passage of the infant. This can compromise the woman's lower-body strength starting in the second trimester.

Labor and delivery

Labor is initiated by complex hormonal signaling between the maternal and fetal bodies. In the ideal labor and delivery process, powerful contractions of the smooth muscle of the uterus push the fully mature fetus (infant) past the cervix and down the birth canal without undue trauma to either the mother or the infant. See the "Prenatal Development" color plate in the middle of the book to see the stages of fetal development.

Stage 1

Stage 1 labor begins with regular, effective uterine contractions. The uterine cervix begin to *efface* (become thinner and wider; see Figure 14-6). The *amniotic membrane* ruptures.

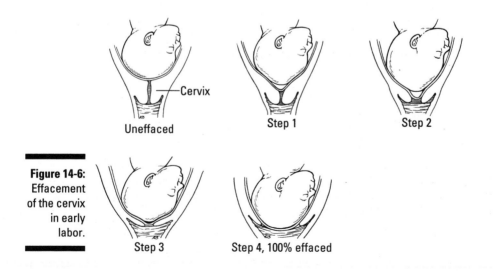

Figure 14-6: Effacement of the cervix in early labor.

Uneffaced — Cervix

Step 1

Step 2

Step 3

Step 4, 100% effaced

Stage 2

Uterine contractions continue pushing the fetus through the birth canal. The cervix becomes fully effaced. The mother is aware of the passage of the fetus and may feel a strong urge to bear down and push the infant along.

At the *transition phase,* the infant emerges head first from the birth canal. This is the end of stage 2 of labor.

Just after delivery, the infant's umbilical cord is cut and tied off. The infant is now totally separated from the mother and will soon have a stylish belly-button.

Stage 3

After the baby is delivered, uterine contractions continue so that the placenta separates from the uterus wall. About 15 minutes after the baby is born, the placenta passes through the birth canal. Uterine contractions continue, during which the uterus contracts, returning eventually to near prepregnant size. Figure 14-7 shows an overview of delivery.

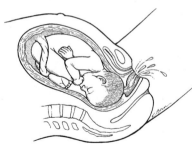
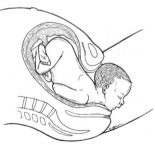

a. Dilation of the cervix and breaking of amniotic sac   b. Delivery of the head

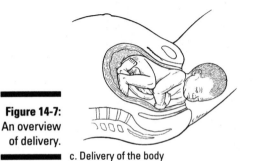
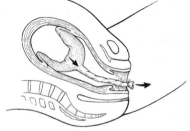

Figure 14-7:
An overview
of delivery.

c. Delivery of the body   d. Delivery of the placenta

Pathophysiology of the Reproductive System

Reproduction is a dangerous business for all animals. The investment of energy is huge, the risks are just as great, and the rewards are distant. But an organism without two generations of descendants has failed in evolutionary terms. All humans are the descendants of organisms that took their chances and survived all the risks.

Reproduction is an intricate interaction among many structures and processes besides those explicitly identified with the reproductive system. Structural defects, hormonal problems, genetic abnormalities, and cancers can all cause problems in reproduction. Here are some of the problems that can affect the reproductive system.

Infertility

Infertility is the inability to fertilize or be fertilized. Infertility may be due to the failure to generate viable gametes, blockages in the "travel routes" of the gametes or the pre-embryo, or damage to or disease in the endocrine glands that control reproduction. Certain bacterial or viral infections, such as mumps, can result in *orchitis* (inflammation of the testes) that can affect fertility. Fertility decreases with age, ending abruptly at menopause in females and declining more gradually in males. Turn to Chapter 15 for more information on aging.

Cancer

Cancers affecting tissues and organs of the reproductive system are among the most prevalent cancers in both men and women worldwide. In men, the most common types of cancer in the reproductive system are prostate cancer and testicular cancer. In women, these are breast cancer, ovarian cancer, and cervical cancer.

Sexually transmitted diseases

Some microbial diseases are transmitted via sexual contact. These *sexually transmitted diseases* (STDs) are endemic in human populations (that is, they exist in all human populations to one extent or another and have since for-ever) because bacteria and viruses can easily be spread from one person to another via the organs and secretions of the reproductive system.

Sexually transmitted infections are similar in most ways to other microbial infections. All the major groups of microbes (bacteria, fungi, protists, and viruses) have evolved some ability to propagate themselves in the very hos-pitable environment of the human reproductive tract. They cause problems in all the same ways: by inducing inflammation, over-activating the immune response, and destroying cells.

In the very special case of *HIV,* the infectious organism (an exotic creature called a *retrovirus*) destroys structures of the immune system, which leaves the body vulnerable to attack by other microbes.

The infectious bacterium *Chlamydia trachomatis* is transmitted by other means as well as sexually; it infects the human eye, joints, and lymph nodes, and also takes up residence in arteries. Chronic, low-level inflammation of the arteries is a major underlying cause of much cardiovascular disease.

Premenstrual syndromes

Up to 80 percent of menstruating women experience physical changes just before the onset of menstrual blood flow. The type and severity of symp-toms varies greatly from one woman to another, but the symptoms tend to remain stable through her reproductive life. *Premenstrual syndrome* (PMS) is a term used to describe a medical syndrome of mood swings, mild *edema* (fluid retention in the tissues), irritability, fatigue, food cravings, and uterine spasms (cramps) that affect an estimated 20 percent to 40 percent of women. Another more severe form, called *premenstrual dysphoric disorder* (PMDD), affects 2 percent to 10 percent of menstruating women, has much in common with mood disorders, and can result in severe disruption of daily activities. Physiological, psychological, environmental, and social factors all seem to play a part in the development of these disorders.

Endometriosis

The *endometrium* is the lining of the uterus, which is shed during menstrua-tion. In *endometriosis,* endometrial tissue grows in or on organs of the body other than the uterus — usually organs in the pelvic cavity, such as the blad-der, ovary, or large intestine. At the end of the uterine cycle, when the hor-mone levels decline, endometrial tissue, whether in the uterus or elsewhere, disintegrates, sometimes causing extreme pain.

Cryptorchidism

During fetal development, the testes are located inside the pelvic cavity, but around the time of birth, the testes descend into the scrotum. Failure of the testes to descend is called *cryptorchidism.* Unless corrected by surgery, cryptorchidism results in sterility.

Hypogonadism

Problems with the pituitary gland, such as injury or tumors, can cause *hypogonadism,* a decline in the function of ovaries or testes. The pituitary gland secretes follicle-stimulating hormone (FSH), which normally spurs on maturation of the oocyte or spermatocyte and the subsequent release of estrogen or testosterone. Symptoms in women include *amenorrhea* (absence of menstruation) and infertility. In men, the symptoms of hypogonadism are impotence and infertility.

Klinefelter's syndrome

The genetic disorder *Klinefelter's syndrome* causes men to have an extra X chromosome in their cells (XXY). Affected boys develop normally, but when they get to puberty, the testes fail. A decline in testosterone production leads to a wide variety of effects: development of breast tissue, impotence, bone loss, and infertility.

Erectile dysfunction

Erectile dysfunction (ED), also called *impotence,* is a condition in which penile erection doesn't follow sexual stimulation. ED has a variety of possible causes, including damaged blood vessels, sometimes because of diseases such as diabetes; psychological factors such as stress and fear; and nerve damage. Some degree of ED is considered a normal part of aging.

Pathophysiology of pregnancy

Even a healthy, young-adult woman carrying a pregnancy under ideal social and economic conditions is at risk for pathophysiology in many organ systems. Carrying a pregnancy and giving birth are major contributors to disability, disease, and death in adult human females. And her existing dependent children experience follow-on consequences — a child whose mother has died or been disabled is likely to have impaired nutrition and ineffective parental protection, which jeopardizes the child's survival to reproductive age.

During childbirth in humans, a relatively large and fragile infant is propelled by powerful uterine contractions through a pelvic girdle that has recently (in evolutionary time) undergone significant adaptation to the upright posture and the bipedal gate. Midwifery is surely the oldest profession and may have contributed profoundly to the survival of early human tribes. Some problems are still being worked out.

Childbirth has other risks — the wounds incurred expose a woman to some kinds of infections, and significant blood loss is quite common. Some other pregnancy-related disorders are

- ✔ **Ectopic pregnancy:** An *ectopic pregnancy* is an abnormal pregnancy where the pre-embryo implants outside the uterus, most commonly in the fallopian tubes. An ectopic pregnancy is often caused by a condition that blocks or slows the movement of a pre-embryo through the fallopian tube to the uterus. This may be caused by a physical blockage in the tube by hormonal factors and by other factors, such as smoking. The fetus can't survive and often stops developing altogether. Ectopic pregnancy is a life-threatening condition.

- ✔ **Gestational diabetes:** Gestational diabetes is *hyperglycemia* (excess glucose in the blood) that develops during pregnancy and that affects both mother and fetus. Diabetes can complicate delivery and can increase risk for diabetes mellitus after the pregnancy.

- ✔ **Incompetent cervix:** In this condition, the cervix is unable to support the pregnancy. The cervix may have been traumatized during an earlier birth. A woman carrying a multiple pregnancy is at increased risk. The cervix dilates prematurely (before the onset of labor), a serious risk for pregnancy loss. To alleviate the problem, in a procedure called *cerclage,* the cervix can be stitched to give it extra support.

- ✔ **Pre-eclampsia and eclampsia:** Just as urine is checked for glucose to help stave off gestational diabetes, the urine is also checked for protein, diagnostic of risk for pre-eclampsia. *Pre-eclampsia* (high blood pressure during pregnancy) can easily escalate into *eclampsia,* characterized by seizures, and possibly coma or even death.

- ✔ **Placenta previa:** In this condition, the placenta covers the cervix, partially or completely, blocking delivery of the fetus or causing heavy bleeding. Placenta previa in the second or third trimester may result in blood loss from the placenta as the growing fetus presses on it. The bleeding puts the mother at risk for preterm labor and the fetus at risk for premature birth. Pregnancies complicated by placenta previa are treated by cesarean delivery.

✔ **Placental abruption:** In this condition, part of the placenta peels away from the uterine wall before the fetus is ready to be delivered. Such an event can deprive the fetus of oxygen and nutrients and, depending on the extent of the tear, can bring on preterm labor and cause life-threatening hemorrhaging in the mother. Maternal hypertension, physical trauma to the mother and fetus (such as a car accident), and a short umbilical cord are among the most common causes of placental abruption.

✔ **Fetal distress:** Labor and delivery are stressful for the fetus. Fetal distress can sometimes complicate the delivery.

Pregnancy loss

Pregnancy loss, also called *spontaneous abortion* and *miscarriage,* is the death of an embryo or fetus without apparent cause within the first 20 weeks of gestation. From 10 percent to 25 percent of all clinically recognized pregnancies end in miscarriage. Many more pregnancies end shortly after implantation, often without the woman realizing that she was pregnant. The bleeding that results may begin around the expected time of her menstrual period.

Miscarriage has many causes. Clinicians speculate that many may be caused by abnormalities in the embryo and not because of any disorder of the woman's reproductive system. Many women experience miscarriage and go on to have a normal pregnancy with a normal outcome (a cute baby). Repeated miscarriage, however, may indicate some form of disorder in either the male or female parent.

Chapter 15

Change and Development over the Life Span

In This Chapter

▶ Unfolding in real time: the drama of development

▶ Summarizing the miracle of reproduction

▶ Changing through life

*I*n the context of anatomy and physiology, *development* means the pattern of change through an organism's lifetime. Development is closely related to the specialized branch of biology called *ontogeny,* which studies an organism's history within its lifetime. Human development has been a subject of much study for thousands of years, especially for parents and grandparents of young human children. It's a good thing for babies that people find them so fascinating: The human baby takes a lot of effort to maintain and a long time to mature.

In this chapter, we take a look at human development, from the creation of the zygote through old age. Although you've already experienced some of this development, you get a glimpse of some of the changes your body will go through as you age.

Programming Development

One way to think of development is as the unfolding in real time and space of a program for generating a unique biological organism. The program is launched when a new *zygote* comes into existence. All zygotes are created the same way and then proceed down the path of development encoded in their own species-specific and individual-specific DNA. (Flip back to Chapter 14 if you need a refresher on the zygote.)

The totality of the DNA of a zygote — that is, its *genome* — comes into existence at the time of fertilization. The DNA in the zygote's nucleus comprises genes (specific DNA sequences) from both its parents, 50-50, but this particular combination of genes has never been seen before and will never be again. A few new genomes achieve evolutionary success; that is, a few new individuals succeed in throwing some of their genes forward to two generations of descendants.

Most genomes, including all human genomes, have aging and death built into the program. All die sooner or later. A few survive until their program has fully unfolded and reached its end.

Stages of development

Development begins in the zygote and continues until death. There's no universally agreed-upon definition of the development stages (although two milestone events — birth and, for females, the onset of menstruation — are universally acknowledged), and the age range at which a person passes from one stage to the next is wide. Change is more or less continuous through life, and different organ systems undergo significant changes on their own development timetable. However, conventionally in human biology, the development milestones that mark the stages are based on developments in the nervous and reproductive systems.

Dimensions of development

The structural and physiological changes that happen during human development include an increase in size, the acquisition of some specialized abilities, and the loss of some other specialized abilities continuously throughout life. When all goes well, *senescence* (aging) is the final stage of development.

The following sections assume an organism for whom all is going well, biologically speaking: no fatal errors in the genome itself and adequate resources to sustain nutrition, thermoregulation, and all the rest of the life-maintaining physiological reactions.

Growth

As in many other species, part of human development involves an increase in size. Increased size is primarily accomplished by the growth of organs that exist in some form in the embryo: The heart grows larger, the brain grows larger, and the bones get longer and heavier. The organs grow by building more of their own tissues, and tissues get bigger by adding cells or increasing cell size. Everything (well, almost everything — there are always exceptions in biology!) grows together, mostly by adding cells.

However, not everything grows equally. Different stages of development are characterized by different proportions of tissue types. For example, both the brain and the skeletal muscle increase in size from infancy to adulthood, but the proportion of muscle tissue to brain tissue is much higher in adulthood.

When a three-dimensional object such as a living body increases in size, the *surface-to-volume ratio* decreases. (Or, to put it another way, the *volume-to-surface ratio* increases — more of your inside parts are dependent on fewer of your outside parts to interact directly with your environment.) The size of a human body strongly influences thermoregulation, fluid balance, and other key aspects of homeostasis.

Differentiation

For humans, the acquisition of new abilities or improvements in existing abilities is part of development. New physiological abilities come about usually because of cell and tissue *differentiation* (function specialization). Tissue specialization begins in the pre-embryonic stage, as we discuss in the "Development before Birth" section later in this chapter. A newborn has some version of more or less all the cell types and tissue types, but many fully differentiated cells must be generated and integrated functionally into tissues at appropriate stages of development. See Chapter 4 for a brief description of the development of a *long bone* (an organ of the skeletal system) by cell growth and differentiation.

Lots of human body functions aren't "learned" but "developed." The ability to digest starch, for example, is acquired during the first year of life, when the body starts producing the necessary enzymes — not when someone teaches a baby how. Toilet training is more about the maturity of the nervous system than the diligence of the parents. The acquisition of a new skill, structure, or process is sometimes accompanied by the loss of existing abilities. A young adult is better at planning than a teenager but has most likely lost some stamina for all-nighters, parties, and road trips. The stages of human development can be characterized by these abilities gained and lost.

Among many aspects of development research, brain research has yielded very interesting data in recent years, aided by advanced imaging technology (see Chapter 1). In the late 1990s, the decades-old doctrine that humans don't generate any new brain cells after birth was definitively shown to be false. Data from many different kinds of studies since then have indicated that the human brain is *plastic* (capable of change and development) well into old age. Turn to Chapter 7 for more about brain development.

Senescence

According to recent theories, age-related decline in specialized and even basic physiological functions is built into new genomes right at the start. Structures at the ends of the chromosomes called *telomeres,* which get shorter and shorter as a genome ages, control the number of times the genome can replicate. Gradually, cells lose the ability to divide. The number of aged and dying cells in a tissue eventually exceeds the number of new cells of their type being made to replace them and the number of cells whose function is to remove these dead and dying cells. The tissue loses its ability to function, which impairs the organism's survival.

The aging processes are an active area of research in anatomy and physiology. In recent decades, therapies and devices to counter aging's effects have dominated the medical products marketplace worldwide.

Development before Birth

Human birth is a commonplace miracle: from a single infinitesimal cell to a human baby in less than nine months. The following sections give a brief overview of how it happens.

If you want more detailed information on pregnancy, check out *Pregnancy For Dummies,* 3rd Edition, by Joanne Stone, Keith Eddleman, and Mary Duenwald (Wiley).

Free-floating zygote to protected embryo

Chapter 14 covers the events leading to the fertilization of a secondary oocyte and implantation of the blastocyst in the uterus from the point of view of female reproductive anatomy and physiology. This section covers those same events from the zygote's point of view, from the fusion of the haploid genomes of the parent gametes to implantation in the uterus. (See Figure 15-1.)

Beginning it all

Fertilization, which takes about a day, begins when a sperm penetrates a *secondary oocyte* (an egg). One sperm binds with the receptors in the *zona pellucida* and enzymatically dissolves membrane proteins at the attachment point, allowing entry to the sperm nucleus only, resulting in the completion of meiosis II and the formation of a second polar body and an ovum. The fusion of the ovum and sperm nuclei, still within the zona pellucida, marks the creation of the zygote and the end of fertilization.

A dangerous journey

The zygote undergoes *cleavage* (mitotic division) immediately. Over the next few days, the daughter cells (called *blastomeres*) divide twice more, to a total of 16 blastomeres, all within the rigid wall of the zona pellucida with no increase in overall size.

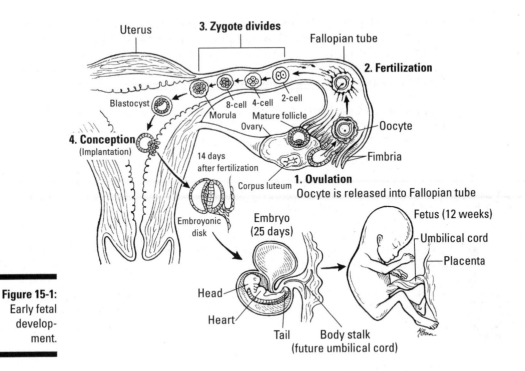

Figure 15-1:
Early fetal develop-
ment.

The mass, now called a *morula* (mulberry-shaped), leaves the fallopian tube and enters the uterine cavity. Cell division continues, still confined within the zona pellucida, and a cavity known as a *blastocele* forms in the morula's center. Around the sixth day after fertilization, the hollow structure, now called a *blastocyst,* "hatches" from the slowly eroded zona pellucida within the uterine cavity. The outer layer of blastocyst cells secretes an enzyme that facilitates implantation in the endometrium. *Angiogenesis* (building of blood vessels) begins in the uterus, and diffusion between mother and blastocyst begins. When this diffusion is established, implantation is complete and the pregnancy is established.

The new genome has survived a very dangerous stage of development. Biologists estimate that up to one-half of blastocysts fail to implant, and they die. But the new genome still has challenges ahead.

The embryonic stage

Weeks three through eight after implantation are called the *embryonic stage*. During these weeks, the embryo's cells begin to differentiate and specialize.

The transition from blastocyst to *embryo* begins when the implanted blasto-cyst develops into a two-layer disc. The top layer of cells *(epiblast)* becomes the embryo and amniotic cavity; the lower layer of cells *(hypoblast)* becomes the yolk sac that nourishes the embryo. A narrow line of cells on the epiblast, called the *primitive streak,* signals *gastrulation* — cells migrate from the epiblast's outer edges into the primitive streak and downward, creating a new, middle layer. By 14 days or so after fertilization, the embryo has *ectoderm, mesoderm,* and *endoderm* layers — the very beginning of tissue formation.

Forming the placenta

Immediately after implantation, the blastocyst initiates the formation of the *placenta,* a special organ that exists only during pregnancy that's made of the mother's cells in the outer layers and the fetus's cells in the inner layer. The placenta serves to support the sharing of physiological functions between the mother and the fetus: nourishment (provision of energy and nutrients), gas exchange (a fetus must take in oxygen and eliminate carbon dioxide before birth), and the elimination of metabolic waste. The placenta allows some substances to enter the fetal body and blocks others. It does a good job of delivering nutrients and maintaining fluid balance, but it's permeable to alcohol, many drugs, and some toxic substances.

In humans, the placenta is a dark red disc of tissue about 9 inches (23 centimeters) in diameter and 1 inch (2.5 centimeters) thick in the center, and it weighs about a pound (roughly half a kilogram). It connects to the fetus by an *umbilical cord* of approximately 22 to 24 inches (56 to 61 centimeters) in length that contains two arteries and one vein. The placenta grows along with the fetus.

Nutrients and oxygen diffuse through the placenta, and the fetal blood picks them up and carries them through the umbilical cord. Then, the wastes that result from the fetus metabolizing the nutrients and oxygen are carried back out through the umbilical cord and diffused into the placenta. The mother's blood picks up the wastes from the placenta, and her body excretes them. Geez, moms start cleaning up after their kids before they're even born!

Both the fetus and the placenta are enclosed within the *amniotic sac,* a double-membrane structure filled with a fluid matrix called *amniotic fluid.* The fluid keeps the temperature constant for the developing fetus, allows for movement, and absorbs the shock from the mother's movements.

Humans are like all animals in that they require a watery environment for early development. Evolution has found ways for animals to breathe air at later stages of development, but not the gestational stages. The amniotic sac is the mammals' way to gestate their young in a watery fluid while remaining on land. The placenta evolved in early mammals and is a defining characteristic of all mammals since then. Other land vertebrates (birds and reptiles) evolved other structures in response to this evolutionary imperative.

Dividing development into trimesters

Officially, by convention, Day 1 of a pregnancy is the first day of the woman's previous menstrual period. Obviously, she wasn't pregnant on that day, nor for numerous days thereafter. But it's easier to be sure about the start date of a menstrual period than about the day of ovulation or fertilization or implantation, so that's the custom that doctors follow. Then, by convention, doctors count ahead 280 days to arrive at the *due date,* the date on which, if all pregnancies and all babies were alike, the birth would take place. The 280 days, usually expressed as 40 weeks, are the human *gestational period* (length of pregnancy). This period is divided, again by convention, into three trimesters, though nothing specific marks the transition from one to the next.

The following sections provide an overview of the development of a fetus's organs through the three stages of pregnancy. See the "Prenatal Development" color plate in the center of the book to get an idea of what a developing fetus looks like.

The first trimester

All the body's organs begin development in the first trimester. The circulatory system forms from small vessels in the placenta three weeks after fertilization. The heart begins to beat.

During the second month, the organ systems continue to develop, and the limbs, fingers, and toes begin to form. The embryo starts to move at the end of the second month, although it's still too small for the mother to feel its movements. Also during the second month, ears, eyes, and genitalia appear, and the embryo loses its tail and begins to look less like a sea horse and more like a human.

At the end of the first trimester, the fetus is about 4 inches long (10 centimeters) and weighs about an ounce (28 grams). The head is large, and hair has begun to grow. The intestines are inside the abdomen, and the urinary system (kidneys and bladder) starts to work.

 If you're counting weeks and feel you're losing track, remember that the trimesters of pregnancy are measured from Day 1 of the mother's last menstrual period. This date is around two weeks earlier than the date of fertilization. The embryonic stage is the second and third month of pregnancy.

The second trimester

The fetus, with all its systems in place, continues programmed development in the second trimester. *Ultrasound imaging* shows the skeleton, head details, and external genitalia. Bone begins to replace the cartilage that formed during the embryonic stage. At the end of the second trimester, the fetus is about 12 to 14 inches (30 to 36 centimeters) long and weighs about 3 pounds (1.4 kilograms).

The third trimester

The fetal development program speeds up in the third trimester. The fetus, with its systems developed, continues to grow in size. Subcutaneous fat is deposited, which serves as a critical energy reserve for brain and nervous system development.

Near the end of the third trimester, the fetus positions itself for birth, turns its head down, and aims for the exit. When the fetus's head reaches the *ischial spines* of the pelvic bones (see Chapter 4), the fetus is said to be *engaged* for birth. (See Figure 15-2.)

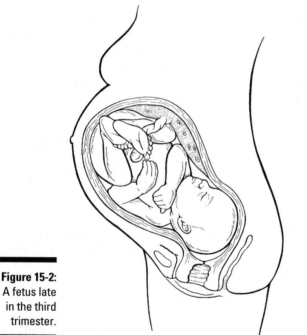

Figure 15-2:
A fetus late
in the third
trimester.

The Human Life Span

Mammals have a general pattern of development from birth to senescence, and humans closely follow that pattern in most ways. In addition to following a typical sequence of events (live birth of dependent young, late development of the reproductive system, hairiness increasing with age, and so on), the mammalian pattern has some "rules" that tie the size of the animal with the pace of development. Generally, the larger the mammal (typical adult size range), the longer the development period. Humans are right where you would expect them to be on this curve.

Every species of mammal has a species-specific version of the pattern, of course, encoded in the species genome. The human species genome encodes a long infant dependency period and, for females, an extraordinary prolongation of life beyond her reproductive years.

Changes at birth

The timing of birth is a compromise between the anatomical needs of a large brain and those of a bipedal gait. The fetus is born a little earlier than would probably be ideal from the point of view of its development, but the pelvis of the adult female has grown narrower and less flexible to support a redistribution of weight and bipedal mobility. Evolution has favored a compromise: a period of a few weeks when the baby, though fully separate from the mother, is still developing in ways that other mammals have completed before birth. Evolution continues to support this compromise relentlessly.

The newborn undergoes a number of changes at birth to allow it to survive outside the womb and adapt to life in a new cold, dry, and terribly immediate environment.

- **The first breath:** The fetus exchanges oxygen and carbon dioxide through the placenta. At birth, the newborn's lungs are filled with amniotic fluid and are not inflated. Within about ten seconds after delivery, the newborn's central nervous system reacts to the sudden change in temperature and environment by stimulating the first breath. The lungs inflate and begin working on their own.

- **Thermoregulation:** Almost as quickly as the lungs start functioning, temperature receptors on the newborn's skin mediate the generation of metabolic heat by muscle action (shivering) and the burning of stores of *brown fat,* a type of fat found only in fetuses and newborns.

- **Digestive system:** The newborn's digestive system starts to work in a limited way immediately after birth. A newborn can digest colostrum and breast milk. Even so, the digestive system can take several weeks to settle down to efficient functioning.

  In the fetus, the liver acts as a storage site for sugar (glycogen) and iron. After birth, the liver begins to takes on its other functions. It begins breaking down waste products such as excess red blood cells.

- **Urinary system:** The fetus's kidneys begin producing urine by the end of the first trimester of pregnancy. The newborn usually urinates within the first 24 hours. The capabilities of the kidneys increase sharply through the first two weeks after birth. The kidneys gradually become able to maintain the body's fluid and electrolyte balance.

- **Immune system:** The immune system begins to develop in the fetus and continues to mature through the child's first few years of life. The newborn is fully exposed for the first time to pathogens in the environment — the placenta doesn't block all infectious organisms or toxins. Some of the mother's antibodies, too, reach the fetal circulation. Colostrum and breast milk contain components that support the development of the infant's immune system.

Infancy and childhood

The long human infancy and childhood is one of the wonders of the biological world.

All organ systems grow and develop in infancy almost as rapidly as during fetal development. (All except the reproductive system. See the upcoming "Adolescence" section.) The physical development milestones in the first year alone would take a full book this size to describe. Just looking at a year-old infant and then at the infant's photos at birth shows you a lot.

Most infants double or triple their birth weight in the first year. Besides adding size to all tissues and organs, the new cells differentiate elaborately and add functionality, according to the individual's genomic scheme. The baby's skeleton changes size, proportion, and composition and takes on all the attributes of bipedalism. The mouth acquires the subtle muscle control to form words and kisses. The baby starts applying the characteristically human opposable thumb to everyday tasks (picking up toys and throwing them down again) in the second half of the first year. The brain grows in size and, just as important, the number and complexity of connections grow astronomically.

A toddler child (between ages 1 and 3) develops sphincter control. Social life begins. Vigorous play coordinates the development of the musculoskeletal and nervous systems. The mechanisms of homeostasis gradually strengthen.

Overall, from infancy to adolescence, the human child becomes bigger, stronger, and smarter every day. The child steadily gains control of the body on a conscious and physiological level. The child is typically fluent in at least one spoken language by age 6. A fully human degree of hypersociability is frequently evident in preadolescents (about age 10 to puberty).

And yet the juvenile's dependency goes on and on. Although most children are weaned before age 2 and can walk fairly long distances by age 10, they are, for the most part, hopeless at obtaining food and shelter. Their caregivers do that for them.

Their physical and mental development is devoted, instead, to mastering the unique aspect of human physiology called "culture." That takes these brilliant human children about 20 years of intense study. During the period of human evolution, the survival of any individual was dependent primarily on the survival of the individual's kin group. Participating effectively in the culture has always been the best way for humans to increase the likelihood of their own survival and the survival of those who carry their genes.

Adolescence

One organ system doesn't undergo much development in infancy and childhood: the reproductive system, which remains in a state of suspended animation from early in fetal development until *puberty,* the first part of the development stage called *adolescence* (*adol-* means "adult").

During puberty, the reproductive systems of males and females emerge from the suspended state. This usually happens sometime between age 11 and 14 for girls; in boys, it begins a couple of years later. Puberty ends when the reproductive system is mature enough to produce viable gametes. (That is, reproduction becomes physically possible.) Hormones play a very large part in this complex process, and getting them all to function smoothly together typically takes a few years.

The hormones produced during adolescence cause physical and neurological changes in female and male bodies. The frequent hormone surges cause acne, among other miseries, and they often trigger emotionally uncomfortable mood swings. The brain's "executive" functions (judgment, impulse control, and risk-assessment) are frequently impaired.

Growth spurts are common during puberty, and growth continues through adolescence. Bones lengthen, muscle mass increases, and all organs reach near-adult size. The primary and secondary sex organs grow and mature. Fat and muscle are redistributed. Adolescents can maintain high levels of physical activity, fueled by the output of millions of new mitochondria. Adolescent sleep-wake cycles may be strikingly different from those of children and adults.

Female puberty

In females, the ovarian and uterine cycles begin (see Chapter 14), which becomes evident when menstruation starts. After the female is ovulating, pregnancy is possible. The female breast develops during puberty.

Other changes that occur in females during puberty include growth of hair in the *axillary* (armpit) and *pubic regions,* and the development of the female fat distribution pattern: more on the hips, thighs, and breasts.

Male puberty

Hormones from the *anterior pituitary* in boys allow them to produce testosterone, and as a result, to begin to develop sperm regularly. Testosterone has certain effects, such as initiating the growth of facial and chest hair, building lean muscle, causing hair to develop in the axillary and pubic regions, and making hair on arms and legs dark and coarse. The vocal cords lengthen, which causes deepening of the voice. The penis and testes enlarge. Males develop broader shoulders and narrower hips than females.

Young adulthood

Typically, young adulthood is a time of good health and resilience. Many people complete their parents' bid for evolutionary success during this stage. (That is, they become parents themselves.) Energy levels can remain high all through the 30s, 40s, and 50s for some people. However, after growth is completed, energy requirements decrease, and adults must decrease their caloric intake to avoid accumulating body fat that can put long-term stress on several organ systems.

The gradual physical decline of senescence actually begins during these years. Influenced by genetic and environmental factors, arteries begin to accumulate damage, slightly more bone is lost than is made, and the same thing happens with the structural proteins of the skin. Muscle mass declines slowly but not imperceptibly. Damage accumulates from repetitive injuries, bad habits, and bad genes.

Anatomy and culture

Just about everything about human development is more or less "typical" for mammals: The young are born alive, ready to breathe air and suck milk from their mothers' mammary glands. As for size, human infants are within the very wide range of normal for a mammalian infant (between a shrew and a whale). Some mammalian infants are ready to run with their herd at one day of age, but lots of mammal babies are pretty helpless.

Only one thing is very odd about human development *(ontogeny)*, and that's the exceptionally long duration of immaturity. Compared with any other known species, even humans' closest evolutionary relatives, human babies take a long time to grow up. The evolution of this remarkable trait has been the subject of much scientific research and speculation at least since Darwin in the fields of anthropology, evolutionary biology, psychology, and genetics.

The consensus is that the long human childhood extends the brain's development, giving it time to absorb the culture's complexities. Humans are a hypersocial species, and learning culture is an absolute requirement for survival. The period between weaning and adulthood is strongly devoted to learning spoken language, nonverbal communication, and other aspects of culture. As we discuss in Chapter 9, the very food that humans eat must be prepared within the social group.

Anatomy and culture have been developing together for a few million years. All organ systems have adapted to culture, and the human animal is fully hypersocial. Humans have never been independent of the social group.

Middle age

For most people, young adulthood ends sometime in the 40s or 50s. The cellular cycles that replace cells and repair tissues slow down. Loss of bone and muscle mass accelerates. However, most of the time, these losses aren't critical and can be mitigated by medical therapies or lifestyle adjustments (diet, exercise, sunscreen, and so on).

For men, reproductive capacity diminishes. For women, it disappears altogether at *menopause,* usually around age 50. Production of some hormones diminishes, triggering anatomical and physiological changes great and small.

The brain, however, continues to develop, cognitively and in many other ways. Some recent brain research shows that an older brain thinks better about some things than a younger brain does, including making financial decisions, exercising social judgment (intuitive judgments about whom to trust), and recognizing categories. The older brain is better at seeing the proverbial forest and following the gist of an argument. These are typically the peak years of professional or occupational achievement. In addition, across all occupations and ethnicities, a sense of well-being peaks as people reach middle age.

An extraordinary aspect of human development is the length of this period. Human females commonly outlive their fertility by three decades — a third of the life span in a stage of life that almost no other animal experiences! Many researchers see a matrix of evolutionary cause-and-effect conjoining natural selection and cultural evolution in the very existence of your grandma.

Growing creaky

Life expectancy beyond the reproductive years is dependent to a great extent on genes.

The developments of *senescence* (growing old) are gradual and diffuse. Body parts don't work quite as well as they used to, and they just keep getting worse. For some people, this is a brief stage of life after a long, healthy middle age. Others aren't so lucky.

In particular, the immune system doesn't work as well as it used to, and malformed cells that would have been immediately eliminated at age 35 now may evade immune surveillance and become cancerous.

Age-related changes in the large arteries, as well as cumulative damage to the smaller vessels, can bring about problems with blood pressure.

Inactivity and chronic overconsumption of calories have their worst effects now, in all the major systems.

Still, the brain continues to develop. Research has been shown repeatedly that, under the right conditions, the brain continues to produce new cells and make new connections among neurons in adults as old as 100.

Table 15-1 lists some age-related changes to the body's systems.

Table 15-1   Age-Related Changes to the Body's Systems and Associated Health Implications

Body System	Change	Implications
Circulatory System (see Chapter 9)	Heart increases in size.	Increased risk of thrombosis and heart attack.
	Fat is deposited in and around the heart muscle.	Varicose veins develop.
	Heart valves thicken and stiffen.	
	Resting heart rate decreases.	
	Maximum heart rate decreases.	
	Pumping capacity declines.	
	Arterial walls fill with plaque and don't stretch well.	
Digestive System (see Chapter 11)	Loss of teeth.	Increased risk of hiatal hernia, heartburn, peptic ulcers, constipation, hemorrhoids, and gallstones.
	Peristalsis slows.	
	Diverticulosis.	Colon cancer and pancreatic cancer increased in elderly.
	Liver requires more time to metabolize alcohol and drugs.	

Body System	Change	Implications
Endocrine System (see Chapter 8)	Glands shrink with age.	Thyroid disorders and diabetes can occur.
Immune System (see Chapter 13)	Thymus gland shrinks with age.	Cancer risk increases.
	Number and effectiveness of T lymphocytes decrease with age.	Infections more common in elderly.
		Autoimmune diseases (such as arthritis) increase.
Integumentary System (see Chapter 6)	Epidermal cells are replaced less frequently.	
	Fibers in the dermis thicken, and less collagen is produced, reducing elasticity.	Skin loosens and wrinkles.
	Adipose tissue in face and hands decreases.	Sensitivity to cold increases.
	Fewer blood vessels and sweat glands.	Body is less able to adjust to increased temperature.
	Melanocytes decrease.	Hair grays, skin becomes more pale.
	Number of hair follicles decreases.	Hair thins.
Muscular System (see Chapter 5)	Mass and strength decrease.	
	Muscle tissue deteriorates and is replaced by connective tissue or fat.	
	Fewer mitochondria in muscle cells.	Endurance decreases due to fewer mitochondria.
	Cardiovascular and/or nervous system changes.	Function can decrease due to cardiovascular or nervous system changes.

(continued)

Table 15-1 *(continued)*

Body System	Change	Implications
Nervous System (see Chapter 7)	Brain cells die and are not replaced.	Learning, memory, and reasoning decrease.
	Cerebral cortex of brain shrinks.	Reflexes slow.
	Decreased production of neurotransmitters.	Alzheimer's disease occurs in elderly people.
Reproductive System (see Chapter 14)	*Females:* Menopause occurs between 45 and 55 years of age and causes cessation of ovarian and uterine cycles, so eggs are no longer released, and hormones such as estrogen and progesterone are no longer produced.	Osteoporosis, wrinkling of skin, and increased risk of heart attack
	Males: Possible decline in testosterone level after age 50; enlarged prostate gland; decreased sperm production.	Vascular (or other) problems can cause impotence in males.
Respiratory System (see Chapter 10)	Breathing capacity declines.	Risk of infections such as pneumonia increases.
	Gas exchange and lung volume decreases due to thickened capillaries, loss of elasticity in muscles of rib cage.	
Skeletal System (see Chapter 4)	Cartilage calcifies, becoming hard and brittle.	Osteoporosis risk increases.
	Bone resorption occurs faster than creation of new bone.	More time is required for bones to heal if they break.
Urinary System (see Chapter 12)	Kidney size and function decreases.	Kidney stone risk increases.
	Decreased bladder capacity.	Incontinence.
	Enlarged prostate gland in men.	More frequent urination urges.
		Urinary tract infections more likely.

Part V
The Part of Tens

The 5th Wave · By Rich Tennant

"I'm really pumped for this test. I can feel the watchamacallit flowing through those little round tube things in my body."

In this part . . .

As in many aspects of life, you can't study anatomy and physiology without knowing something about related science concepts. To help in that regard, Chapter 16 focuses on some important chemistry and physics concepts that are tied to the body and its processes.

Chapter 17 offers you some fascinating details about the human body, both in how it functions and in how it differs from other species.

Chapter 16

Nearly Ten Chemistry Concepts Related to Anatomy and Physiology

In This Chapter

▶ Understanding the nature of energy

▶ Getting a handle on fluid properties, osmosis, and polarity

▶ Transferring electrons with redox reactions

*B*iology is a very special application of the laws of chemistry and physics. Biology follows and never violates the laws of the physical sciences, but this fact can sometimes be obscured in the complexity and other special characteristics of biological chemistry and physics.

This chapter contains a review of some of the principles of chemistry and physics that have special application in anatomy and physiology. Some of these principles overlap — for example, probability is one factor that drives the process of osmosis. Although what follows are oversimplified explanations of very profound and complex matters, we hope they help you better understand anatomy and physiology.

Energy Can Neither Be Created Nor Destroyed

The *first law of thermodynamics* is that energy can be neither created nor destroyed — it can only change form. Throughout any process, the total energy in the system remains the same. This law is one of the fundamental concepts in physics, chemistry, and biology.

Energy is the ability to bring about change or to do work. It exists in many forms, such as heat, light, chemical energy, and electrical energy. Light energy can be captured in chemical bonds, such as in the process of photosynthesis. In physiological processes, the energy in the chemical bonds of ATP is transformed into work when the chemical bonds are broken — to move things, for example, and to generate heat. (And where did the energy in ATP come from in the first place? Photosynthesis.)

Although the total energy in a system always remains the same, the energy available for biological processes may not remain the same. Biological processes can use energy only in certain specific forms. A physiological process that uses ATP doesn't use all the energy stored in those chemical bonds, but the leftover energy isn't in a form that can be used in another physiological process. It is "lost" to physiology, mostly as heat flowing out into the surrounding environment.

Everything Falls Apart

In our universe, energy is required to create "order" — for example, to build the atomic and molecular aggregations we call "matter" or "stuff." Without continuous input of energy (maintenance), stuff falls apart. No news here for dwellers in the real world. As a physicist might put it, all systems tend toward increasing *entropy* (disorder). This is the *second law of thermodynamics*.

Energy always moves from a point of higher concentration to a point of lower concentration, never the reverse. For example, where two adjacent objects are of different temperatures, heat flows only from the warmer object (higher energy) to the cooler object (lower energy). A state of order contains more energy than a state of disorder because of the energy that went into building the state of order. Energy flows outward into the relative chaos of disorder.

Because living systems are highly ordered, the implications of the second law of thermodynamics are profound for physiology. The law means that *physiological homeostasis* (the maintenance of order) is an active process that requires energy. The energy that must be applied to drive any physiological process comes from releasing the chemical bonds in ATP. It means that physiological reactions proceed in only one direction — they aren't reversible (unlike, say, sodium and chlorine ions that go into solution in water and then reorganize back into salt crystals spontaneously when the water is removed).

The ultimate physiological implication of the second law is the inevitability of death.

Everything's in Motion

Particles in a solution fly around constantly and collide with one another all the time. The higher the temperature, the more frequent and harder the collisions. Chemists and physicists call this *Brownian motion*. It's the reason why any reaction that *can* happen *will* happen, because (most of) the particles required for the reaction will collide sooner or later. (But see the "Probability Rules" section that follows.)

Brownian motion is also a mechanism of entropy. Each of the molecular collisions converts energy in the molecules to heat, in which form the energy is transferred to the surroundings.

Probability Rules

Everything that can happen will happen — *some of the time.* Other times, it won't. The proportion of times it *does* happen depends on a lot of factors. If a solution contains large numbers of each of two molecules required for a reaction, the different types will collide frequently. The higher the solution's temperature, the more frequently those collisions will be hard (energetic) enough to facilitate the reaction. But almost never will every possible reaction actually happen. Just by chance, some of these molecules won't meet up with their counterpart molecule. That's life. The chance, or randomness, can be quantified as *probability.* As with this hypothetical reaction, so with everything else related to biology and physiology: Probability, not certainty, rules.

By the way, the existence of life itself is highly improbable. And the probability of the existence of the uniqueness that is you is more improbable still.

Polarity Charges Life

A molecule is said to be *polar* when the positive and negative electrical charges are separated between one side of the molecule and the other. For example, a molecule of water is polar because the unequal sharing of electrons between the hydrogen and oxygen atoms concentrates the negative charge on the hydrogen atoms. So the water molecule has a positive charge at one end and a negative charge at the other, like a magnet. It attracts and holds other polar molecules. Methane is *nonpolar* because the carbon shares the electrons with the four hydrogen atoms uniformly.

Polarity underlies a number of physical properties of a substance, including surface tension, solubility, and melting and boiling points. In physiology, polarity strongly determines which molecules form bonds and which don't.

The strong polarity of water is a significant factor in physiology. It accounts for the ability of so many substances to dissolve in water. One could argue that the strong polarity of the water molecule, especially in the liquid phase, is a factor underlying the existence of life on earth.

Oil and Water Don't Mix

More specifically for the study of physiology, lipid and water don't mix. Living cells use this principle to control the flow of substances in and out.

Lipids are a large and varied group of organic compounds, including fats and oils. All lipids have *hydrophobic* portions to them — that is, they don't mix with water. Why not? Because a lipid is nonpolar, so it can't form bonds with water. Water molecules push nonpolar molecules aside to get closer to other polar molecules.

Visualize a party where some people gather around the TV to watch a game and others congregate in the kitchen. The game-watchers are the polar entities, supporters of one team or the other, and the other folks are the nonpolar entities, sharing an interest in nonpolar subjects like biological development and the effect of heat on complex organic substances. To belabor the analogy: The polar entities, after they've taken their positions close to other polar entities, tend to maintain their state and position relative to the entities they've bonded with, while vibrating in place. The nonpolar entities move around relative to one another, and they hold and release nonpolar side-chains (children) easily and often. A different set of polar entities (the teenagers) conduct different physiological processes in seclusion from both the nonpolar and the other polar entities.

The cell membrane is made up of lipid macromolecules that have nonpolar (hydrophobic) subunits at one end and polar (hydrophilic) subunits at the other. These lipid macromolecules spontaneously line up, some with their polar ends "facing" the watery cytoplasm inside the cell and some facing the watery fluid outside the cell; their nonpolar ends line up in between. The double layer of lipids, called the *lipid bilayer,* blocks polar solutes (for example, amino acids, nucleic acids, carbohydrates, proteins, and ions) from diffusing across the membrane but generally allows for the diffusion of hydrophobic molecules.

Fluids and Solids

Physiological processes, generally speaking, take place in fluids, and the properties of fluids are very important in these processes.

In everyday conversation, "fluid" means "liquid," something that's usually water-based, like juice, broth, or tea. In physics and chemistry, a watery solution is one kind of fluid, whether it's one you'd care to drink or not. Air is another kind of fluid. Fats are fluids, even when they're solid: Butter is exactly the same substance whether cold or warm, and so is every other form of fat. Technically speaking, glass and pure metals are fluids!

Salt, in contrast, is a solid. Salt (NaCl) crystals flow out of their containers in every kitchen and dining room, yes, but that doesn't make salt a fluid. It's got to do with the molecular structure. In solids, atoms are tightly packed together in a geometrically precise formation called a *crystalline lattice.* Sodium chloride is the model for this: Equal numbers of sodium and chlorine ions, each linked to six other ions, all pull each other in as tightly as the forces of polarity (electrical charge) require and allow. Solids are rigid at the molecular level; once bound together in a crystalline lattice, every atom in the molecule remains in place relative to its surrounding molecules. (That is, until the moment the forces of polarity pull them apart and into solution.)

In fluids, things move around more. Components come together in various ways — carbon dioxide and molecular oxygen dissolve from air into water and back into air. Fluids take the shape of their container. Air flows into and fills your alveoli. A watery mass in your stomach changes shape with every churning contraction. Gaseous fluids can be compressed. However, the compressibility of liquids is very limited because the forces of polarity are already pulling the water molecules together just about as tightly as they can be made to go.

Diffusion and Osmosis

Besides recycling energy back into the universe, Brownian motion (see the "Everything's in Motion" section earlier in the chapter) distributes molecules of various substances throughout the universe, and within each tiny subunit of the universe called a living cell. The cell has adapted two physical processes, called *diffusion* and *osmosis,* to regulate the entry and exit of solute and solvent (water) molecules, respectively, and thus control the composition of the cytoplasm. Note that both diffusion and osmosis are chemical processes, not biological ones, although they are exploited by every cell of every organism in existence.

Diffusion is a chemical process in which a solute molecule moves *passively* (without input of energy) through a solution. In chemistry terms, molecules of solute move through the solution (Brownian motion), resulting (by probability) in a net migration of molecules from a region of high concentration to a region of low concentration. This, of course, induces continuous change in the concentration gradient. When the concentration of the solute is equal throughout the solution, Brownian motion continues but there is no net change in the distribution of solute molecules. Diffusion has brought about equilibrium in that solution with respect to that solute.

If two solutions are separated by a membrane that is permeable to a particular solute, the system reaches equilibrium with respect to that solute when its concentration is equal throughout both solutions. Essentially, with respect to the solute, there is no separation between the solutions.

Osmosis is similar to diffusion, except that there is *always* a semipermeable membrane involved, and it is solvent molecules that cross the membrane and consequently equalize their concentration on both sides. In physiological processes, that means water molecules move in and out of cells to maintain an equilibrium between the cell and its surroundings. Osmosis is important in biological systems because the movement of some types of molecules, but not others, across the cell membrane is key to many physiological processes. In general, biological membranes, including the cell membrane, are impermeable to large organic solutes such as polysaccharides, but permeable to water and small, uncharged solutes.

Redox Reactions Transfer Electrons

The concept of *reduction-oxidation* (or *redox*) reactions is basically this: An electron is transferred from one chemical entity (atom or molecule) to another. The entity that receives the electron is said to be *REDuced*. The entity that releases the electron is said to be *OXidized*. In a redox reaction, the reduction of one entity is always balanced by the oxidation of another. The entities are called a *redox pair*. The redox reaction changes the *oxidation state* of both entities. In some cases, the oxidized entity undergoes another reaction to acquire another electron. Note that this isn't a simple reversal of a redox reaction but a new reaction that involves another electron "donor" and frequently requires an enzyme catalyst.

Here's a clever mnemonic to help with the terminology: OIL RIG — Oxidation Is Losing, Reduction Is Gaining (electrons, that is).

In biological systems, redox reactions are highly controlled and very important. Chemical energy is stored in electron bonds and released (made available for work) by redox reactions. Redox reactions are commonly part of *signaling pathways*. A change in the oxidative state of some molecules carries information. A change in the oxidation state of an entity can affect its polarity, which, in turn, affects its solubility in water and thus its ability to enter or leave a cell through the cell membrane. An entity that becomes more soluble can also become more available metabolically, which can be very important for some metal ions like iron and calcium.

Redox reactions play a crucial role in both of the most important reactions in biology: photosynthesis and cellular respiration. *Photosynthesis* is the reduction of carbohydrate to glucose and the oxidation of water molecules to molecular oxygen, using light energy. (Molecular oxygen is O_2 — the oxygen atoms from two water molecules joined together.) In *cellular respiration,* glucose is oxidized to CO_2 and O_2 is reduced to water.

Chapter 17

Ten Phabulous Physiology Phacts

This merest smattering of the everyday miracles of the anatomy and physiology of this one species inspires awe at the power of evolution's forces.

Unique to You: Hand, Hand, Fingers, Thumb

There they are, down at the end of your arms, one on each side, a matching pair, unique to you in their details. Other humans have them, too, but no other animal has them. Your hands are certainly far removed from the front paws more or less typical of mammals, and they're greatly specialized compared even to those of other primates, including man's closest evolutionary relatives.

One specialization is the *opposable thumb,* which is a thumb that can touch each finger on the same hand. (Go ahead — try it now.) Along with that, the human thumb is *prehensile,* meaning capable of grasping. This anatomy underlies the development of manual dexterity and fine motor skills in humans. The prehensile, opposable thumb makes possible tool making, hunting and gathering, textile and metal crafts, art, writing, cooking, and possibly the very existence of human culture.

Nothing's Better than Mother's Milk

All the copious research that has been done on the topic over many years has pointed in the same direction: The best nutrition for a human baby is human milk. Human milk is a complex mixture of over 200 different components, and no other substance produced in another animal or yet in a laboratory matches its ability to meet the needs of a human infant. A baby doesn't necessarily need its *own* mother's milk, though. The composition of milk is remarkably consistent — the age, health status, diet, or geographic location of the mother notwithstanding.

Like all foods, the core components of milk are carbohydrates, proteins, and fats. The proportions of these components in the watery matrix and the specific carbohydrate, protein, and fat molecules in a species' milk are precisely adapted to that species' infants' needs. Human milk is the ideal nourishment for a slow-growing, warm-blooded newborn ape: low in protein (rat's milk has 12 times as much), high in lactose, a sugar with twice the energy content of glucose, and high in the essential fatty acids required for neural development. (As we discuss in Chapter 15, the configuration of the human female pelvis drives the birth of a baby with a relatively undeveloped brain.)

Human milk contains many other substances that affect nutrition and development in different ways. Milk and its precursor, *colostrum,* essentially lend the baby part of the mother's immune system until it can make its own: B cells, T cells, neutrophils, macrophages, and antibodies (see Chapter 13). Lactoferrin and iron-binding protein cause the scant iron in milk to be fully absorbed across the digestive membrane. Milk also has human hormones and growth factors that are believed by some to be required to optimize the development of the brain and other organs.

It's Apparent: Your Hair Is Different

Along with milk, hair is a defining characteristic of class Mammalia. This class has found hair to be an adaptable accessory and put it to many uses: mechanical protection, UV protection, thermoregulation, sexual selection, social signaling, and waterproofing, among others.

Among the great apes, the genus Homo is distinguished by an apparent lack of hair. Evolutionary theorists suggest that the ape forebears of Homo were about as hairy as gorillas, who put their hair to use in all the aforementioned ways. What could have driven so drastic a change in so useful an accessory?

Anatomists note that humans haven't "lost" their hair — the skin is covered with hair follicles at about the same density as other apes. But the hair itself is different. Most of it is short and fine. Head hair is longer and coarser than body hair. Head hair and body hair may be curly. (No other ape has curly hair anywhere.) It may be lightly pigmented or apigmented. How does Homo escape a predator or thermoregulate under that? Maybe he runs away on his long, hairless legs, cooled by a steady stream of water from newly-evolved glands on his hairless chest and arms. Homo evolved evaporative cooling for a hunter's life on the hot, dry, equatorial savannah.

A typical mammal has more or less dense hair covering the epidermis. As far as thermoregulation goes, wearing a warm blanket is good for retaining heat but bad for dissipating it. Many mammals, including many large predators, rely on panting and certain patterns of behavior to thermoregulate. For example, on hot days, some mammals lie in the shade near the watering hole. The hunter who could be active when others were torpid escaped predation and ate well.

But what about the cold nights? An agile thinker could put his prey's skin and fur to its rightful purpose, keeping a mammalian body warm and safe from abrasion and mechanical shock.

The Almonds of Emotion

The *amygdalae* are paired structures in the middle brain almost exactly the size and shape of almonds, whence they get their name. They've been attracting attention in neuropsychiatry for 60 years — for pretty much the entire history of neuropsychiatry.

Early research found that neural circuits through the amygdalae connected the midbrain, among the most primitive brain structures, to the frontal cortex, the most recent and advanced. These circuits are part of the limbic system and are thought to be critical in regulating emotion and in guiding emotion-related behaviors.

The amygdalae have been associated clinically with a range of mental and emotional conditions, including religious rapture, depression, autism, and even "normalcy." Physicians have widely and publicly discussed one case in particular — a woman whose amygdalae are partly nonfunctional. This patient is incapable of experiencing the emotion of fear. The doctors have tried everything, not just for research purposes but because a total lack of fear is a maladaptive trait; it threatens her well-being and survival. This patient has been injured and victimized in situations that normal, healthy fear would have kept her far away from.

Some neuropsychiatric researchers theorize that the amygdalae evolved as part of a protective mechanism. Probably very early on in the evolution of vertebrates, the amygdalae reacted to change in the chemical environment, moving the organism away from toxic substances. New organisms adapted this functionality to perceive and respond to new stimuli in the environment. Stepping back from a chemical spill is still a model of appropriate, fear-based reflexive behavior. The circuit between the frontal lobes and the amygdalae is what generates such survival-enhancing behavior.

You Smell Well!

It is often said that, compared with other animals, the human nose is poor at gathering the information available from volatile molecules in the environment. Just how does the human olfactory sense compare with that of other animals? Take a look at the evidence.

As with other mammals, the human olfactory structures are located at the interface of the brain and the airway. Specialized neurons called *olfactory neurons,* actually protuberances from the brain, sit right on the border of the nasal passages, behind and slightly above the nostrils.

An olfactory neuron bears olfactory receptors on its plasma membrane. Compared with the strict antigen-specificity of the receptors on lymphocytes, olfactory receptors operate in a fuzzier way. An olfactory receptor recognizes a certain chemical feature of an odor molecule, but that feature is present in numerous kinds of odor molecules. The receptor can bind any odor molecules that have that feature. Thus, humans don't have a single receptor for "coffee" or "lavender" or "wet dog." They have many receptors for many kinds of molecules released into the air and drawn into the nose. The brain assembles its olfactory perception of the environment by aggregating the signals from the various receptors. The process is similar to that of vision. Odor recognition is like object recognition based on aggregating many different impulses from the retina.

Molecular biology experiments in the early 1990s led to the identification and cloning for research purposes of a large family of olfactory receptors. The experiments showed that the family of genes coding these receptors is the largest in the mammalian genome. Some animals have more than 1,000 different receptors; humans have about 450. One out of every 50 human genes is for an odor receptor!

Based on information from the complex set of olfactory receptors, the brain can determine the concentration of an environmental odor and distinguish a new odor signal from the background odor noise.

The 450 olfactory receptors give humans their very sophisticated palates, allowing them to enjoy the possibly hundreds of different flavors of a savory meal. Apart from entertainment, this faculty may have helped humans discover new food sources as they moved into new climates and environments.

What a Small Mouth You Have, Grandma

Compared with other apes, the human *buccal* apparatus (mouth) is puny. The capacity of the male adult human mouth is about the same as that of the male adult chimp, who is only about two-thirds the man's size. The chimp can open twice as wide, showing teeth about twice as big. The man has nothing like the chimp's enormous, muscular lips for expressing juice from fibrous foods. The human *temporalis* and *masseter* muscles are small and weak compared with the chimp's powerful jaw muscles. Yet, with all this extra power, the chimp spends about six hours each day masticating (chewing food), whereas the human hunter-gatherer spends less than one hour. Yes, the puny mouth of a human chews food much more effectively than the mighty mouth of a chimp.

The pattern continues with the rest of the human digestive apparatus. Compared with other primates and adjusted for size, the human digestive system's structures are smaller, proportionately, than that of 97 percent of primates. In particular, the human colon is only 60 percent as large as expected for a primate of human size.

Yet here you are, extracting ample fuel and nutrition through massively scaled-down anatomical structures and strikingly abbreviated physiological processes. How does that work? Some evolutionary theorists answer that question in two words: cooked food.

Cooking makes food more nutritious — higher in physiologically available calories, carbohydrates, and protein. Moreover, and possibly more significant, the processes of digesting cooked food are very energy-efficient. It takes perhaps 10 percent less energy to masticate and digest cooked versus uncooked food. This energy bonus is certainly enough to give an individual and a species a survival advantage in many environments. Under this selection pressure, a group of apes evolved into a genus that is absolutely dependent for survival on cooked food.

Microbes: We Are Their World

For thousands of millions of tiny creatures, your gut is the only universe they know. They live and die in that warm, moist, nutrient-rich, immune-protected environment. They work almost every minute of their lives, providing a service to their community and their universe and abiding by the laws of thermodynamics. These good citizens of the gut are adapted specifically to that environment, in the way of symbiotic organisms, and can survive nowhere else.

The internal tissues — blood, bone, muscle, and the others — are normally free of microbes. But the surface tissues — the skin, the digestive and respiratory tracts, and the female urogenital tract — have distinctive colonies of symbiotic microorganisms. The term *symbiosis* (adjectival form, *symbiotic*) describes a more or less cooperative and reciprocal relationship between organisms and species. Symbioses are, by definition, good for all. That brings us back to your supporting role as Mr. or Ms. Universe.

The microbe colonies derive from the "host" (a particular human body) a steady supply of nutrients, a stable environment, and protection. The host gets help with some tricky digestive tasks, stimulation of the development and activity of the immune system, and protection against colonization by other, less-well-adapted (pathogenic) microbes. The number and species in a microbe colony is influenced by various characteristics of the host, including age, sex, diet, and genetic makeup.

More than 200 species of bacteria are commonly found in one or another of the human symbiotic colonies. The human body, consisting of about 100 trillion eukaryotic "self" cells, carries about ten times as many prokaryotic (microbe) cells in the intestines alone. Because of their small size, the prokaryotic cells make up only about 1 percent to 2 percent of the body mass.

A question to ponder: If these microbes are, literally, vital to your survival and you are, just as literally, vital to their survival, does that change your understanding of "you" and "them"?

Oxygen Habitually Overreacts

It was the biggest environmental disaster ever on the planet. After life forms had gotten along for a billion years or so, obtaining energy from chemosynthetic processes, some bacterial forms evolved the ability to capture the energy of light itself. This was a huge advantage for them, but unfortunately for all life forms then in existence, photosynthesis has a toxic byproduct: molecular oxygen. Slowly, as photosynthesizing bacteria prospered, overly reactive molecular oxygen replaced moderately reactive methane in the earth's atmosphere.

We can only speculate about what happened next, and some cell biologists have speculated very imaginatively. They started by noticing the strong similarities between certain organelles and some bacteria (between mitochondria and rod-shaped bacteria, for example). Then came the stunning discovery that mitochondria and *chloroplasts* (the photosynthesizing organelles of plant cells) had their own DNA, completely separate from the genomic DNA in the nucleus, and that this DNA was bacterial in character. Gradually, the view has come to be accepted that eukaryotic cells began as symbiotic communities of different types of bacteria sheltering from the damaging effects of atmospheric oxygen inside a bubble of phospholipid bilayer. That was the beginning of a whole new book of life.

Talkin' about Breath Control

You don't have to think about breathing. The steady in-and-out continues while you sleep and go about your daily business. The depth and rhythm adjust to your level of effort. Just climb those stairs; breathing will take care of itself. Many of us would have died young if breathing required constant attention.

Humans are capable of controlling their breath, though. Cetaceans (whales and dolphins) can, too; must, in fact, and some also use breath control to sing. Other animals can't — or at least don't appear to. Canines in chorus are not really controlling their breath.

Humans use breath control to generate speech. Humans can make finely controlled exhalation pass over the vocal cords while the length and thickness of the cords is changing to generate different frequencies of sound. The lips, tongue, glottis, and other structures shape the vibration, allowing you to make those finely distinguished sound symbols called "words" and "syllables." Singing, closely related to talking, requires even finer breath control. We doubt that such a hypersocial, hypercommunicative species as Homo sapiens could have evolved as far as it has without speech and singing.

Various religious practices and breathing disciplines take breath control in another direction. Conventional physiological thinking is skeptical of the idea that the conscious brain can reach "down" and exert control over the quintessentially autonomic processes of breathing. However, brain imaging studies and other experimental results have shown that some people with a long history of regular meditation practice have important differences in their neural systems. Some people believe that systematic breath control exercises can provide benefits to many organ systems: the circulatory, digestive, neurological, and endocrine systems for a start.

Hanging Out with Hemoglobin

Hemoglobin is the predominant protein in the red blood cells (RBCs). Specialized for the transport of the blood gases, hemoglobin has the capacity to transport molecular oxygen (attached to the heme group) and carbon dioxide (attached to the globin portion) simultaneously.

In its oxygenated state, it's called *oxyhemoglobin* and is bright red. In the reduced state it's called *deoxyhemoglobin* and is dark red. Thus the colors of oxygenated arterial and deoxygenated venous blood. (Venous blood isn't blue as it looks through the skin, especially fair skin.)

A single hemoglobin molecule is a complex made up of four ring-structured heme groups attached together by the polypeptide chains of a globin. At the center of each heme ring is a single atom of iron, which forms an unstable, reversible bond with oxygen. The flexibility of the attachment between the heme groups facilitates both the binding of oxygen in the oxygen-rich environment of the pulmonary alveoli and its release in the oxygen-depleted environment of the capillary bed.

A typical RBC contains almost nothing but hemoglobin molecules floating in cytoplasm. Hemoglobin makes up about 97 percent of the dry weight of RBCs, and around 35 percent of the total content, including water. The hemoglobin molecules are put together in RBCs as they mature, before the nucleus and other organelles die off. The heme subunits mostly stay together during the four-month life span of the fully differentiated RBC. When the cell dies, the complex is released and broken up in the liver. The iron is salvaged and recycled in new hemoglobin molecules. The rest of the hemoglobin becomes a chemical called *bilirubin,* which is secreted through the bile and into the large intestine, where it gives feces their characteristic yellow-brown color.

Index

• A •

abdominal muscles, 110–112
abdomino-pelvic cavity, 21, 22
accessory organs to digestion, 222–226, 230–232
acetyl coenzyme A, 31
acetylcholine, 150
Achilles tendon, 115, 116
acidosis, 244
acquired immune deficiency syndrome (AIDS), 267
actin, 105, 106
action potential of neurons, 148, 149
active transport mechanism, 50–51
adenosine diphosphate, 31, 32
adenosine triphosphate. *See* ATP
ADH (antidiuretic hormone)
  in alcohol hangover, 246
  in blood volume, 194, 244
  description of, 160, 164
  in diabetes insipidus, 173
  as peptide hormone, 158
  in water reabsorption, 242
adipose tissue, 67, 125, 236
adolescence, 305–306
adrenal glands, 158, 166–168
adrenaline, 228. *See also* epinephrine
adult stem cells, 62
aerobic pathway, 30, 31
AIDS (acquired immune deficiency syndrome), 267
air, components of, 208
airway disorders, 210–211
albumin, 187
alcohol hangover, 246
alimentary canal, structures of, 214–221
alkalosis, 244
allergy, 265–266
alopecia (baldness), 126, 131–132
alveoli, 204
amine hormones, 159
amino acids, 57–58

amniotic sac and fluid, 300
amphiarthroses, 79
amygdalae, 321–322
anabolic reactions, 28–29, 38
anabolites, 51
anaerobic pathway, 30, 33
anal fistulas, 230
anaphylaxis, 266
anatomical planes, 17, 18
anatomical position, 15, 16
anatomy
  definition of, 9
  of plants and animals, 10
  in science, 10–11
  subsets of, 11
  technical word fragments, 13–14
androgen insensitivity syndrome, 175
anemia, 197, 230
angina, 196
angiogenesis, 299
angiotensin II, 243–244
angular movements of joints, 81
animals, 10, 34–35, 124. *See also* mammals
ankles, 92
ankylosing spondylitis, 95
antagonistic muscles, 104
anterior, definition of, 15
anterior pituitary glands, 163–164
antibodies, 187, 203, 231, 259–260
antibody-mediated immunity, 263
antidiuretic hormone. *See* ADH
antigen, 259
aorta, 182, 186
apocrine sweat glands, 128
apoptosis, 188, 257
appendages, 90–92, 113–116
appendicitis, 228
appendicular body, 17, 20
appendicular skeleton, 88–92
arches of feet, 92
areolar tissue, 67
arms, 90–91, 113–114
arrector pili muscle of hair follicles, 128

arrhythmia, 192, 195
arteries, 182–183
arteriosclerosis, 99–100
arterioventricular node, 190, 191
arthritis, 94–95
asthma, 211, 265
atherosclerosis, 194–195, 196
atlas, 86
ATP (adenosine triphosphate)
  cellular respiration, 29, 31, 32, 33
  at death, 107
  production of, 45
atrial natriuretic hormone, 244
atrioventricular bundle, 190, 191
autodigestion, 54
autoimmune diseases, 94, 265
autonomic system, 141–142, 208
autonomous contraction, 104
autotrophs, 28
axial body, 17, 19–20
axial skeleton
  description of, 81–82
  rib cage, 86–88
  skull, 82–84
  spinal column, 84–86, 93
axis, 86
axon of neurons, 137, 138

• *B* •

B cells, 255, 256–257
back, muscles of, 112
bacteria, 48
balance, sense of, 153
baldness (alopecia), 126, 131–132
basal cell carcinoma, 131
basophils, 255, 261
biceps muscle, 113
bile, 224, 226
bilirubin, 188, 226, 326
binding sites of antibodies, 259–260
biology
  as application of laws of chemistry, 313
  definition of, 9
  specializations within, 11
birth, changes at, 303–304
bladder, 237–238
blastocele, 299

blastocysts, 286, 299
blockage of urethra, 247
blood. *See also* blood flow; blood pressure;
    blood vessels
  circuits of through heart and body, 192–193
  clotting cascade, 40–41, 194–195
  as connective tissue, 186–187
  disorders of, 197–198
  filtering of, 240
  glucose in, 36, 170–171
  pH of, 200, 240, 244–245
  plasma, 187
  platelets, 189
  red blood cells, 39, 187–188, 198, 210,
    325–326
  white blood cells, 189, 203, 254, 256–258
blood flow
  to heart, 180–181
  to kidneys, 236
  in thermoregulation, 35
  through heart, 192–193
blood glucose concentration, 36
blood pressure
  capillaries, 185
  description of, 193–194
  hypertension, 196, 244
blood vessels
  arteries, 182–183
  capillaries, 183–185
  in circulatory system, 181–182
  in dermis, 125, 129
  disorders of, 196–197
  veins, 185–186
blood volume, regulation of, 243–244
blood-brain barrier, 147–148
body temperature, regulation of. *See*
    thermoregulation
bolus, 217
bone
  cells of, 39, 67
  regulation of growth of, 234
  structure of, 76–77
  types of, 78
  vitamin D, 130
bone marrow, 77
bone tissue, 74–75
bowel syndromes, 229–230
brachiocephelic veins, 185
bradykinin, 262

brain
  development of, 304, 306, 307
  in nervous system, 139–140, 142–148
  plasticity of, 297
brain cells, 39
brain stem, 143, 145–146
breasts, 278–279, 287, 320
breathing
  controlled, 208–209, 325
  description of, 205
  first breath of newborns, 303
  measurements of, 206
  normal, 206–207
  under stress, 207–208
broad ligaments, 276
bronchial trees, 204
bronchitis, 211
Brownian motion, 314–315, 317
buccal membrane, 216, 217
buffers, 244
bulbourethral glands, 285

• C •

calcaneus, 92
calcitriol, 234
cancer, 264, 290
capillaries, 50, 182, 183–185, 204
capillary exchange, 183, 184
capsule of kidney, 236, 237
carbon dioxide, 208
carbonic-acid-bicarbonate buffer, 245
carboxyhemoglobin, 188
carcolemma, 103
cardiac cycle, 190–192
cardiac disorders, 195–196
cardiac muscle
  cells of, 39
  description of, 104
  function of, 99–100
cardiac muscle tissue, 102
cardiac veins, 181
cardiovascular system. See circulatory
    system
cartilage, 67, 75
CAT (computed axial tomography), 18–19
catabolic reactions, 28–29, 38
catabolites, 51

catalysis, enzymes in, 58
catecholamines, 167
caudal, definition of, 15
cavities in body, 21–22. See also specific
    cavities
cell biology
  cell cycle, 62–66
  eukaryotic cells, 46–54
  functions of cells, 43–46
  genes and genetic material, 58–61
  macromolecules, 54–58
  tissue, 66–69
cell body of neurons, 137, 138
cell cleavage, 62
cell cycle
  division, 62–63
  DNA replication, 63–65
  interphase, 63
  mitosis, 44, 65–66, 272–273
cell membrane, 48–51
cell walls, 48
cell-mediated immunity, 263
cells. See also cell biology; cell cycle;
    eukaryotic cells; stem cells; tissues
  B, 255, 256–257
  bone, 39, 67
  brain, 39
  cardiac muscle, 39
  central nervous system, 41
  chemoreceptor, 207
  diploid, 272
  endothelial, 147
  epidermis, 123
  functions of, 43–46
  haploid, 62, 272
  of immune system, 254–258
  Langerhans, 123
  liver, 40
  malignant, 264
  mast, 255, 262
  membranes of, 52
  multinucleate, 101, 102
  muscle, 40, 101–102
  neuroglial, 137
  neurons, 136–137
  NK, 255, 256–257
  nucleus of, 46, 51
  olfactory, 154–155
  prokaryotic, 48

cells *(continued)*
  red blood, 39, 187–188, 198, 210, 325–326
  replacement of, 38–39
  skin, 119, 124
  somatic, 62, 63, 272
  syncytium, 51
  T, 254, 255, 256–257, 266
  white blood, 189, 203, 254, 256–258
cellular level of organization, 24
cellular respiration, 29–33, 318
central, definition of, 16
central nervous system, 41, 139–140
cerebellum, 143, 144
cerebrospinal fluid (CSF), 146–147
cerebrum, 143–144, 147
cervical vertebrae, 85, 86
cervix, uterine, 277, 287, 293
chambers of heart, 178
chemistry, laws of, 313–318
chemoreceptor cells, 207
chewing, muscles used in, 108
chickenpox, 267
childhood, 304–305
chitin, 87
Chlamydia trachomatis, 291
chloroplasts, 324
cholecystokinin, 220, 226
chromosomes, 276, 292, 297–298
chronic inflammation, 266
chronic obstructive pulmonary disorder, 212
chronic pain syndrome, 156
chyme, 219
circular movements of joints, 81
circulatory system
  age-related changes to, 308
  blood, 186–189
  blood vessels, 181–186
  functions of, 177
  heart, 178–181
  heartbeat and blood flow, 189–195
  lymphatic system in, 251–254
  pathophysiology of, 195–198
  waste substances in, 234
circumcision, 284
citric acid, 31
clavicles, 87, 88, 89
cleft palates, 94

clinical medicine, definition of, 11
clitoris, 278
clotting cascade, 40–41, 194–195, 231
coccygeal vertebrae, 85, 86
colon, 221
communication
  by cells, 46
  between self and environment, 135
  using scientific jargon, 12–15
compact bone, 77
comparative anatomy, 11
complement system proteins, 258–259
complementary pairs, 57, 60
computed axial tomography (CAT or CT), 18–19
concentration gradient, 49
connective tissue
  blood as, 186–187
  functions and types of, 67
  skeletal system, 74–75
constipation, 227
control of breathing, 208–209, 325
cooking of food, 323
coronary arteries, 180–181
cortex, 139
corticosteriods, 166
cortisol, 167
cosmetics, use of, 122
costal cartilages, 87, 88
coughing, 208
cramps in muscles, 118
cranial cavity, 21, 22
cranial nerves, 140
cranium, 82, 83
creation of products by cells, 45–46
Crohn's disease, 229–230
cross sections, 17
crossing-over stage of meiosis, 273
cryptorchidism, 292
crystalline lattice, 317
CSF (cerebrospinal fluid), 146–147
culture, socialization of child into, 304–305, 306
curvature of spine, 85, 93
cytokinesis, 65
cytoplasm, 31, 46, 52

• D •

Darwin, Charles (scientist), 18
daughter cells, 44
DCT (distal convoluted tubule), 236, 242
dead space, 206
deep, definition of, 16
degranulation, 261
deltoid muscle, 112
dendrites of neurons, 137, 138
deoxyribonucleic acid. *See* DNA
depolarization of neurons, 148, 149
dermatitis, 131
dermis, 124–125
development
  before birth, 298–302
  dimensions of, 296–298
  genome in, 295–296
  stages of, 296
  through life span, 302–310
developmental anatomy, 11
diabetes, 36, 172–173
diaphragm
  in coughing or sneezing, 208
  functions of, 100, 205
  hiccups, 118
  in normal breathing, 206–207
  in relation to lungs, 202
diarrhea, 227–228
diarthroses, 79–80
diencephalon, 147, 163
differentiation, 44–45, 51, 297
diffusion, 49, 183, 317
digestion, 214
digestive smooth muscle, 100
digestive system
  accessory organs of, 222–226, 230–232
  age-related changes to, 308
  cooking of food, 323
  energy exchange in, 213
  functions of, 213–214
  mucous membranes in, 203, 215–216, 225
  of newborns, 303
  pathophysiology of, 227–232
  structures of, 214–221
digital imaging technology, 19

diploid, definition of, 44, 272
direct secretion process, 241
diseases. *See also* pathophysiology
  of accessory organs to digestion, 230–232
  autoimmune, 94, 265
  bowel, 229–230
  enzymes, 58
  of heart, 195–196
  immune-mediated, 265–266
  infectious, 266–267
  of liver, 231
  of lungs, 211–212
  of oral cavity, 227
  sexually transmitted, 290–291
  X-linked, 117
distal, definition of, 15, 16
distal convoluted tubule (DCT), 236, 242
division of cells, 62–63
DNA (deoxyribonucleic acid). *See also*
    genome
  in cell nucleus, 51
  description of, 55–57
  in diploid cells, 272
  double helix, 59–60
  in mitochondra, 53
  replication of in meiosis, 273
  of zygotes, 295–296
DNA replication, 63–65, 66
dopamine, 150
dorsal, definition of, 15, 16
dorsal cavity, 21–22
Duchenne muscular dystrophy, 117
duodenal ulcers, 228–229
duodenum, 219, 220

• E •

ear wax, 128
ears, 152–153
eccrine sweat glands, 128
eclampsia, 293
ectopic pregnancy, 293
edematous pancreatitis, 232
ejaculation, 285
electrocardiograms, 191–192
electrolytes, 166, 243

electron transport chain, 32–33
electrons, 318
elimination of digestive waste, 214, 221.
    See also excretion; urinary system
embolus, 195, 196
embryo, 62, 299–300
emotion and amygdalae, 321–322
emphysema, 212
endocardium, 180
endochondral ossification, 77
endocrine glands, 159, 162, 220
endocrine organs, kidneys as, 234–235
endocrine system
  age-related changes to, 309
  as diffuse, 163
  glands, 162–172
  in homeostasis, 37
  hormones, 157–162
  pathophysiology of, 172–175
endogenous substances, hormones as, 157
endomembrane system, 52
endometriosis, 291
endometrium, 277, 281
endomysium, 103
endoplasmic reticulum, 54
endorphin, 150
endoskeleton, 87
endothelial cells, 147
endothelium, 68
energy, laws of, 28, 313–314
enteric endocrine, 170–171
enteric nervous system, 141
entropy, 314, 315
environment, skin as mediating interaction
    with, 120
enzymes, 45, 57, 58, 225, 226
eosinophils, 255, 261
epicardium, 180
epidermis, 39, 67, 121, 123–124
epididymis, 284
epiglottis, 201–202, 217, 218
epimysium, 103
epinephrine, 150, 159, 160, 167
epiphysis, 77
epithelial tissue, 67–69
erectile dysfunction, 292
erythropoietin, 234
esophagus, 217–218

estrogen, 168, 169, 280–282
ethmoid bone, 83
eukaryotic cells
  cytoplasm, 52
  description of, 46
  internal membranes, 52
  lysosomes, 54
  membrane of, 48–51
  mitochondria, 52–53
  nucleus, 51
  organelles of, 47
  origins of, 324–325
  protein construction, 53–54
  structure of, 47
eukaryotic organisms, 208
evolution
  of mammary glands, 279
  of respiratory system, 199
evolutionary biology, 11
evolutionary success, 283
evolutionary theory, 10–11
excitatory neurotransmitters, 150
excretion, 23, 235
exercise, weight-bearing, 89
exfoliants, 122
exocrine glands, 162
exocytosis, 161
exoskeleton, 87
expiration, 207
expiratory reserve volume, 206
external urethral sphincter, 238
extracellular fluid, 234, 243, 251
eyes, 250

• *F* •

facial bones, 82, 83–84, 94
facial expression, muscles used in, 108
FAD (flavinadenine dinucleotide), 32, 33
fallopian tubes, 277
fascicles, 103
fat digestion, 220–221
feces, 221
feedback, 37
feet, 92
femur, 75, 76, 91, 92
fertilization, 286, 298

fetus
  bone development in, 77
  early development of, 298–299
  as embryo, 62, 299–300
  passive immunity conferred on, 263
  placenta, 300
  trimesters of pregnancy, 301–302
  vernix caseosa around, 128
fibers, 103
fibrinogen, 187
fibrocartilage, 75
fibromyalgia, 118
fibrous connective tissue, 75
fibula, 91
fight-or-flight response, 167, 207–208
filtrate, 240–241
filtration, 50
fingernails, 127, 132
fingerprints, 124
fingers, 75, 76, 91, 114, 319
first law of thermodynamics, 28, 313–314
first trimester of pregnancy, 301
flat bone, 78
flat feet, 92
flavinadenine dinucleotide (FAD), 32, 33
floating ribs, 87, 88
fluid balance, 35–36
fluid-mosaic model of cell membrane, 48, 49
fluids, 316–317
fMRI (functional magnetic resonance imaging), 19
food, cooking of, 323
foramen magnum, 84
foreskin, 284
frontal bone, 83
frontal plane, 17, 18
functional magnetic resonance imaging (fMRI), 19

• G •

GABA, 150
gallbladder, 223, 224
gallstones, 232
gametes, 41, 44, 62, 63, 271–276
ganglia, 138
gastric juice, 218, 226, 250
gastric ulcers, 228–229

gastrins, 171, 226
gastrocnemius muscle, 115, 116, 118
gastrointestinal tract, structures of, 214–221
gastrulation, 300
gene expression, 51, 58, 60
genes, 58–61
genome, 51, 58, 60, 295–296. See also DNA
gestational diabetes, 173, 293
gestational period, 301
gingivitis, 227, 266
girdles, 88–89
glands
  adrenal, 158, 166–168
  bulbourethral, 285
  endocrine, 159, 162, 220
  exocrine, 162
  gonads, 168
  hypothalamus, 163–164
  lacrimal, 201
  mammary, 279
  pancreas, 170–171
  parathyroid, 171
  pineal, 172
  pituitary, 37, 158, 163–164, 244
  prostate, 284, 285
  salivary, 217
  sebaceous, 120, 128, 162
  in skin, 127–128
  stomach as, 171
  sudoriferous, 128
  sweat, 120, 128, 129, 162
  thymus, 172
  thyroid, 158, 165
glomerular capsule, 236, 237, 240
glomerulus, 236
glucose in blood, 36, 170–171
glutamate, 150
gluteus maximus muscle, 115
gluteus medius muscle, 115
glycolysis, 30, 31, 32
glycoprotein hormones, 158
Golgi body, 54
gonadocorticoids, 167
gonads, 168
goose bumps, 128
gout, 95, 245
graded potential of neurons, 148
granular, definition of, 257

granules, 261
gravity, muscles as fighting, 98
gray matter, 139, 145
gross anatomy, 11. *See also* skeletal system
growth in human development, 296–297
gums, 216–217, 227

• *H* •

hair and hair follicles, 126, 128, 131–132, 320–321
hamstrings, 115, 116
haploid cells, 62, 272
hard palates, 201, 218
head, muscles of, 108–110
healing of wounds, 40–41
hearing, receptors for, 152–153
heart
    blood flow through, 192–193
    blood supply to, 180–181
    cardiac cycle, 190–192
    diseases of, 195–196
    pulse, 189–190
    structure of, 178–179
    tissues of, 179–180
hematopoiesis, 77, 188
hemoglobin, 187–188, 325–326
hemoglobinopathy, 197–198
hemostasis, 194
hepatic portal system, 185, 186
hepatitis, viral, 231
herpes viruses, 266, 267
heterotrophs, 28
hiccups, 118
hip fractures, 94
hipbones, 89
hippuric acid, 240
histamine, 258, 261, 262
histologic anatomy, 11
HIV, 266, 291
holding breath, 208
homeostasis. *See also* thermoregulation
    blood glucose concentration, 36
    blood pressure, 194
    fluid balance, 35–36
    hormones, 158
    measurement of variables in, 37
    muscular system contribution to, 99
    physiological, 314

process of, 27
    skin, 129
    urinary system in, 242–245
homeothermy, 34–35
Homo sapiens, 9, 12
hormone replacement therapy, 169
hormones. *See also* glands
    to control blood volume, 243–244
    description of, 157–158
    in digestive system, 226
    in menstrual cycle, 280–281
    receptors, 161–162
    sources and functions of, 159–161
    types of, 158–159
human milk, 279, 320
humerus, 75, 90
humoral immunity, 263
hunchback, 93
hyaline cartilage, 75, 77
hyoid bone, 85, 201–202
hyperpolarization of neurons, 148, 149
hypertension, 196, 244
hyperthyroid disorders, 174
hypodermis, 125
hypogonadism, 292
hypothalamus, 37, 143, 147, 163, 164
hypothyroid disorders, 173–174
hypoxemia, 210, 212
hypoxia, 210

• *I* •

IgG antibodies, 260
ileum, 220
iliac veins, 185
ilium, 89
immovable joints, 79
immune system
    age-related changes to, 309
    cells of, 254–258
    functions of, 249–250
    lymphatic system, 251–254
    mechanisms of, 260–262
    molecules of, 258–260
    of newborns, 303–304
    pathophysiology of, 264–267
    pus, 261
immune-mediated diseases, 265–266
immunity, 262–264

immunization, 263–264
immunoglobulins, 187, 203, 231, 259–260
implantation of blastocyst in uterus, 286
impotence, 292
impulses of nervous system, 135, 148–151
incompetent cervix, 293
incontinence, 247
infancy, 304
infections of urinary tract, 246–247
infectious diseases, 266–267
inferior, definition of, 16
inferior vena cava, 185, 186
infertility, 290
inflammation, 231, 262, 266
inflammatory bowel disease, 229
ingestion, 213–214
inhibitory neurotransmitters, 150
inoculation, 263–264
inspiration, 206, 207
inspiratory reserve volume, 206
insulation to hold temperature constant, 35
insulin, 36, 170–171
insulin metabolism, abnormalities in,
    172–173
integument
  age-related changes to, 309
  dermis, 124–125
  description of, 119
  epidermis, 39, 67, 121, 123–124
  functions of, 120–121
  hypodermis, 125
  pathophysiology of, 130–132
  proteins of, 57
  structure of, 121–125
  in thermoregulation, 120, 129
internal intercostal muscles, 110
internal urethral sphincter, 238
interneurons, 137, 145
interphase of cell cycle, 62, 63
interstitial fluid. See extracellular fluid
intervertebral disks, 79
intestinal lining, 39
intestines
  in digestive system, 219–221
  disorders of, 227–229
  enteric endocrine, 171
Invertebrata classification, 73
involuntary muscle cells, 101
irregular bone, 78

irritable bowel syndrome, 220
ischemia, 196
ischium, 89

• J •

jargon, 12–15
jaundice, 231
jejunum, 220
joints
  arthritis in, 94–95
  cartilage as component of, 75, 76
  movement of, 80–81
  sacroiliac, 89
  types of, 79–80
jugular veins, 185
juvenile arthritis, 95

• K •

Kaposi sarcoma, 264
keratins, 120, 123, 124
ketone bodies, 240
kidney stones, 245–246
kidneys
  adrenal glands, 166–168
  blood supply, 236
  description of, 235–236
  as endocrine organs, 234
  endocrine system, 243–245
  nephron, 236
  pathologies of, 245–246
  tissues of, 236
  urine composition, 36, 239–240
kingdoms, 10, 12
Klinefelter's syndrome, 292
Krebs cycle, 31, 32–33
kyphosis, 93

• L •

labia of vulva, 278
labor and delivery, 288–289
lacrimal bones, 83
lacrimal glands, 201
lactation, 279
Langerhans cells, 123
large intestine, 221

laryngopharynx, 202
larynx, 218
lateral, definition of, 15
latissimus dorsi muscle, 112
legs, 91–92, 114–116
leukocytes, 254–258, 256. *See also* white
    blood cells
levels of organization, 22–25, 43
life span
    adolescence, 305–306
    age-related changes to body systems,
        308–310
    changes at birth, 303–304
    description of, 302–303
    infancy and childhood, 304–305
    middle age, 307
    senescence, 307–308
    young adulthood, 306
ligaments, 75, 79
limbic system, 147
limbs, 90–92, 113–116
lipid hormones, 158
lipids, 123, 316
liver
    as accessory organ of digestion, 222–224
    bile production in, 224
    functions of, 224
    regeneration of, 223–224
    structures of, 222–223
    symptoms of disease of, 231
    veins of, 185, 186
liver cells, 40
lobes
    of cerebral hemispheres, 143, 144
    of liver, 222
lobules, liver, 222, 223
long bone, 76–78, 297
loop of Henle, 236, 241
lordosis, 93
lumbar vertebrae, 85, 86
lumen
    of arteries, 182, 183
    of digestive tract, 214, 215
lungs, 202–204, 211–212
lymph nodes, 252–253
lymphatic ducts, 252
lymphatic system, 251–254
lymphatic vessels, 252

lymphocytes, 253, 256–257
lysosomes, 54

• *M* •

macromolecules
    description of, 54–55
    nucleic acids and nucleotides, 55–57
    polysaccharides, 57
    proteins, 57–58
    structure of, 55
macrophages, 253, 255, 257–258
macular degeneration, 156
malaria, 198
malignant cells, 264
malignant melanoma, 131
mammals
    amniotic sac of, 300
    bladder of, 237
    hair of, 320–321
    marine, 34
    pattern of development of, 302–303
    respiratory system of, 199
    thermoregulation in, 129
mandible, 83, 108, 218
manubrium, 87, 88, 89
marine mammals, 34
masseter muscle, 323
mast cells, 255, 262
matrix, 67
maxillae, 84
medial/median, definition of, 15, 16
medical imaging, 18–19
medulla, 236, 242
medulla oblongata, 146
medullary cavity, 77
meiosis, 62, 272–274
melanin, 123, 124, 126, 131
melatonin, 172
men. *See also* sperm
    puberty in, 305–306
    reproductive system of, 282–285
    urethra of, 239, 284
menarche, 277
menopause, 169, 277, 282, 307
menstrual cycle, 41, 277, 280–282
messenger RNA (mRNA), 60
metabolic syndrome, 172

metabolism
  anabolic and catabolic reactions of, 28–29
  gene expression, 51
  process of, 27, 29–33
  purpose of, 29
  urination as final step in, 233
  waste products of, 234
microbes, 323–324
microvilli, 241
midbrain, 145
middle age, 307
milk production in breasts, 279, 320
miscarriage, 294
mitochondra, 31, 52–53, 97
mitosis, 44, 65–66, 272–273
moisturizers, 122
molecules of immune system, 258–260
monocytes, 255, 257
monomers, 55
morula, 299
motion, laws of, 314–315
motor nerve fibers, 141
motor neurons, 137, 138, 145
mouth
  in digestive system, 216–217, 218, 323
  diseases of, 227
  saliva, 225, 250
mouth feel, 214
movement and muscular system, 98
mRNA (messenger RNA), 60
mucous membranes (mucosa)
  of bladder, 237
  in digestive system, 215–216, 219, 225
  in immune system, 250
  of urethra, 238
mucus in respiratory system, 203
multinucleate cells, 101, 102
multiple sclerosis, 156
muscle cells, characteristics of, 101–102
muscle contraction
  to hold temperature constant, 35
  to move blood through veins, 186
  for movement, 98
muscle spasms, 117–118
muscle spindles, 99, 103
muscle tissue
  cardiac, 104
  cells of, 101–102
  description of, 97

skeletal, 102–104
smooth, 105
types of, 69, 100–105
muscle tone, 98–99
muscular dystrophy, 117
muscular system. *See also specific muscles*
  age-related changes to, 309
  functions of, 98–100
  names of skeletal muscles, 107–116
  overview of, 97
  pathophysiology of, 116–118
  sliding filament model, 105–107
  tissue types, 100–105
musculoskeletal system and integument, 121
myelin, 137, 156
myocardial infarctions, 195–196
myocardium, 180
myofibrils, 103, 105
myosin, 105, 106
myotonic muscular dystrophy, 117
myxedema, 174

• *N* •

NAD+ (nicotinamide adenine dinucleotide),
    31, 33
nails, 127, 132
nasal bones, 84
nasopharynx, 201
neck, muscles of, 109, 110
necrotizing pancreatitis, 232
negative feedback mechanism, 37, 171
nephron, 236
nerve fibers, 138, 141, 152
nerves, 138, 140–141
nervous system
  age-related changes to, 310
  brain, 139–140, 142–148
  central nervous system, 41, 135, 139–140
  as electrical communications network,
      135–136
  endocrine system compared to, 157
  enteric, 141
  functions of, 136
  impulses of, 148–151
  muscle spindles, 99
  neural tissues, 136–139
  parasympathetic, 141

nervous system *(continued)*
  pathophysiology of, 156
  peripheral, 136, 140–142
  sensory receptors in dermis, 125, 130
  sensory system, 151–155
  somatic, 141
  sympathetic, 141, 142
nervous tissue, 69
neural tissues, 136–139
neuroglial cells, 137
neurons
  description of, 136–137
  hormones, 159
  motor, 145
  sensory, 145, 322
  transmission of impulses across, 148–149
neurotransmitters, 149–151, 159
neutrophils, 255, 257, 261
nicotinamide adenine dinucleotide (NAD+), 31, 33
nitrogen, molecular, 208
nitrogenous components of urine, 239
NK cells, 255, 256–257
node, definition of, 191
nomenclature in book, 2
noninflammatory bowel disease, 229
norepinephrine, 150, 160, 167–168
nose
  receptors in, 154–155, 214, 322
  in respiratory system, 200–201
nuclear envelope, 51
nuclear medicine, 19
nucleic acids, 55–57
nucleotides, 55
nucleus of cells, 46, 51

• *O* •

obstructions of urinary tract, 246
occipital bone, 83
olfaction, 154–155, 214, 322
olfactory neurons, 322
one gene, one protein model, 59
ontogeny, 295, 306
oogenesis, 274, 277
opposable thumbs, 14, 319
oral cavity
  in digestive system, 216–217, 218
  diseases of, 227

organ level of organization, 24–25
organ system
  culture, 306
  role in excretion, 234
  skin as, 119
  terminology for, 13–14
organ system level of organization, 25
organ transplants, 265
organelles, 46, 47, 52
organism
  cells and, 43
  definition of, 22, 23, 27
  eukaryotic, 208
organism, functions of
  growing, replacing, and renewing, 38–41
  homeostasis, 34–37
  metabolism, 28–33
  transferring energy, 28
organism level of organization, 25
organs. *See also* skin; *specific organs*
  definition of, 24
  of female reproductive system, 276–279
  hormones, 159, 163
  of male reproductive system, 283–284
  sense organs, receptors in, 151
osmosis, 49, 317, 318
osopharynx, 201–202
osteoarthritis, 94
osteoblasts, 89, 93
osteoporosis, 89, 93–94
ova, 41, 44, 274–275
ovarian cycle, 280–281
ovaries, 277, 287
overflow incontinence, 247
ovulation, 277, 281
oxaloacetic acid, 31
oxidation, 32, 318
oxidative phosphorylation, 30, 32–33
oxygen, molecular, 208
oxytocin, 279

• *P* •

palatine bones, 84, 94
pancreas
  as accessory organ of digestion, 224–225
  functions of, 170–171
  inflammation of, 232
pancreatic juice, 225, 226

papillary region of dermis, 124–125
parasites in blood cells, 198
parasympathetic nervous system, 141
parathyroid glands, 171
parietal bones, 83
parietal layer of serous pericardium, 180
passive transport mechanism, 49–50
pathogens, 130
pathophysiology
   of circulatory system, 195–198
   definition of, 11
   of digestive system, 227–232
   of endocrine system, 172–175
   of immune system, 264–267
   of integument, 130–132
   of muscular system, 116–118
   of nervous system, 156
   of pregnancy, 292–294
   of reproductive system, 290–294
   of respiratory system, 210–212
   of skeletal system, 93–95
   of urinary system, 245–247
PCT (proximal convoluted tubule), 236, 241
pectoral girdle, 88–89
pectoralis major muscles, 110, 111
pelvic cavity, 276
pelvic girdle, 88–89
penis, 284
peptide hormones, 158
pericardial cavity, 180
perimysium, 103
periosteum, 75
peripheral, definition of, 16
peripheral nerve cells, 41
peripheral nervous system, 136, 140–142
peristalsis, 215, 217, 221
peritoneum, 219, 236
peritubular capillaries, 236, 241
peritubular capillary endothelial cells, 234
permeability of cell membranes, 48–51
pernicious anemia, 230
pH of blood, 200, 240, 244–245
phagocytes (phagocytizing leukocytes),
      188, 257–258
phagocytosis, 260
phalanges, 91
pharynx, 201–202, 217–218
phenylketonuria (PKU), 58
pheromones, 128

phospholipid bilayer, 48, 52, 208, 316
phospholipid molecules, 48
photosynthesis, 208, 213, 318
phrenic nerves, 205, 206
physiology, overview of, 9–11
pineal glands, 172
piriformis muscle, 117
pituitary gland, 37, 158, 163–164, 244
PKU (phenylketonuria), 58
placenta, 287, 289, 300
placenta previa, 293
placental abruption, 294
plane, definition of, 17
planes of body, 17, 18
plants, 10, 28. *See also* photosynthesis
plaque on teeth, 227
plasma, 187, 251
platelets, 40, 189
pleural sac, 202–203
plexus, 139
pluripotent stem cells, 62
pneumonia, 211–212
poikilothermy, 35
polarity, 48, 315
polarization of neurons, 148
polycythemia, 210
polymers, 55
polysaccharides, 55, 57
pons, 146
positive feedback mechanism, 37
positron emission tomography, 19
posterior, definition of, 15
posterior pituitary glands, 164
post-herpetic neuralgia, 267
precapillary sphincters, 183, 184–185
pre-eclampsia, 293
pre-embryos, 286
pregnancy
   adaptation to, 286–288
   development before birth, 298–302
   fertilization, 286
   implantation, 286
   labor and delivery, 89–90, 288–289
   pathophysiology of, 292–294
   placenta, 287, 289, 300
   trimesters of, 301–302
prehensile thumbs, 114, 319
premenstrual syndromes, 291
primary bronchi, 202, 204

primary oocytes, 274
primary response, 263
probability, laws of, 315
products, creation and transport of by cells, 45–46
progesterone, 168, 281–282
prokaryotic cells, 48
prone, definition of, 15
prostaglandins, 285
prostate gland, 284, 285
protein construction in cells, 53–54
protein synthesis, 60–61
proteins, 55, 57–58
protist, 10, 46
proximal, definition of, 15, 16
proximal convoluted tubule (PCT), 236, 241
puberty, 305–306
pubis, 89
pulmonary circulation path, 192–193
pulmonary veins, 185
pulse, 189–190
Purkinje fibers, 190, 191
pus, 261
pyloric sphincter, 220
pylorus, 219
pyruvate molecules, 52
pyruvic acid, 31

• Q •

quadriceps femoris muscles, 115

• R •

radiopharmaceuticals, 19
radius, 76, 90, 91
reabsorption of water from filtrate, 240–242
receptors for hormones, 161–162. See also sensory receptors
rectum, 221
rectus abdominis muscle, 111
red blood cell production, regulation of, 234
red blood cells, 39, 187–188, 198, 210, 325–326
red pulp of spleen, 254
reduction, definition of, 32
reduction-oxidation (redox), 318
reflex arcs, 145
regeneration of liver, 223–224
regions of body surface, 17, 19–20
renal artery, 236
renin-antiotensin system, 243–244
repair of tissue, 40
replication fork, 63, 64
repolarization of neurons, 148, 149
reproductive system
  in adolescence, 305–306
  age-related changes to, 310
  determining sex, 276
  functions of, 271–272
  meiosis, 272–274
  of men, 282–285
  mucus in, 203
  ova, 41, 44, 274–275
  pathophysiology of, 290–294
  pregnancy, 285–289, 298–302
  sperm, 44, 275–276
  of women, 276–282
residual volume, 206
respiratory chain, 32–33
respiratory membrane, 39, 204–205
respiratory system. See also diaphragm
  age-related changes to, 310
  breathing, 205–209, 303, 325
  functions of, 199–200
  lungs, 202–204
  mucus in, 203
  nose, 200–201
  pathophysiology of, 210–212
  pharynx, 201–202, 217–218
  respiratory membrane, 204–205
  trachea, 202, 218
respiratory tract, 200–205
resting potential of neurons, 148, 149
reticular region of dermis, 125
reticular tissue, 67
rheumatoid arthritis, 94
rib cage, 86–88
ribosomes, 54
rickets, 130
rigor mortis, 107
RNA (ribonucleic acid), 55–57
Roentgen, Wilhelm Conrad (scientist), 18
rotator cuff muscles, 112

• S •

sacral vertebrae, 85, 86
sacroiliac joint, 89
saggital plane, 17, 18
saliva, 225, 250
salivary glands, 217
sarcomeres, 97, 103, 105–106
sartorius muscle, 115, 116
scapulae, 89
scar tissue, 41
science, anatomy and physiology in, 10–11
scoliosis, 93
scrotum, 284
sebaceous glands, 120, 128, 162
sebum, 128
second law of thermodynamics, 314
second trimester of pregnancy, 301, 302
secondary oocytes, 274–275, 286, 298
secondary response, 263
secretin, 220, 226
seminal vesicles and fluid, 285
semipermeable cell membranes, 48
senescence, 296, 297–298, 307–308
sensory nerve fibers, 141
sensory neurons, 137, 138, 145
sensory receptors
  in dermis, 125, 130, 152
  function of, 136
  in sense organs, 151–155, 214, 322
sensory system, 151–155
septicemia, 247
serotonin, 150
sex cells, 62, 63. *See also* gametes
sex hormones, 168
sexually transmitted diseases, 290–291
shaving, 122
shingles, 267
short bone, 78
sickle cell disease, 197–198
sight, receptors for, 153–154
signaling pathways, 318
singing and breath control, 208, 325
sinoatrial node, 190
sinuses, 82, 84, 201
sitz bones, 89

skeletal muscle
  cells of, 40
  functions of, 98–100
  of head, 108–110
  of lower limbs, 114–116
  names of, 107–108
  synergistic and antagonistic, 104
  of torso, 110–112
  of upper limbs, 113–114
skeletal muscle tissue, 102–104
skeletal system. *See also* appendicular
    skeleton; axial skeleton
  age-related changes to, 310
  bone structure, 76–77
  connective tissue, 74–75
  description of, 73
  development of, 77–78
  joints, 79–81
  pathophysiology of, 93–95
  structural functions of, 74
skin. *See also* integument
  as barrier to invaders, 250
  cancers of, 131
  glands in, 127–128
  nerve endings in, 130
  size and weight of, 119
skin fibroblasts, 40, 41
skull, 82–84
sliding filament model, 101, 105–106
small intestine, 220–221
smell, sense of, 154–155, 214
smooth muscle tissue, 40, 102, 105
sneezing, 208
sodium bicarbonate, 226
soft palates, 201, 218
solids, 317
solutes, 35
somatic cells, 62, 63, 272
somatic nervous system, 141
somatic reflex arc, 102
spasms in muscles, 117–118
speaking and breath control, 208, 325
species, definition of, 10
sperm, 44, 275–276
spermatogenesis, 39, 275, 276, 284
sphenoid bone, 83
sphincter muscles, 100

spinal cavity, 21, 22
spinal column, 84–86, 93
spinal cord, 84, 139–140, 143, 145
spinal nerves, 141
spinal tap, 147
spleen, 254
spongy bone, 77
spontaneous abortions, 294
squamous cell carcinoma, 131
squamous epithelium, 204
stem cells, 38, 44, 62, 63, 188
sternocleidomastoid muscles, 109, 110
sternum, 87, 89
stomach
  in digestive system, 218–219
  disorders of, 227–229
  as gland, 171
stratum corneum, 121, 122, 123
stratum germinativum, 124
stratum granulosum, 123
stratum lucidum, 123
stratum spinosum, 123
stress
  breathing under, 207–208
  as cause of diarrhea, 228
stress incontinence, 247
striation of muscle cells, 101, 102
strokes, 196–197
structural proteins, 45
students, uses of book by, 2, 3, 6
subcutaneous layer of skin, 125
sudoriferous glands, 128
superficial, definition of, 16
superior, definition of, 16
superior vena cava, 185
supine, definition of, 156
sutures, 79, 82, 83
swayback, 93
sweat glands, 120, 128, 129, 162
sweating to hold temperature constant, 35
swelling of lymph nodes, 253
symbiosis, 324
sympathetic nervous system, 141, 142
synapses, 149–151
synarthroses, 79
syncytium cells, 51
syndrome, definition of, 198, 229

synergistic muscles, 104
synovial joints, 79–80
systemic circulation path, 192, 193

● *T* ●

T cells, 254, 255, 256–257, 266
talus, 92
tanning, 126
taste, sense of, 155
taxonomy, 12
teeth, 216–217, 227
telomeres, 296, 297–298
temporal bones, 83
temporalis muscle, 323
tendons, 75, 103
terminology, 12–15
testes, 275, 284
testosterone, 168
thalamus, 143, 147
thermodynamics, laws of, 28, 313–314
thermoregulation
  capillaries, 184
  description of, 34–35
  in newborns, 303
  skin, 120, 129
thirst reflex, 36
thoracic cage, 86–88
thoracic cavity, 21, 22
thoracic vertebrae, 85, 86
thrombocytes, 194
thumbs, prehensile and opposable, 114, 319
thymus, 172, 254
thyroid disorders, 173–174
thyroid gland, 158, 165
thyroid hormones, 165
tibia, 91, 92
tidal volume, 206
tissue level of organization, 24
tissues. *See also* connective tissue; muscle
    tissue
  adipose, 67, 125, 236
  bone, 74–75
  building, 44–45
  definition of, 66
  epithelial, 67–69
  of heart, 179–180

of kidneys, 236
nervous, 69
repair of, 40
toenails, 127
toes, 91, 92, 116
tongue, 155, 217
torso, muscles of, 110–112
total incontinence, 247
touch, receptors for, 152
trabeculae, 77
trachea, 202, 218
traits, 59
transcription, 60, 61
transitional epithelium tissue of bladder, 237–238
translation, 60, 61
transport of products by cells, 45–46
transverse plane, 17, 18
trapezius muscle, 109, 110, 112
triceps muscle, 113
trimesters of pregnancy, 301–302
tuberculosis, 211–212
tunica externa, tunica media, and tunica interna
of arteries, 182, 183
of veins, 185

● *U* ●

ulcerative colitis, 230
ulcers, gastric and duodenal, 228–229
ulna, 90, 91
ultrasound imaging, 19
ultraviolet (UV) radiation, 123, 124, 126, 130
umbilical cord, 300
umbilical region, 16
ureter, 236, 237
urethra, 238–239, 247, 284
urge incontinence, 247
urinalysis, 241
urinary system
age-related changes to, 310
bladder, 237–238
description of, 233
functions of, 233–234
in homeostasis, 242–245
kidneys, 235–237

of newborns, 303
pathophysiology of, 245–247
structures of, 235–239
urethra, 238–239
urine, 239–242
urine
color of, 241
composition of, 36, 239–240
description of, 234
expelling, 242
urobilinogen, 240
uterine cervix, 277, 287, 293
uterine cycle, 281–282
uterus, 277, 287
UV (ultraviolet) radiation, 123, 124, 126, 130

● *V* ●

vaccination, 263–264
vagina, 277–278
vaginal delivery of infants, 89–90
vagus nerves, 140
valves
of heart, 179
in veins, 186
Varicella-Zoster virus, 267
vas deferens, 284
vascular disorders, 196–197
vascular endothelium, 182, 183
veins, 185–186
ventilation, 199–200
ventral, definition of, 15, 16
ventral cavity, 21, 22
ventricles, 146
vertebrae, 84, 85–86
Vertebrata classification, 73
villi, 219
viral hepatitis, 231
viruses, 267
vision, 153–154
vital capacity, 206
vitamin D, 126, 130
vitiligo, 265
vocal cords, 201–202
voluntary muscle cells, 101–102
vomer bone, 84
vulva, 278

• W •

walls of digestive tract, 215–216
warm-blooded animals. *See* animals; mammals
water
  in body, 234
  lipids, 316
  poisoning by, 241
  polarity of, 315
  reabsorption of from filtrate, 240–242
weight-bearing exercise, 89
white blood cells, 189, 203, 254, 256–258
white matter, 139, 145
white pulp of spleen, 254
wisdom teeth, 217
women. *See also* menopause; menstrual cycle; ova
  pregnancy, 285–289, 298–302
  puberty in, 305
  reproductive system of, 276–282
  urethra of, 238
word fragments for precise terminology, 13–14
wounds, healing, 40–41
wrists, 91, 114

• X •

X chromosomes, 276, 292
xiphoid process, 87, 88
X-linked diseases, 117
X-ray images, 18

• Y •

Y chromosomes, 276
young adulthood, 306

• Z •

Z lines, 106
zona pellucida, 274, 298
zygomatic bones, 84, 108
zygotes, 44, 62, 63, 272, 295–296